Flavor of Meat and Meat Products

Flavor of Meat and Meat Products

Edited by

FEREIDOON SHAHIDI
Departments of Biochemistry and Chemistry
Memorial University of Newfoundland

SPRINGER-SCIENCE+BUSINESS MEDIA, B.V.

First edition 1994

Originally published by Chapman & Hall in 1994
Softcover reprint of the hardcover 1st edition 1994

Typeset in 10/12pt Times by Acorn Bookwork, Salisbury, Wiltshire

ISBN 978-1-4613-5911-1 ISBN 978-1-4615-2177-8 (eBook)
DOI 10.1007/978-1-4615-2177-8

A catalogue record for this book is available from the British Library

Library of Congress Catalog Card Number: 94–71018

∞ Printed on acid-free text paper, manufactured in accordance with ANSI/NISO Z39.48-1992 (Permanence of Paper)

Preface

Flavour is an important sensory aspect of the overall acceptability of meat products. Whether we accept or reject a food depends primarily on its flavour. Both desirable and undesirable flavour effects are contemplated. Furthermore, threshold values of different flavour-active compounds have an important effect on the cumulative sensory properties of all foods.

Meat from different species constitutes a major source of protein for most people. Although raw meat has little flavour and only a blood-like taste, it is a rich reservoir of non-volatile compounds with taste-tactile properties as well as flavour enhancers and aroma precursors. Non-volatile water-soluble precursors and lipids influence the flavour of meat from different species. In addition, mode of heat processing and the nature of additives used may have a profound effect on the flavour of prepared meats.

This book reports the latest advancements in meat flavour research. Following a brief overview, chapters 2 to 5 discuss flavours from different species of meat, namely beef, pork, poultry and mutton. In chapters 6 to 12 the role of meat constituents and processing on flavour are described. The final section of the book (chapters 13 to 15) summarizes analytical methodologies for assessing the flavour quality of meats.

I wish to thank all the authors for their cooperative efforts and commendable contributions which have made this publication possible.

Fereidoon Shahidi

Series foreword

The 8th World Congress of Food Science and Technology, held in Toronto, Canada, in 1991 attracted 1400 delegates representing 76 countries and all five continents. By a special arrangement made by the organizers, many participants from developing countries were able to attend. The congress was therefore a most important international assembly and probably the most representative food science and technology event in that respect ever held. There were over 400 poster presentations in the scientific programme and a high degree of excellence was achieved. As in previous congresses much of the work reported covered recent research and this will since have been published elsewhere in the scientific literature.

In addition to presentations by individual researchers, a further major part of the scientific programme consisted of invited papers, presented as plenary lectures by some of the leading figures in international food science and technology. They addressed many of the key food issues of the day including advances in food science knowledge and its application in food processing technology. Important aspects of consumer interest and of the environment in terms of a sustainable food industry were also thoroughly covered. The role of food science and technology in helping to bring about progress in the food industries of developing countries was highlighted.

This book is part of a series arising from the congress and including bibliographical details. The series editors are Professor Marvin Tung of the Technical University of Nova Scotia, Halifax, Nova Scotia, Canada; and Dr. Gordon Timbers of Agriculture and Agri-Food Canada, Ottawa, Ontario, Canada. The book presents some of the most significant ideas which will carry food science and technology through the nineties and into the new millennium. It is therefore essential reading for anyone interested in the subject, including specialists, students and general readers. IUFoST is extremely grateful to the organizers from the Canadian Institute of Food Science and Technology for putting together a first class scientific programme and we welcome the publication of this book as a permanent record of the keynote papers presented at the congress.

Dr. D.E. Hood
(President, International Union of Food Science & Technology)

Contributors

M. Bae-Lee	Yukong Ltd, Naugatuck, CT 06770, USA
M.E. Bailey	University of Missouri–Columbia, Department of Food Science and Human Nutrition, 21 Agriculture Building, Columbia, Missouri 65211, USA
K.L. Bett	US Department of Agriculture, Agricultural Research Service, Southern Regional Research Center, 1100 Robert E. Lee Boulevard, New Orleans, Louisiana 70124, USA
T.J. Braggins	Meat Industry Research Institute of New Zealand (Inc.), PO Box 617, Hamilton, New Zealand
T. Cheraghi	Wageningen Agricultural University, Department of Food Science, Wageningen, The Netherlands
J.I. Gray	Department of Food Science and Human Nutrition, Michigan State University, East Lansing, MI 48824, USA
C.-T. Ho	Department of Food Science, Rutgers University, New Brunswick, NJ 08903, USA
C.M. Hollenbeck	Red Arrow Products Company Inc., Manitowoc, WI, USA
A.J. MacLeod	Department of Chemistry, King's College London, Strand, London, WC2R 2LS, UK
G. MacLeod	Department of Food Science, King's College London, Campden Hill Road, London, W8 7AH, UK
J.A. Maga	Department of Food Science and Human Nutrition, Colorado State University, Fort Collins, Colorado 80523, USA
D.S. Mottram	University of Reading, Department of Food Science and Technology, Whiteknights, Reading, RG6 2AP, UK
Y.-C. Oh	Department of Food Science, Rutgers University, New Brunswick, NJ 08903, USA

A.M. Pearson Department of Animal Science, Oregon State University, Corvallis, OR 97331, USA

N. Ramarathnam Japan Institute for the Control of Aging, Division of Nikken Foods Co. Ltd., 710-1 Haruoka, Fukuroi-City, Shizuoka, Japan 437-01

D.H. Reid Meat Industry Research Institute of New Zealand (Inc.), PO Box 617, Hamilton, New Zealand

J.P. Roozen Wageningen Agricultural University, Department of Food Science, Wageningen, The Netherlands

L.J. Rubin Department of Chemical Engineering and Applied Chemistry, University of Toronto, Toronto, Ontario, Canada, M5S 1A4

F. Shahidi Departments of Biochemistry and Chemistry, Memorial University of Newfoundland, St. John's, Newfoundland, Canada, A1B 3X9

H. Shi Scientific Research Institute of the Food and Fermentation Industry, Ministry of Light Industry, Beijing, People's Republic of China

M.E. Smith Meat Industry Research Institute of New Zealand (Inc.), PO Box 617, Hamilton, New Zealand

A.J. St. Angelo US Department of Agriculture, Agricultural Research Service, Southern Regional Research Center, 1100 Robert E. Lee Boulevard, New Orleans, Louisiana 70124, USA

B.T. Vinyard US Department of Agriculture, Agricultural Research Service, Southern Regional Research Center, 1100 Robert E. Lee Boulevard, New Orleans, Louisiana 70124, USA

O.A. Young Meat Industry Research Institute of New Zealand (Inc.), PO Box 617, Hamilton, New Zealand

Contents

1 Flavor of meat and meat products—an overview

F. SHAHIDI

1.1 Introduction

Flavour is an important sensory aspect of the overall acceptability of meat products. The overwhelming effect of flavour volatiles has a tremendous influence on the sensory quality of muscle foods. However, the taste properties of high molecular weight components and contribution of non-volatile precursors to the flavour of meat should also be considered.

Although raw meat has little aroma and only a blood-like taste, it is a rich reservoir of compounds with taste tactile properties as well as aroma precursors and flavour enhancers (Crocker, 1948; Bender and Ballance, 1961). Non-volatile precursors of meat flavour include amino acids, peptides, reducing sugars, vitamins and nucleotides. Interaction of these components and/or their breakdown products produces a large number of intermediates and/or volatiles which contribute to meat flavour development and aroma generation during heat processing. Lipids also play an important role in the overall flavour of meat which is distinct and species dependent (Mottram *et al.*, 1982; Mottram and Edwards, 1983).

Dietary regime, metabolic pathway, and species of animals under investigation may have an effect on the flavour quality of meat. For example, branched fatty acids such as 4-methyloctanoic and 4-methylnonanoic acids are mutton-specific and a swine sex odour compound is associated with boars (Wong *et al.*, 1975; Gower *et al.*, 1981).

1.2 Meat flavour volatiles

Nearly 1000 compounds have so far been identified in the volatile constituents of meat from beef, chicken, pork, and sheep (Shahidi *et al.*, 1986; Shahidi, 1989). These volatiles were representative of most classes of organic compounds such as hydrocarbons, alcohols, aldehydes, ketones, carboxylic acids, esters, lactones, ethers, furans, pyridines, pyrazines, pyrroles, oxazoles and oxazolines, thiazoles and thiazolines, thiophenes and other sulphur- and halogen-containing substances. It is believed that the predominant contribution to aroma is made by sulphurous- and carbonyl-containing volatiles (Shahidi, 1989).

Although the chemical nature of many flavour volatiles of meat from different species is similar qualitatively, there are quantitative differences. For example, it has been reported that mutton aromas contain a higher concentration of 3,5-dimethyl-1,2,4-trithiolane and 2,4,6-trimethylperhydro-1,3,5-dithiazine (thialdine) as compared to those of other species. Other sulphur-containing compounds were also present in high concentration and were attributed to the high content of sulphurous amino acids in mutton as compared with those of beef and pork. Similarly a higher concentration of alkyl-substituted heterocyclics was noted in mutton volatiles (Buttery *et al.*, 1977). Mercaptothiophenes and mercaptofurans were significant contributors to beef aroma (Macleod, 1986).

Compared to the total number of volatile compounds identified in meat from different species, only a small fraction of them have been reported to possess meaty aroma characteristics (Shahidi, 1989) and these are mainly sulphur-containing in nature. While most of the sulphurous volatiles of meat exhibit a pleasant meaty aroma at concentrations present in meat, at high levels their odour is objectionable. Therefore, both qualitative and quantitative aspects of volatiles have to be considered when assessing the flavour quality of muscle foods. In addition, possible synergisms between various aroma constituents have to be considered.

Finally, in the evaluation of flavour quality of meat, the contribution to taste by amino acids, peptides and nucleotides must be considered. These compounds not only interact with other components to produce flavour volatiles, they also contribute to sweet, salty, bitter, sour and umami sensation of muscle foods. In the production of soups and gravies, proteins are partially hydrolysed to enhance taste sensation of the molecules. Therefore, studies in this area would allow us to optimize conditions to yield products with a maximum level of acceptability.

1.3 Impact of processing and storage on meat flavour

Processing of meat such as curing (Shahidi, 1992) and/or smoking (Maga, 1987) brings about a characteristic flavour in the products. Interaction of nitrite with meat constituents retards the formation of off-flavour volatiles which may mask the natural flavour of meat (Shahidi, 1992). On the other hand, smoking of meat may produce/deposit a number of new compounds in the products. Therefore, process flavours contribute greatly to the availability of a wide range of well-loved products.

While cured products retain their flavour and do not undergo oxidative changes for reasonably long storage periods, cooked meats, as such, are highly prone to oxidation. Progression of oxidation and meat flavour deterioration is dependent primarily on the species of meat and its lipid content. Furthermore, interaction of oxidative products with muscle food components may in turn bring about changes in the colour, texture and nutritional value of meats (Spanier *et al.*, 1992). Therefore, control of oxidative processes and methods to quantify these changes have received considerable attention over the last few decades.

As analytical methodologies have improved, the identification of new flavour-active compounds contributing to flavour at low threshold values has become possible. Use of results from fundamental research to improve the quality of meat products will continue to have an impact in new industrial developments. As more data become available, better understanding of the mechanisms of flavour perception becomes possible.

References

Bender, A.E. and Ballance, P.E. (1961). A preliminary examination of the flavour of meat extract. *J. Sci. Food Agric.* **12**, 683–687.

Buttery, R.G., Ling, L.C., Teranishi, R. and Mon, T.R. (1977). Roasted lamb fat: Basic volatile components. *J. Agric. Food Chem.* **25**, 1227–1229.

Crocker, E.C. (1948). The flavor of meat. *Food Res.* **13**, 179–183.

Gower, D.E., Hancok, M.R. and Bannister, L.H. (1981). In *Biochemistry of Taste and Olfaction*, eds. Cagan, R.H. and Kave, M.R. Academic Press, New York, pp. 7–31.

MacLeod, G. (1986). The Scientific and Technological Basis of Meat Flavours. In *Development in Food Flavours*, eds. Birch, G.G., Lindley, M.G. Elsevier Applied Science, London. pp. 191–223.

Maga, J.A. (1987). The flavor chemistry of wood smoke. *Food Reviews International* **3**, 139–183.

Mottram, D.S. and Edwards, R.A. (1983). The role of triglycerides and phospholipids in the aroma of cooked beef. *J. Sci. Food Agric.* **34**, 517–522.

Mottram, D.S., Edwards, R.A. and MacFie, H.J.H. (1982). A comparison of the flavor volatiles from cooked beef and pork meat systems. *J. Sci. Food Agric.* **33**, 934–944.

Shahidi, F. (1989). Flavour of cooked meats. In *Flavour Chemistry: Trends and Developments*, eds. Teranishi, R., Buttery, R.E. and Shahidi, F. ACS Symposium Series 388, American Chemical Society, Washington, D.C. pp. 188–201.

Shahidi, F. (1992). Prevention of lipid oxidation in muscle foods by nitrite and nitrite-free compositions. In *Lipid Oxidation in Food*, ed. St. Angelo, A.J. ACS Symposium Series 500, American Chemical Society, Washington, D.C. pp. 161–182.

Shahidi, F., Rubin, L.J. and D'Souza, L.A. (1986). Meat flavor volatiles: A review of the composition, techniques of analysis, and sensory evaluation. *CRC Crit. Rev. Food Sci. Nutr.* **24**, 141–243.

Spanier, A.M., Miller, J.A. and Bland, J.M. (1992). Lipid oxidation: Effect on meat proteins. In *Lipid Oxidation in Food*, ed. St. Angelo, A.J. ACS Symposium Series 500, American Chemical Society, Washington, D.C. pp. 161–182.

Wong, E., Nixon, L.N. and Johnson, C.B. (1975). Volatile medium chain fatty acids and mutton flavor. *J. Agric. Food Chem.* **23**, 495–498.

2 The flavour of beef

G. MACLEOD

2.1 Introduction

The flavour of beef has been investigated more extensively than any other meat flavour, probably because of its greater consumer popularity, and hence its commercial significance in the creation of successful simulated meat flavourings. Literature reports over the past 30 years show that the flavour of beef is highly complex. In its simplest format, it consists of taste-active compounds, flavour enhancers and aroma components.

2.2 Taste-active compounds

With regard to taste (MacLeod and Seyyedain-Ardebili, 1981; MacLeod, 1986; Kuninaka, 1981; Haefel and Glaser, 1990), sweetness has been associated with glucose, fructose, ribose and several L-amino acids such as glycine, alanine, serine, threonine, lysine, cysteine, methionine, asparagine, glutamine, proline and hydroxyproline. Sourness stems from aspartic acid, glutamic acid, histidine and asparagine, together with succinic, lactic, inosinic, *ortho*-phosphoric and pyrrolidone carboxylic acids. Saltiness is largely due to the presence of inorganic salts and the sodium salts of glutamate and aspartate; while bitterness may be derived from hypoxanthine together with anserine, carnosine and other peptides, and also the L-amino acids histidine, arginine, lysine, methionine, valine, leucine, isoleucine, phenylalanine, tryptophan, tyrosine, asparagine and glutamine.

The umami taste has a characteristic savoury quality and is supplied by glutamic acid, monosodium glutamate (MSG), 5′-inosine monophosphate (IMP), 5′-guanosine monophosphate (GMP) and certain peptides. Although generally speaking, glutamate is the most important contributor, its presence at a lower concentration in beef than in pork or chicken, for example, gives rise to a lower perceived umami taste intensity in beef (Kato and Nishimura, 1989; Kawamura, 1990). Similarly, the effect of conditioning these three species has shown a significant increase in intensity of the savoury, brothy taste in pork and chicken after aging, yet no significant difference in conditioned beef (Nishimura *et al.*, 1988). This disparity was paralleled by the observation that the increased concentration of free amino acids and of oligopeptides on aging was significantly smaller in beef than in pork or chicken (Nishimura *et al.*, 1988). However, subsequent rates of changes in concentration of these non-volatile components and of others on continued heating, and the associated aroma manifestations, could well alter the overall comparative conclusions on flavour improvement, in general, on aging different meat species.

2.3 Flavour enhancers

A more important sensory contribution than the taste characteristics of glutamic acid, MSG, IMP and, to a lesser extent, GMP in beef, is their flavour enhancing property. It has been proposed that our sensory receptors for flavour enhancement are independent and sterically different from the traditional basic taste receptors (Kuninaka, 1981; Kawamura, 1990). The 5′-ribonucleotides have strong flavour potentiating effects individually, but more importantly, they exhibit a potent synergistic effect when present, as in meat, in conjunction with glutamic acid or MSG (Kawamura, 1990). It appears that, in some mammals, this synergy is due to an induced increased strength of binding of glutamate to the receptor protein site, whereas in other mammals, an enhanced amount of glutamate is actually bound (Kawamura, 1990). The 5′-nucleotides are reported to enhance meaty, brothy, MSG-like, mouthfilling, dry and astringent qualities; they suppress sulphurous and HVP-like notes, while sweet, sour, oily/fatty, starchy and burnt qualities remain unchanged (Kuninaka, 1981).

Thermal decomposition of both classes of flavour enhancers may occur, with a resultant loss of activity. For example, at 121°C and a pH of 4.5–6.5 (e.g. during canning), an initial loss of the phosphate group from IMP and GMP, converting the nucleotide into the corresponding nucleoside, is followed by slow hydrolysis releasing the base—either hypoxanthine (from IMP) or guanine (from GMP) (Shaoul and Sporns, 1987). The first triggering reaction of the phosphate loss is depressed in the presence of certain divalent metals, e.g. calcium ions (Kuchiba *et al.*, 1990). Under similar heating conditions (100°C/pH 4–6), glutamic acid and MSG are converted into pyrrolidone carboxylic acid (PCA) (Gayte-Sorbier *et al.*, 1985). The reverse reaction is also possible, but is favoured by extreme pH values of <2.5 or >11 (Airaudo *et al.*, 1987). Not only is there no flavour enhancement activity from any of the decomposition products (Kuninaka, 1981; Gayte-Sorbier, 1985), but a distinct off-flavour is associated with PCA at certain concentrations.

2.4 Aroma components

Aroma components are generated in beef from non-volatile precursors on cooking. Primary reactions occurring are (1) lipid oxidation/degradation;

(2) thermal degradation and inter-reactions of proteins, peptides, amino acids, sugars and ribonucleotides; and (3) thermal degradation of thiamine. But reaction products become reactants, and the end result is a complex and intertwining network of reactions. In consequence, the most recent edition of the now classic TNO–CIVO publication *Volatile Compounds in Food* lists 880 volatile components reported from cooked beef (Maarse and Visscher, 1989). To place this figure in a better perspective, a rough breakdown of the chemical classes represented is shown in Table 2.1 (Maarse and Visscher, 1989).

With such a large number of potential contributors to the sensory perception of cooked beef aroma, the critical question is, 'What is the relative sensory significance of these volatiles?' The answer is not totally clear, but three conclusions can be drawn. First, many are relatively unimportant. Secondly, the term 'meatiness' can be cleanly dissected sensorially into about ten different odour qualities (Galt and MacLeod, 1983), in which

Table 2.1 Chemical classes of aroma components reported from cooked beef

Class of compound	Number of components reported
Aliphatic	
Hydrocarbons	103
Alcohols	70
Aldehydes	55
Ketones	49
Carboxylic acids	24
Esters	56
Ethers	7
Amines	20
Alicyclic	
Hydrocarbons	44
Alcohols	3
Ketones	18
Heterocyclic	
Lactones	38
Furans and derivatives	44
Thiophenes and derivatives	40
Pyrroles and derivatives	20
Pyridines and derivatives	21
Pyrazines and derivatives	54
Oxazol(in)es	13
Thiazol(in)es	29
Other sulphur heterocyclics	13
Benzenoids	80
Sulphur Compounds (not heterocyclic)	72
Miscellaneous	7

From Maarse and Visscher (1989).

case, many of the identified volatiles are acting as 'aroma modifiers' contributing buttery, caramel, roast, burnt, sulphurous, green, fragrant, oily/fatty and nutty qualities. Structure/activity correlations, or at least associations, do exist in the literature for many of these odour qualities. Thirdly, some aroma components do contribute a truly specific 'meaty' odour, and are therefore character impact compounds. Several potent, key and trace meaty compounds are present in natural cooked beef aromas and many remain to be identified. Clearly, the ultimate chemical and sensory resolution of the beef flavour complex relies primarily on defining these particular meaty/beefy compounds and the reactions which generate them.

A search of the literature for individual chemical compounds described as meaty (e.g. MacLeod, 1986; Werkhoff *et al.*, 1989, 1990; Guntert *et al.*, 1990) shows that, of the 880 cooked beef aroma components identified to date (Maarse and Visscher, 1989; Werkhoff *et al.*, 1989, 1990), only 25 have been reported to possess a meaty odour. These are presented in Figure 2.1. The meaty quality of several of these has been under attack by some workers, probably because it is a difficult quality to define with precision. Nevertheless, they are included for completeness. Also, many compounds are meaty only at certain concentrations, usually very low concentrations which are often unspecified. In the discussion below, which considers how various cooked beef aroma components are formed during cooking, the meaty compounds of Figure 2.1 are numbered **1**–**25**. All other compounds are quoted without a numerical code label.

2.4.1 *Effect of heat on sugars and/or amino acids*

To the seasoned flavour chemist, even a cursory glance at Table 2.1 shows that a high proportion of the total number of volatiles identified is derived from reactions which result from the effect of heat on sugars and/or amino acids. Strecker degradations and Maillard reactions are critical contributors, both chemically and sensorially. Furthermore, a host of secondary reactions can occur involving the products of the above reactions (e.g. H_2S, NH_3, thiols and simple carbonyl compounds), thereby increasing quite considerably the variety of compounds which may be formed.

Strecker degradation is depicted in Figure 2.2 (MacLeod and Ames, 1988), and several Strecker aldehydes are well-known cooked beef aroma components, e.g. acetaldehyde (from alanine), methylpropanal (from valine), 2-methylbutanal (from isoleucine), 3-methylbutanal (from leucine), phenylacetaldehyde (from phenylalanine) and methional, which readily decomposes into methanethiol, dimethyl sulphide, dimethyl disulphide and propenal (from methionine).

The Maillard reaction between compounds containing a free amino group (e.g. amino acids, amines, peptides, proteins, ammonia) and

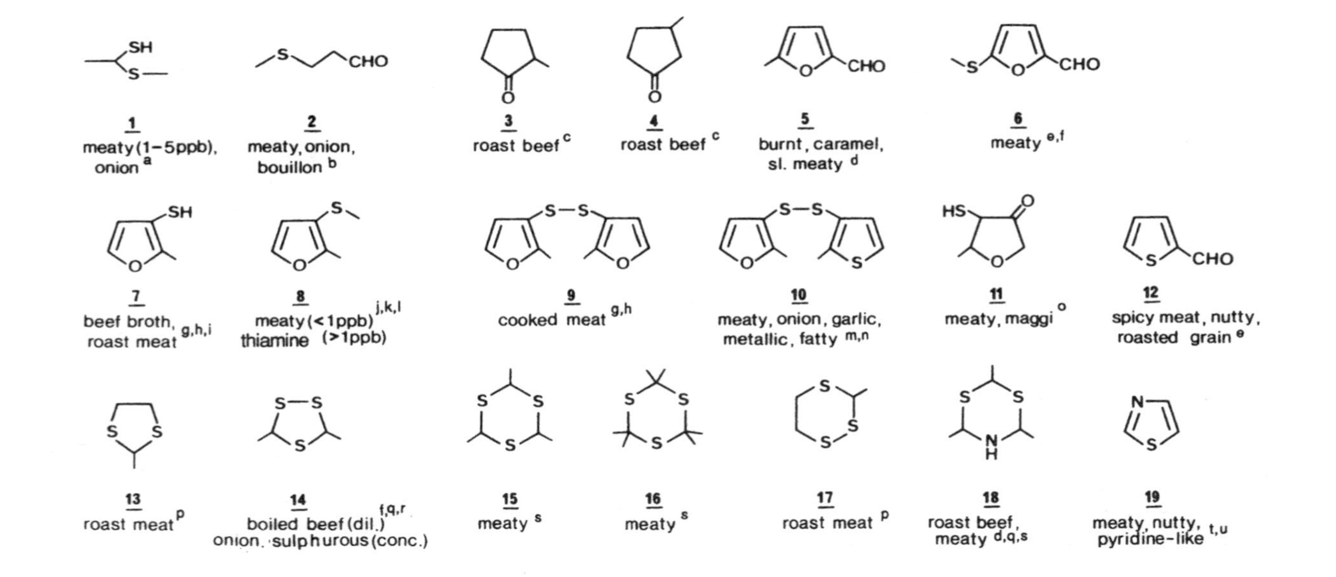

Figure 2.1 Compounds identified from cooked beef aromas (Maarse and Vischer, 1989; Werkhoff *et al.*, 1989, 1990) and reported to possess meaty odour ([b]Arctander, 1969; [t]Baltes, 1979; [a]Brinkman *et al.*, 1972; [h]Evers *et al.*, 1976; [u]Furia and Bellanca, 1975; [l]IFF Inc., 1979; [g]Katz, 1981; [q]Kubota *et al.*, 1980; MacLeod, 1986; [j]MacLeod and Ames, 1986; [x]Mussinan *et al.*, 1976; [c]Nishimura *et al.*, 1980; [d]Ohloff and Flament, 1978; [w]Pittet and Hruza, 1974; [e]Roedel and Kruse, 1980; [r]Self *et al.*, 1963; [f]Shibamoto, 1980; [k]Tressl and Silwar, 1981; [p]Tressl *et al.*, 1983; [o]van den Ouweland and Peer, 1975; [i]van der Linde *et al.*, 1979; [v]Vernin, 1979, [y]1982; [n]Werkoff *et al.*, 1989, [m]1990; [s]Wilson *et al.*, 1974).

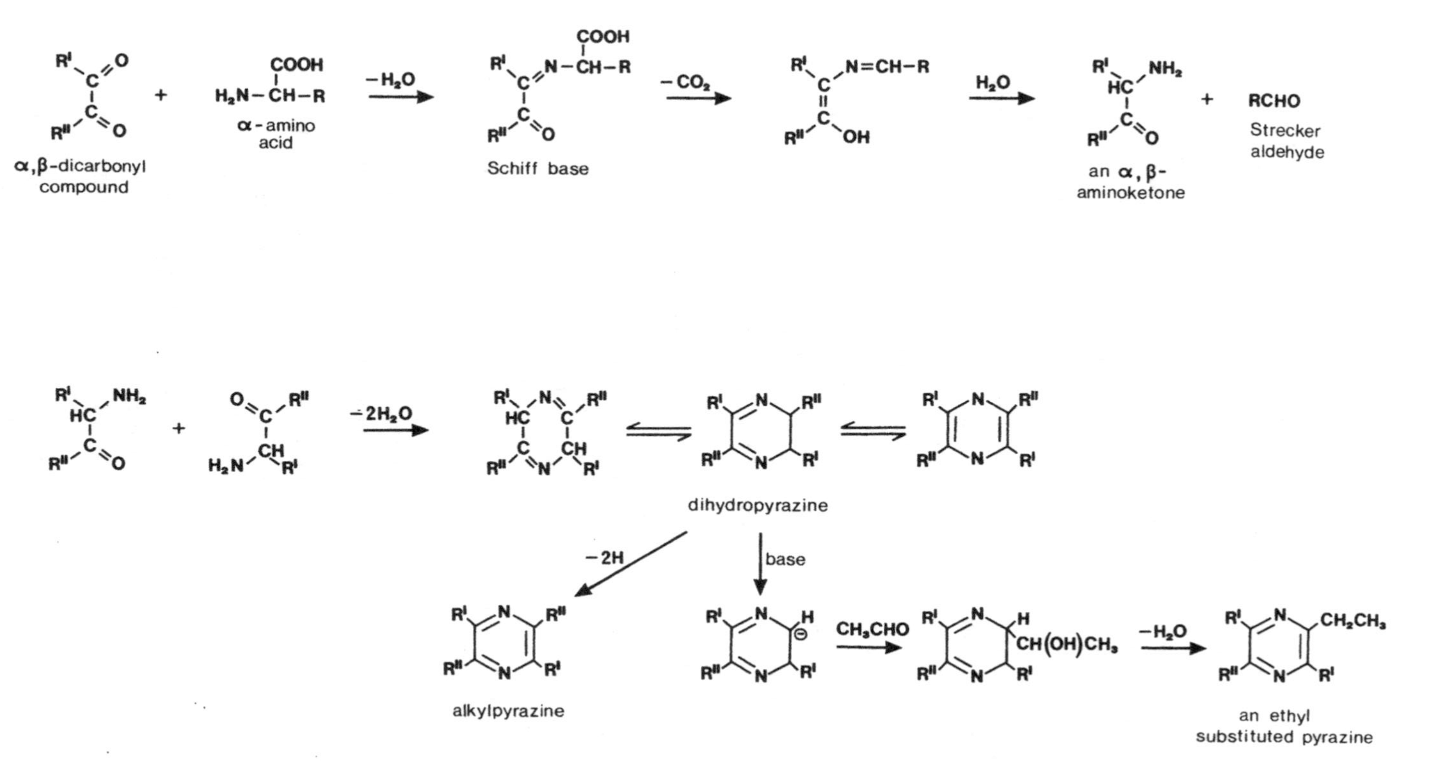

Figure 2.2 The Strecker degradation of α-amino acids and the subsequent formation of alkylpyrazines. (Reprinted with permission from MacLeod, G. and Ames, J.M. (1988) *CRC Crit. Revs Food Sci. and Nutr.*, **27**, 219–400. Copyright CRC Press, Inc., Boca Raton, FL.)

carbonyl compounds (e.g. aldehydes, ketones, reducing sugars) is an open-ended complex reaction consisting of a series of inter-reactions and decompositions and resulting in numerous volatile products. The first stage in the reaction of an α-amino acid with an aldose or ketose is the formation of an Amadori or Heyns compound, respectively. Both are non-volatile but they are heat labile and they decompose thermally. The second gross stage involves rearrangement and subsequent decomposition of the Amadori and Heyns compounds, as exemplified in Figure 2.3 (MacLeod and Ames, 1988). Thus, 2-furaldehyde forms from pentoses, and 5-hydroxymethyl-2-furaldehyde from hexoses. Additionally, a number of dicarbonyl and hydroxycarbonyl fragmentation products can be formed, e.g. glyoxal, glycolaldehyde, glyceraldehyde, pyruvaldehyde, hydroxyacetone, dihydroxyacetone, diacetyl, acetoin and hydroxydiacetyl. These are all highly reactive compounds, and therefore the final global stage of the Maillard reaction can be considered as the decomposition and further inter-reaction of furanoids and aliphatic carbonyl compounds, formed as exemplified in Figure 2.3, with other reactive species present in the system, e.g. H_2S, NH_3, amines and aldehydes. Yayalan has recently proposed that, since the currently accepted mechanisms of 1,2- and 2,3-enolizations of the open chain form of the Amadori compounds, and their subsequent dehydration (as just described), do not adequately account for many products observed in model systems and in foods, alternative mechanisms should be considered, e.g. direct dehydrations from cyclic forms of the Amadori compounds (Yayalan, 1990). By whichever mechanism, the final stage reactions are extremely large in number and varied in nature and cannot be generalized. Examples are as follows.

Furanoids often arise, as shown in Figure 2.3, via the 1,2-enolization pathway. One exception is 2-acetylfuran which probably is derived from a 1-deoxyhexosone intermediate (Tressl *et al.*, 1979). Some other furanoids, e.g. 4-hydroxy-5-methyl-(2*H*)furan-3-one and 2,5-dimethyl-4-hydroxy-(2*H*)furan-3-one, also arise by this 2,3-enolization route. 2-Furaldehyde is an important precursor of other furanoids, and indeed of other heterocyclic compounds too, such as thiophenes and pyrroles. This is because the oxygen of the furan ring, in the presence of H_2S or NH_3, may be substituted by sulphur or nitrogen, forming the corresponding thiophenes or pyrrole derivatives. The formation of such compounds is shown in Figure 2.4 (MacLeod and Ames, 1988).

The most likely pathway for the formation of alkylpyrazines is by self-condensation of the α,β-aminoketones formed during Strecker degradation, as shown in Figure 2.2. While alkylpyrazines are frequently occurring volatiles in many heated foods, the bicyclic pyrazines, namely the alkyl-(5*H*)6,7-dihydrocyclopenta[*b*]pyrazines and the pyrrolo[1,2-*a*]pyrazines, are unique to grilled and roasted beef aromas (Maarse and Visscher, 1989). The former arise from reaction of an alkylhydroxycyclopentenone

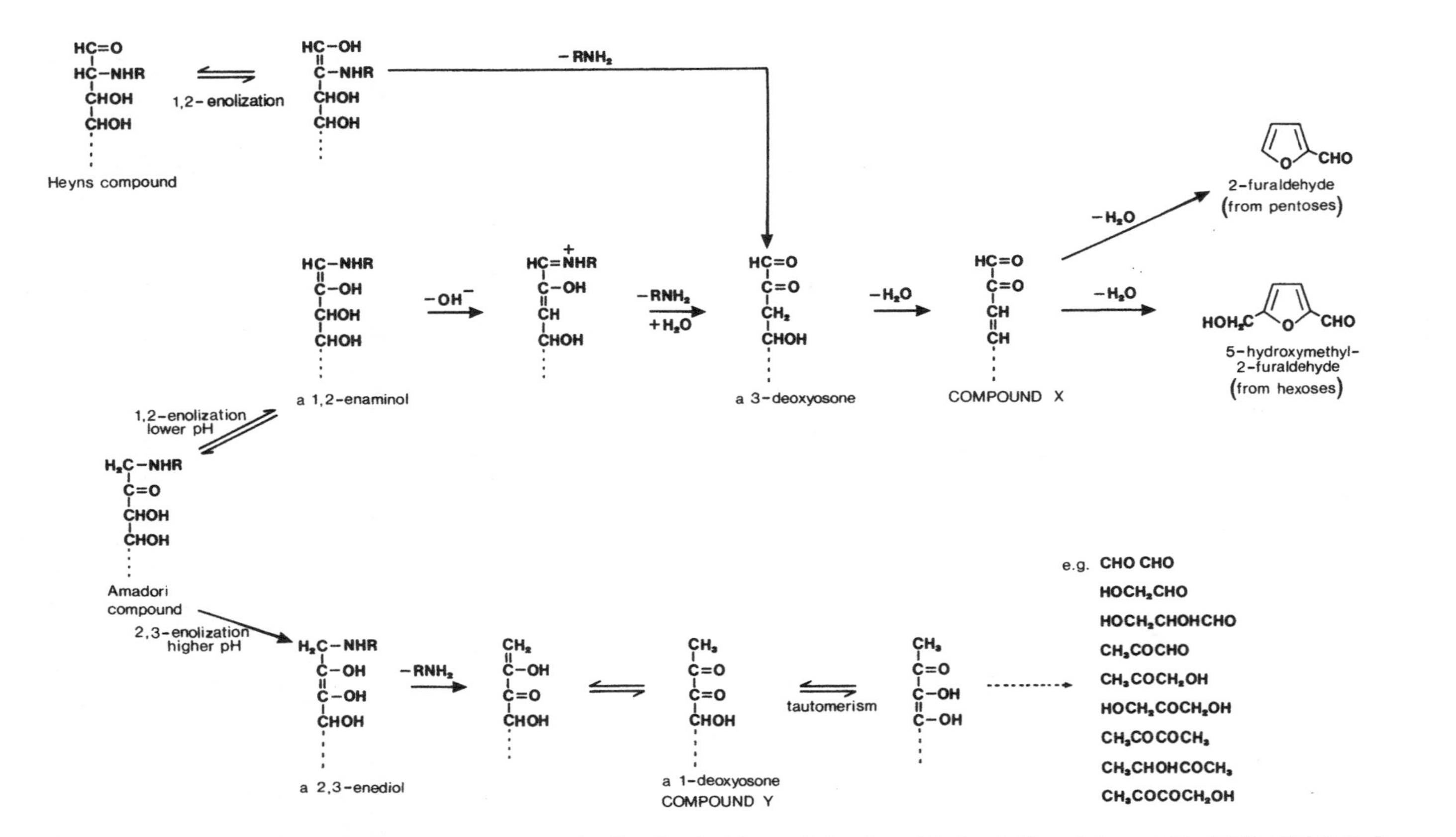

Figure 2.3 Decomposition of Amadori and Heyns compounds. (Reprinted with permission from MacLeod, G. and Ames, J.M. (1988) *CRC Crit. Revs Food Sci. and Nutr.*, **27**, 219–400. Copyright CRC Press, Inc., Boca Raton, FL.)

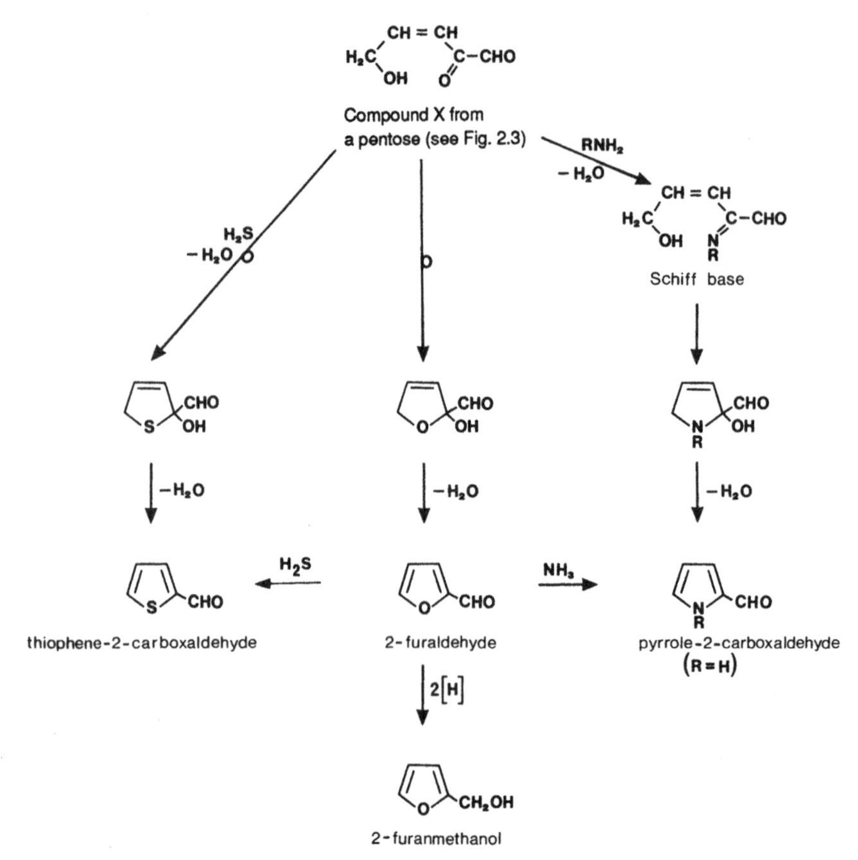

Figure 2.4 Formation of some related furan, thiophene and pyrrole derivatives from Maillard intermediates. (Reprinted with permission from MacLeod, G. and Ames, J.M. (1988) *CRC Crit. Revs Food Sci. and Nutr.*, **27**, 219–400. Copyright CRC Press, Inc., Boca Raton, FL.)

(e.g. cyclotene) with an α,β-dicarbonyl and NH_3 as shown in Figure 2.5 (Flament *et al.*, 1976; MacLeod and Seyyedain-Ardebili, 1981), while the latter derive from condensation of an α,β-aminoketone (from Strecker degradation) with a hydroxy α,β-dicarbonyl compound (formed from reducing sugars) as shown in Figure 2.6 (Flament *et al.*, 1977; MacLeod and Seyyedain-Ardebili, 1981).

Maillard type reactions of cysteine and/or cystine have been extensively studied, and a recent review summarizes the various cysteine-specific Maillard products (Tressl *et al.*, 1989). Reaction products include carbonyl compounds, amines, benzenoids, acids, lactones, thiols, sulphides, furanoids, thiophenoids, thianes, thiolanes, pyrroles, pyridines, pyrazines, thiazoles, thiazolines and thiazolidines. The type of products formed, and their concentrations, are strongly influenced in model systems, by the

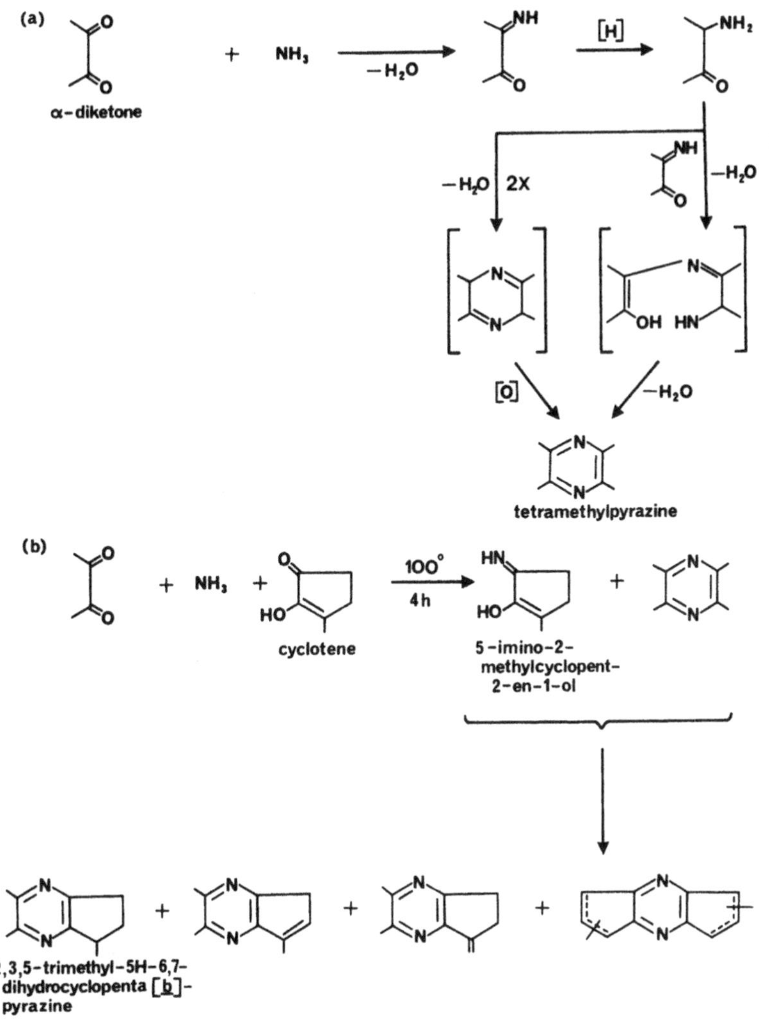

Figure 2.5 Formation of alkyl-(5*H*)6,7-dihydrocyclopenta[*b*]pyrazines by Maillard reaction. (Reprinted with permission from MacLeod, G. and Seyyedain-Ardebili, M. (1981) *CRC Crit. Revs Food Sci. and Nutr.*, **14**, 309–437. Copyright CRC Press, Inc., Boca Raton, FL.)

solvent. For example, reaction of cysteine/dihydroxyacetone in water favours the formation of sulphur products, especially mercaptopropanone and some thiophenes, whereas in triglyceride or glycerine systems, dehydration reactions are favoured producing preferentially various pyrazines and thiazoles (Okumura *et al.*, 1990).

The long list of chemical classes deriving from cysteine/cystine systems, and indeed the identification of several representatives within each class,

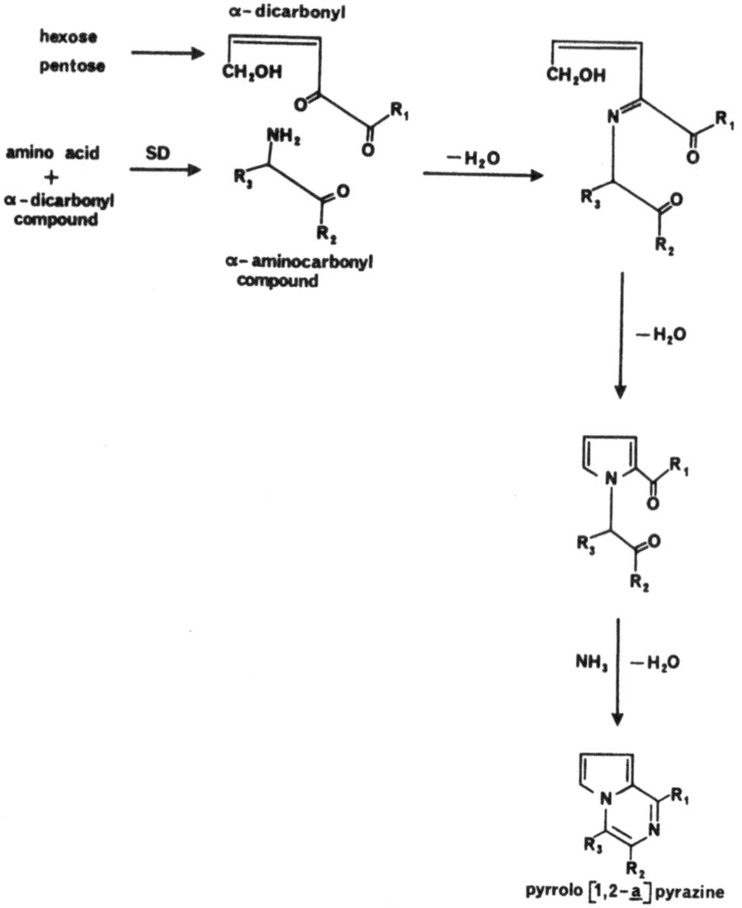

Figure 2.6 Formation of pyrrolo[1,2-*a*]pyrazines by Maillard reaction. (Reprinted with permission from MacLeod, G. and Seyyedain-Ardebili, M. (1981) *CRC Crit. Revs Food Sci. and Nutr.*, **14**, 309–437. Copyright CRC Press, Inc., Boca Raton, FL.)

serves to stress the considerable impact of cysteine (in particular) as a Strecker/Maillard reactant. This significance of cysteine is attributed to the very high reactivity of its initial Strecker degradation products, namely mercaptoacetaldehyde, acetaldehyde, H_2S and NH_3, all of which undergo numerous further reactions. Sheldon *et al.* (1986) recently showed that some Maillard reaction between cysteine and glucose occurs even at room temperature in the dark. At high temperatures approximating roasting conditions (i.e. Shigematsu conditions), the products formed reflect the fact that amino acid pyrolysis (rather than Strecker degradation) is an important reaction (de Rijke *et al.*, 1981), and cysteine is primarily transformed into six products, namely mercaptoacetaldehyde, acetaldehyde, cysteamine, ethane-1,2-dithiol, H_2S and NH_3 (Tressl *et al.*, 1983). These highly reactive compounds trigger a number of aldol and other condensation reactions with sugar degradation products and with each other. For example, the reaction of ethane-1,2-dithiol/acetaldehyde/H_2S gives rise to 2-methyl-1,3-dithiolane (13) and 3-methyl-1,2,4-trithiane (17) (Tressl *et al.*, 1983).

Further examples of Figure 2.1 compounds arising from Maillard type reactions are thiazole (19) from cysteine/pyruvaldehyde (Kato *et al.*, 1973), 2,4,5-trimethyloxazole (24) from cysteine/butanedione (Ho and Hartman, 1982), thiophene-2-carboxaldehyde (12) (Scanlan *et al.*, 1973), 3-methyl-1,2,4-trithiane (17) and thialdine (18) (de Rijke *et al.*, 1981) from cysteine/glucose; 3-methyl-1,2,4-trithiane (17) and 2,4-dimethyl-5-ethylthiazole (21) from cysteine/cystine–ribose (Mulders, 1973), 2-methylfuran-3-thiol (7), 2-methyl-3-(methylthio)furan (8) and 2,4-dimethyl-5-ethylthiazole (21) from cysteine/ribose (Whitfield *et al.*, 1988; Farmer *et al.*, 1989; Mottram and Salter, 1989; Farmer and Mottram, 1990; Mottram and Leseigneur, 1990), methional (2) and 2-methyl-3-(methylthio)furan (8) from methionine/reducing sugar (Tressl *et al.*, 1989) and the 2- and 3-methylcyclopentanones (3, 4) from cyclotene/H_2S/NH_3 reaction (Nishimura *et al.*, 1980).

Frequently the effective sulphur reactant is H_2S as shown, for example, in Figure 2.4, explaining the generation of some thiophenes. Thiazol(in)es derive from the reaction of an α,β-dicarbonyl compound, an aldehyde, H_2S and NH_3 as shown in Figure 2.7 (MacLeod and Ames, 1988; Takken *et al.*, 1976). In the absence of H_2S, parallel reactions yield oxazol(in)es.

On a similar theme, inter-reactions involving acetaldehyde, H_2S and NH_3 explain the formation of several other cyclic sulphur compounds identified in cooked beef aromas, e.g. 3,5-dimethyl-1,2,4-trithiolane (14), trithioacetaldehyde (15) and thialdine (18). These reactions are shown in Figure 2.8 (MacLeod and Seyyedain-Ardebili, 1981; Takken *et al.*, 1976). The thiadiazine is decomposed on storage into thialdine (Kawai *et al.*, 1985). Interestingly, the isomeric trithiolanes (14) are major products (63% of total volatiles) from heated glutathione, whereas H_2S/NH_3 interaction products such as the dithiazine and thiadiazine predominate (70% of total volatiles) from heated cysteine (Zhang *et al.*, 1988). This is explained by the fact that a very much milder degradation occurs with glutathione and it releases H_2S more readily than NH_3 (Zhang *et al.*, 1988). It is generally established that glutathione is the main H_2S precursor in meat during the early stages of cooking, but cysteine takes over this major role on prolonged heating. The formation of another reportedly important beef aroma component is shown in Figure 2.8, i.e. 1-(methylthio)ethanethiol (1) which derives from reaction of acetaldehyde with methanethiol and H_2S.

The reactants required for the reactions just described are liberated on

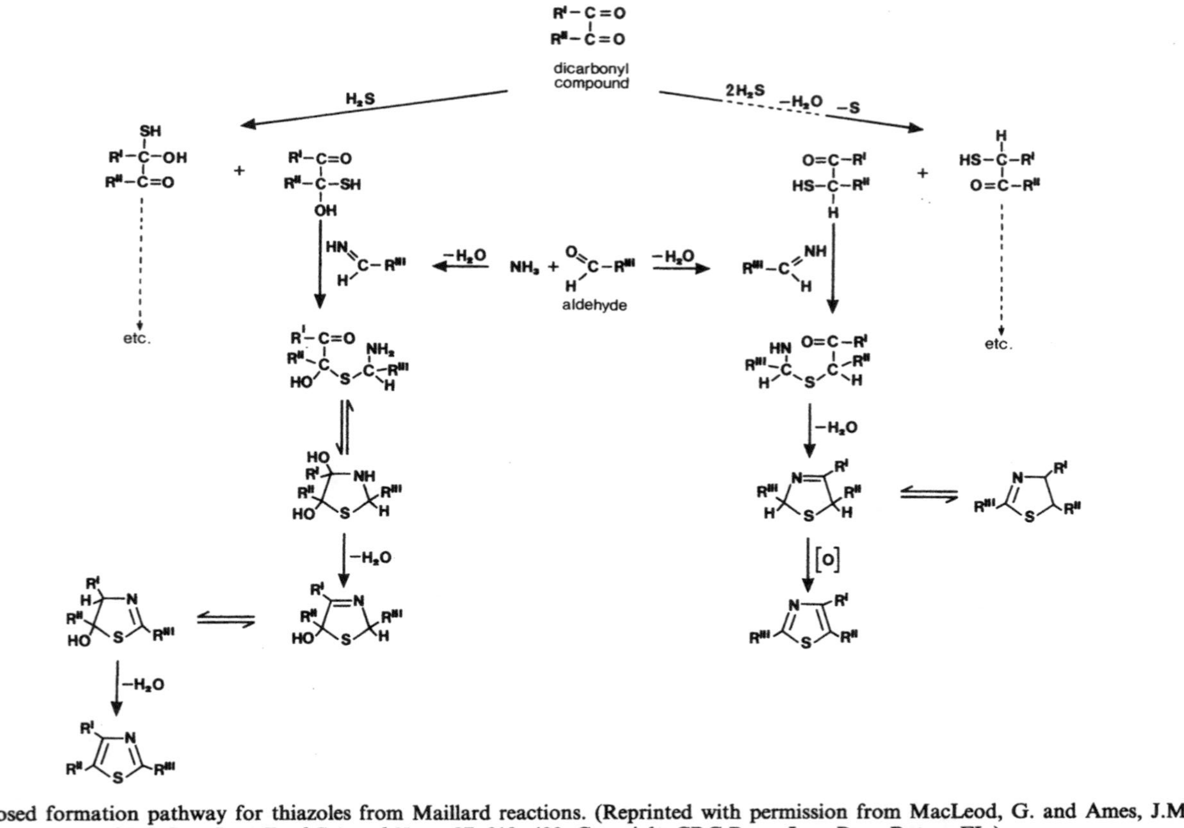

Figure 2.7 Proposed formation pathway for thiazoles from Maillard reactions. (Reprinted with permission from MacLeod, G. and Ames, J.M. (1988) *CRC Crit. Revs Food Sci. and Nutr.*, **27**, 219–400. Copyright CRC Press, Inc., Boca Raton, FL.)

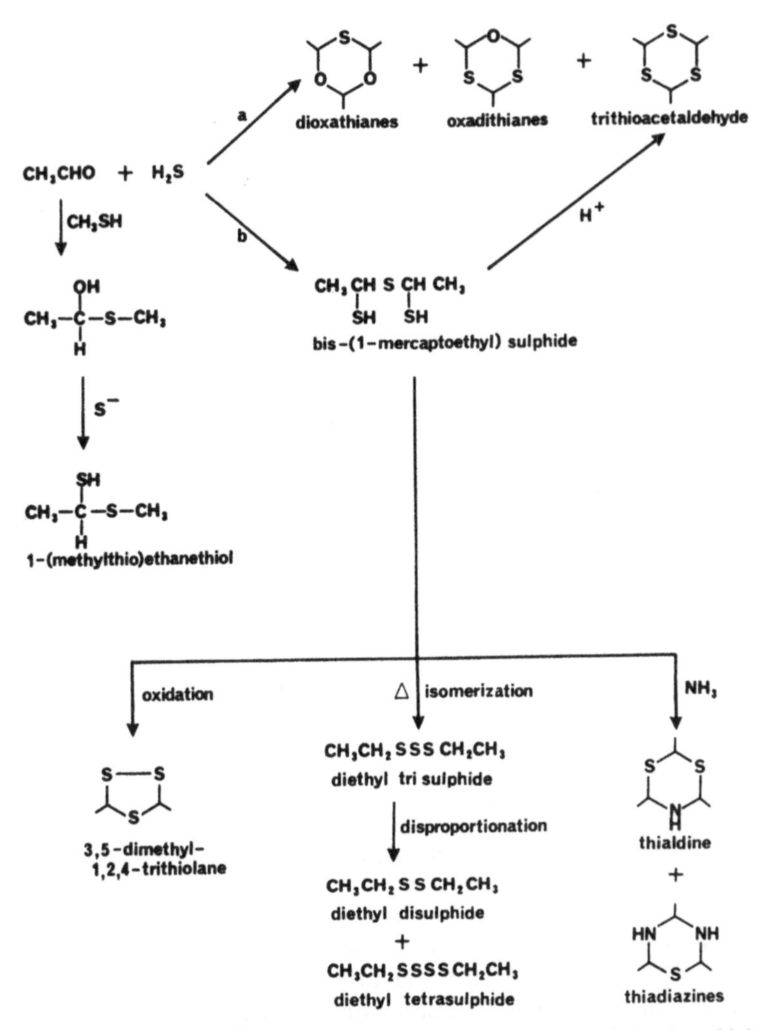

Figure 2.8 Some reactions of aldehydes with hydrogen sulphide/ammonia/methanethiol: a = atmospheric pressure; b = closed vessel + excess H_2S. (Reprinted with permission from MacLeod, G. and Seyyedain-Ardebili, M. (1981) *CRC Crit. Revs Food Sci. and Nutr.*, **14**, 309–437. Copyright CRC Press, Inc., Boca Raton, FL.)

heating either aqueous cysteine (Shu *et al.*, 1985b) or cystine (Shu *et al.*, 1985a) alone. Shu *et al.* (1985b) have reported the thermal degradation products of cysteine in water at 160°C/30 min at different pH values. They identified 26 volatile products of which five have been described as meaty, i.e. 2-methyl-1,3-dithiolane (**13**), 3-methyl-1,2-dithiolane, 3,5-dimethyl-1,2,4-trithiolane (**14**), 3-methyl-1,2,4-trithiane (**17**), thiazole (**19**) and 1,2,3-trithiacyclohept-5-ene. The most vigorous reaction occurred at the

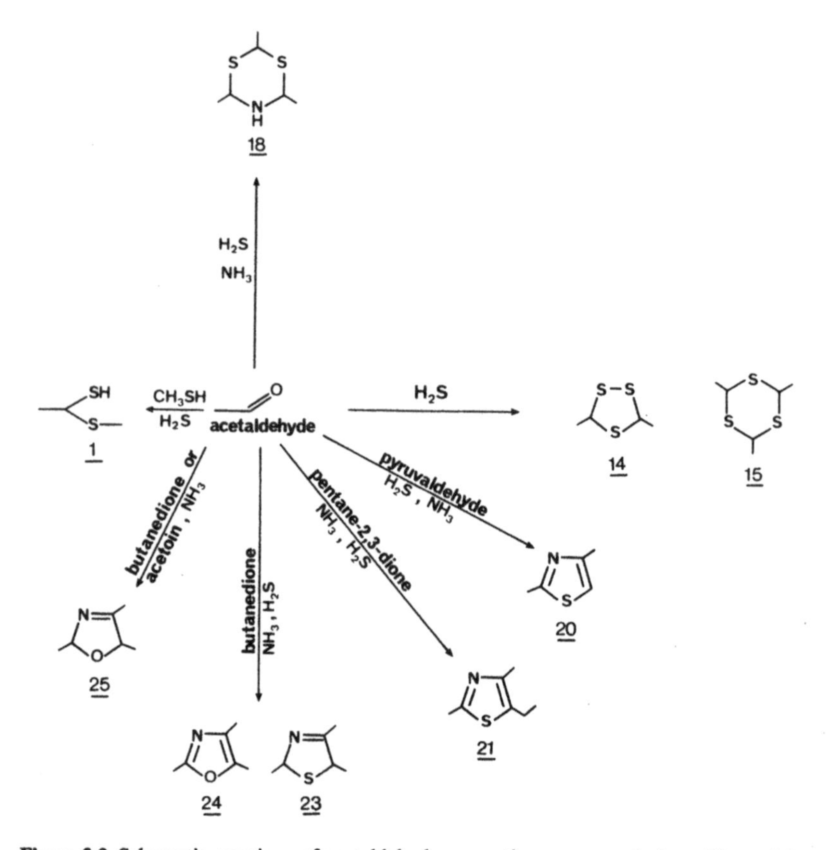

Figure 2.9 Schematic reactions of acetaldehyde-generating compounds from Figure 2.1.

isoelectric point (IEP) of 5.1, which is also the most relevant pH as far as beef is concerned. The volatile mixture was relatively simple and contained 55% acetone. This pH also favoured the formation of the isomeric 3,5-dimethyl-1,2,4-trithiolanes (14). At pH 2.2, the main volatile products were cyclic sulphur compounds containing 5-, 6- and 7-membered rings, together with thiophenes. The major component (34% of the isolate) was the novel meaty compound 1,2,3-trithiacyclohept-5-ene. Only a mild degradation occurred at pH 7.1 and, apart form butanone, only cyclic compounds were identified. Major products were the 3,5-dimethyl-1,2,4-trithiolanes (14), 3-methyl-1,2,4-trithiane (17) and 2-propanoylthiophene (Shu *et al.*, 1985b).

From aqueous cystine treated similarly, 42 compounds were identified; they were mainly thiazoles, aliphatic sulphides, thiolanes and thianes (Shu *et al.*, 1985a). Significantly more thiazoles were present than in the corresponding cysteine system (Shu *et al.*, 1985b). Several compounds reported to be meaty were generated at the pH of meat (pH 5.5). Figure 2.1 compounds produced were 3-methylcyclopentanone (4), 2-methyl-1,3-dithiolane (13), 3,5-dimethyl-1,2,4-trithiolane (14), 3-methyl-1,2,4-trithiane (17), trithioacetaldehyde (15) and thiazole (19) (Shu *et al.*, 1985a). On the whole, cyclic sulphur compounds were generated more readily at pH 2.3, which is the converse of what was reported for the cysteine system where the thiolanes formed preferentially at the IEP rather than below it (Shu *et al.*, 1985b). It was suggested that this was related to the thermal stability of the disulphide bond of cystine at different pH values.

Often an integral part of the degradations just described, but not necessarily so, is a series of reactions of simple low molecular weight compounds, in particular aliphatic aldehydes, alkane-2,3-diones, methanethiol, H_2S and NH_3. Acetaldehyde is involved in many of them and its relevant reactions are schematically represented in Figure 2.9 (Katz, 1981; Mussinan *et al.*, 1976; Takken *et al.*, 1976), explaining the formation of the reportedly meaty 1-(methylthio)ethanethiol (1), 3,5-dimethyl-1,2,4-trithiolane (14), trithioacetaldehyde (15), thialdine (18), 2,4-dimethylthiazole (20), 2,4-dimethyl-5-ethylthiazole (21), 2,4,5-trimethyl-3-thiazoline (23), 2,4,5-trimethyloxazole (24) and 2,4,5-trimethyl-3-oxazoline (25).

2.4.2 *Reactions of hydroxyfuranones*

Two important cooked beef aroma precursors are 4-hydroxy-5-methyl-(2*H*)furan-3-one (HMFone) and 2,5-dimethyl-4-hydroxy-(2*H*)furan-3-one (HDFone). Both have been isolated from beef (Tonsbeek *et al.*, 1968, 1969) but not from any other meats. Natural precursors of HMFone in beef are ribose-5-phosphate and PCA or taurine (to a lesser extent) or both, and the furanone forms on heating at 100°C/2.5 h in a dilute aqueous medium of pH 5.5 (Tonsbeek *et al.*, 1969). PCA itself is readily formed from ammonia and glutamic acid or glutamine on heating (Tonsbeek *et al.*, 1969). The furanone can also be derived from 5′-ribonucleotides via ribose-5-phosphate obtained, for example, by heating at 60°C (Macy *et al.*, 1970) or during autolysis in muscle (Jones, 1969). HMFone is also generated from Maillard reaction of pentoses, e.g. ribose. Precursors of the related HDFone are hexoses, e.g. glucose or fructose under Maillard conditions. HDFone is a relatively unstable compound. Its optimum stability is at pH 4 and its decomposition increases rapidly with temperature (Hirvi *et al.*, 1980). Shu *et al.* thermally degraded HDFone in water at 160°C/30 min at pH 2.2, 5.1 and 7.1 (Shu *et al.*, 1985c). Degradation occurred more readily at the lower pH values. They identified 20 volatile products, most of which were reactive simple aliphatic carbonyls and dicarbonyls; the remainder of the volatile reaction product mixture consisted of alkyl-substituted (2*H*)furan-3-ones, which were favoured at higher pH values. Most of these were shown to derive from pentane-2,3-

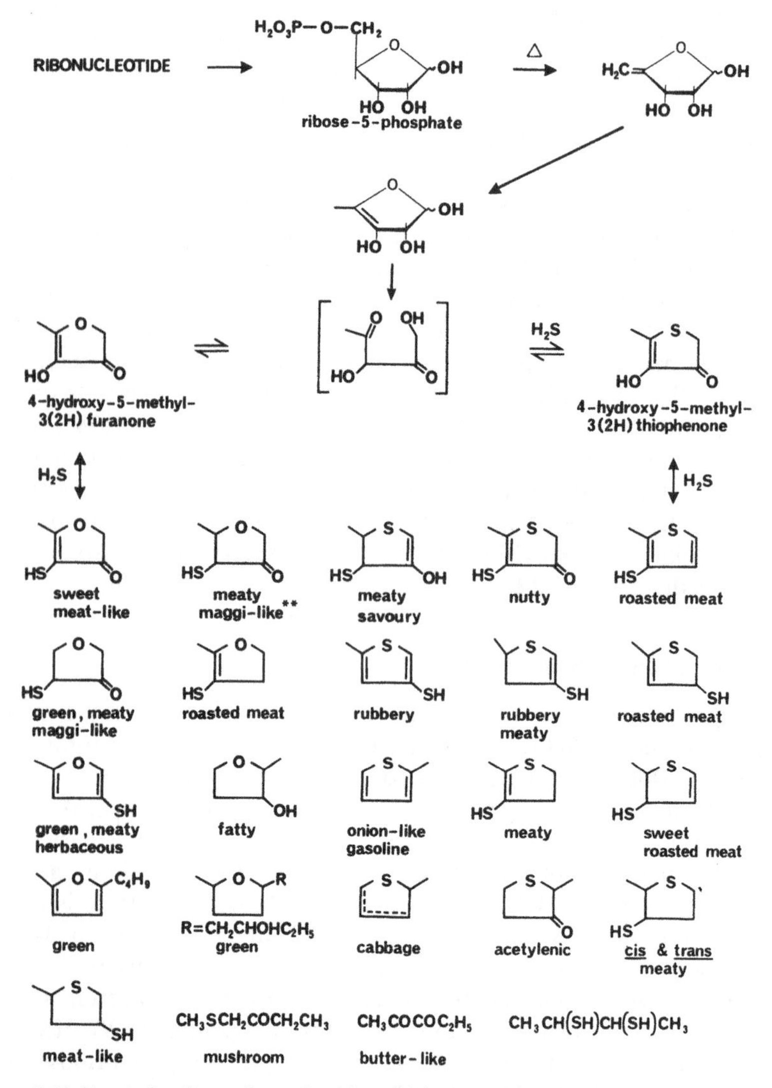

Figure 2.10 Formation from ribonucleotides of 4-hydroxy-5-methyl-(2*H*)furan-3-one, and its reaction with hydrogen sulphide. (Reprinted with permission from MacLeod, G. and Seyyedain-Ardebilli, M. (1981) *CRC Crit. Revs Food Sci. and Nutr.*, **14**, 309–437. Copyright CRC Press, Inc., Boca Raton, FL.)

dione as intermediate, indicating that the HDFone ring opened initially on heating.

The significance of these two hydroxyfuranones in cooked beef aroma is in their reaction with hydrogen sulphide. For HMFone/H_2S, the reaction is summarized in Figure 2.10 (MacLeod and Seyyedain-Ardebili, 1981;

van den Ouweland and Peer, 1975; van den Ouweland *et al.*, 1978). Again, ring opening is indicated and an initial partial substitution of the ring oxygen with sulphur occurs. The reaction product mixture possesses an overall odour of roasted meat and the majority of the reaction products are mercaptofuranoids and mercaptothiophenoids, most of which possess meaty odours (van den Ouweland and Peer, 1975; van den Ouweland *et al.*, 1978). Only one of these (asterisked in Figure 2.10) has so far been identified from natural cooked beef aromas, i.e. 5-methyl-4-mercapto-tetrahydrofuran-3-one (**11**) (Ching, 1979). It is likely however that the remainder are present, since several of them possessed GC retention times and odour port assessments which were very similar to those of trace components detected from a natural beef broth aroma isolate (Ching, 1979). Bodrero *et al.*, using surface response methodology to study the contribution of various volatiles to cooked beef aroma, showed that the HMFone/H_2S reaction product mixture gave the highest predicted score of their entire study (Bodrero, 1981). HDFone reacts with H_2S in a similar manner (van den Ouweland and Peer, 1975; van den Ouweland *et al.*, 1978) producing at least two meaty compounds, i.e. 2,5-dimethylfuran-3-thiol and 2,5-dimethyl-4-hydroxy-(2*H*)thiophene-3-one, neither of which has yet been identified from beef. The chances are that other compounds identified from this reaction are meaty also, but their individual sensory properties have not been reported.

Shu *et al.* have reacted HDFone with cystine at 160°C/30 min in water at pH 2.4 (Shu *et al.*, 1985d). The volatile products changed on storage for two weeks at 4°C and were therefore analysed after holding for two weeks. When compared with the products generated from HDFone (Shu *et al.*, 1985c) and from cystine (Shu *et al.*, 1985a) treated similarly, the major difference was the presence of relatively large concentrations of hexane-2,4-dione (16%) and of three thiophenones (22.5%). One of these was the 2,5-dimethyl-4-hydroxy-(2*H*)thiophene-3-one mentioned above and previously reported from HDFone/H_2S reaction, but the other two were novel, namely 2,5-dimethyl-2-hydroxy-(2*H*)thiophene-3-one (possessing a roasted onion odour) and 2,5-dimethyl-2,4-dihydroxy-(2*H*)thiophene-3-one possessing a meaty, pot roast aroma and taste. Yet again, this meaty compound has eluded identification in natural beef. Optimum conditions for the formation of meaty compounds (including these thiophenones and the isomeric 3,5-dimethyl-1,2,4-trithiolanes (**14**) were 160°C/aqueous medium; 75% water/pH 4.7 (Shu and Ho, 1989). More recently, all three thiophenones have been reported from cysteine/glucose reaction (Tressl *et al.*, 1989).

In a study involving HDFone/cysteine, Shu *et al.* (1986) showed that the two novel thiophenones mentioned above were only trace components. Instead, two novel thiophenes were characterized, namely 3-methyl-2-(2-oxopropyl)thiophene and 2-methyl-3-propanoylthiophene. Their odour properties were not described. Meaty compounds formed preferentially at pH 2.2 and 5.1 rather than at pH 7.1, when various secondary reactions generated a host of different compounds (Shu and Ho, 1986, 1988).

2.4.3 Thermal degradation of thiamine

The thermal degradation of thiamine produces some important compounds (Werkhoff *et al.*, 1989, 1990; Guntert *et al.*, 1990; van der Linde *et al.*, 1979; Hartman *et al.*, 1984; Reineccius and Liardon, 1985; Dwivedi and Arnold, 1973). Van der Linde *et al.* (1979) reported five primary products, the main component being 4-methyl-5-(2-hydroxyethyl)thiazole, which is responsible for the formation of several thiazoles on further degradation. The other sulphur-containing primary product is 5-hydroxy-3-mercaptopentan-2-one, a key intermediate compound giving rise to a number of aliphatic sulphur compounds, furans and thiophenes (van der Linde, 1979). Two of these were also present in the previously discussed reaction properties of HMFone/H_2S (van den Ouweland and Peer, 1975; van den Ouweland *et al.*, 1978), namely 2-methyltetrahydrofuran-3-one, and the meaty 2-methyl-4,5-dihydrofuran-3-thiol. Hartman *et al.* degraded thiamine at 135°C/30 min in water (pH 2.3) and also in propane-1,2-diol (Hartman *et al.*, 1984). Few decomposition products were formed from the diol system but several carbonyls, furanoids, thiophenoids, thiazoles and aliphatic sulphur compounds were isolated from the aqueous reaction. In addition, a novel compound was reported, i.e. 3-methyl-4-oxo-1,2-dithiane, also present (and in enhanced concentration) from a model system of cystine/ascorbic acid/thiamine (Hartman *et al.*, 1984). Reineccius and Liardon studied the volatile products from thiamine at lower temperatures (40°C, 60°C, 90°C) and at pH 5, 7 and 9 respectively (Reineccius and Liardon, 1985). At pH 5 and 7, the meaty 2-methylfuran-3-thiol (7) and bis(2-methyl-3-furyl)disulphide (9), together with various thiophenes were the major products, but at pH 9, the meaty compounds (7) and (9) were not significant and the thiophenes predominated. Apart from compounds (7) and (9), other compounds which are reportedly meaty and which have been identified from thermally degraded thiamine are 3-mercaptopentan-2-one, 2-methyl-4,5-dihydrofuran-3-thiol, 2-methyltetrahydrofuran-3-thiol, 4,5-dimethylthiazole and 2,5-dimethylfuran-3-thiol (van der Linde *et al.*, 1979; Hartman *et al.*, 1984).

In their recent paper reporting the formation of selected sulphur-containing compounds from various meat model systems, Guntert *et al.* (1990) have proposed very comprehensive reaction schemes for the thermal degradation of thiamine. An extract of these schemes, in so far as it relates to the formation of some compounds reported to be meaty, is shown in Figure 2.11 (Guntert *et al.*, 1990). In the following, meaty compounds from thiamine are coded (A)–(U); all other compounds are uncoded. One of the primary degradation products, i.e. 4-methyl-5-(2-hydroxyethyl)thiazole, as mentioned above, is responsible for the subsequent generation of several thiazoles, such as 4,5-dimethylthiazole (**B**) and thiazole itself (**19**). Other primary degradation products are 4-amino-5-(aminomethyl)-2-methylpyrimidine (**a**), formic acid and the key intermediates H_2S and 5-hydroxy-3-mercaptopentan-2-one (**c**). The latter can form seven additional intermediates, only one of which, i.e. 3,5-dimercaptopentan-2-one (**b**), is shown in Figure 2.11. Further reactions of **b** lead to the meaty compounds 3-acetyl-1,2-dithiolane (**A**), 2-methyltetrahydrothiophene-2-thiol (**C**), 2-methyl-4,5-dihydrothiophene-3-thiol (**D**) 2-methylthiophene-3-thiol (**F**) and bis(2-methyl-3-thienyl)disulphide (**G**) as shown, whereas intermediate **c** is responsible for the formation of 2-methyl-4,5-dihydrofuran-3-thiol (**E**), 2-methylfuran-3-thiol (**7**) and bis(2-methyl-3-furyl)disulphide (**9**). Two further meaty compounds, that are not shown in Figure 2.11, were also identified (Guntert *et al.*, 1990). These were mercaptopropanone and tetrahydrothiophene-2-thiol. To date, only compounds **7**, **9** and **19** have been identified from beef; mercaptopropanone has been reported from canned pork (Maarse and Visscher, 1989), and it is interesting to note that pork has a significantly higher thiamine content than beef.

As a sequel to this study just described, Werkhoff *et al.* (1989, 1990) have recently published their excellent work on the identification of interesting volatile sulphur compounds giving meaty notes to a model meat system consisting of cystine/thiamine/glutamate/ascorbic acid/water heated at 120°C/0.5 h at an initial pH of 5.0. They positively characterized 70 sulphur components, of which 19 possessed individual odours described as meaty. These are presented in Figure 2.12. The majority are new to the flavour literature. Four have already been identified from natural cooked beef, namely 2-methylfuran-3-thiol (**7**) (Gasser and Grosch, 1988), bis(2-methyl-3-furyl)disulphide (**9**) (Gasser and Grosch, 1988), 2-methyl-3-[2-methyl-3-thienyl)dithio]furan (**10**) (Werkhoff *et al.*, 1989, 1990) and 1-(methylthio)ethanethiol (**1**) (Maarse and Visscher, 1989). Apart from the latter, the others are all recent identities from beef, the most recent being 2-methyl-3-[(2-methyl-3-thienyl)dithio]furan (**10**), identified by retrospective use of information gained from the above model system in which it was the main component. The disulphides **9**, **10**, **G**, **H** and **J** of Figure 2.12 are derived from oxidation of the corresponding thiols. Even air oxidation of the monomers results in dimerization without effort (Werkhoff *et al.*, 1989, 1990). 1-[(2-Methyl-3-thienyl)thio]ethanethiol (**T**) and 1-[(2-methyl-3-furyl)thio]ethanethiol (**U**) are totally new compounds. They have flavour threshold values in water of $<0.05\ \mu g\ kg^{-1}$ (Werkhoff *et al.*, 1989, 1990). They are derived from the reaction of 2-methylthiophen-3-thiol (**F**) or 2-methylfuran-3-thiol (**7**) with acetaldehyde and H_2S respectively. Formation pathways from thiamine have been proposed for most of the

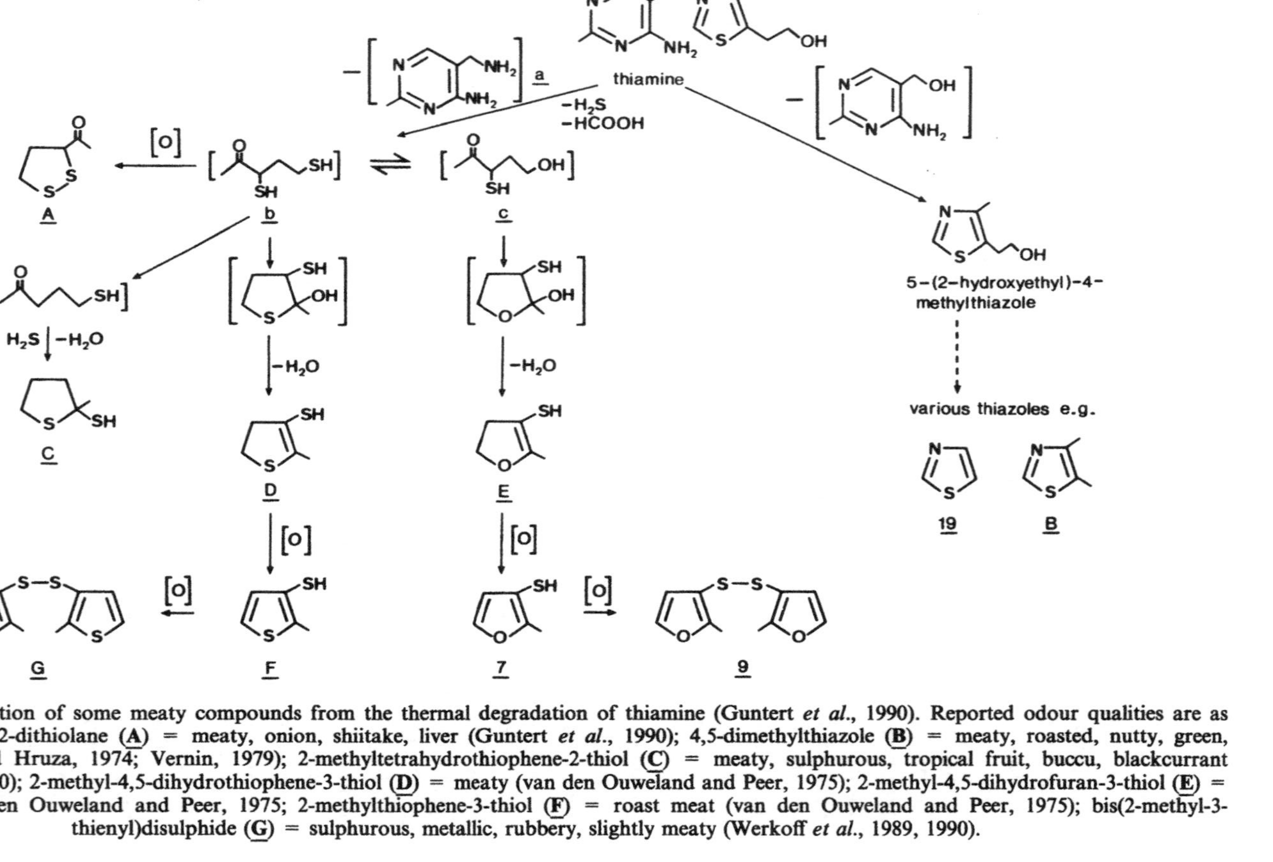

Figure 2.11 Formation of some meaty compounds from the thermal degradation of thiamine (Guntert *et al.*, 1990). Reported odour qualities are as follows: 3-acetyl-1,2-dithiolane (A) = meaty, onion, shiitake, liver (Guntert *et al.*, 1990); 4,5-dimethylthiazole (B) = meaty, roasted, nutty, green, poultry (Pittet and Hruza, 1974; Vernin, 1979); 2-methyltetrahydrothiophene-2-thiol (C) = meaty, sulphurous, tropical fruit, buccu, blackcurrant (Guntert *et al.*, 1990); 2-methyl-4,5-dihydrothiophene-3-thiol (D) = meaty (van den Ouweland and Peer, 1975); 2-methyl-4,5-dihydrofuran-3-thiol (E) = roast meat (van den Ouweland and Peer, 1975; 2-methylthiophene-3-thiol (F) = roast meat (van den Ouweland and Peer, 1975); bis(2-methyl-3-thienyl)disulphide (G) = sulphurous, metallic, rubbery, slightly meaty (Werkoff *et al.*, 1989, 1990).

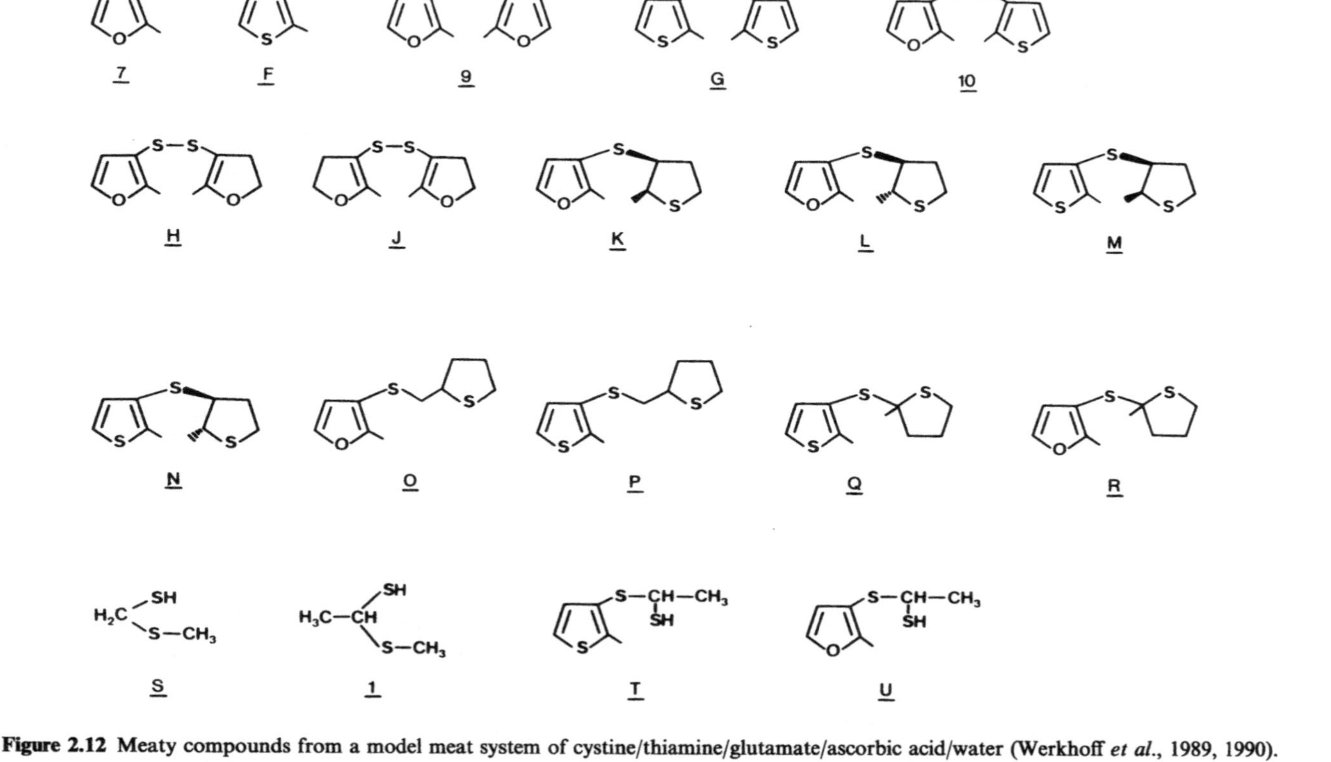

Figure 2.12 Meaty compounds from a model meat system of cystine/thiamine/glutamate/ascorbic acid/water (Werkhoff *et al.*, 1989, 1990).

Figure 2.12 compounds (Werkhoff *et al.*, 1989, 1990), highlighting the important role of thiamine as precursor of meaty compounds.

2.4.4 Lipid oxidation/degradation

Lipid decomposition creates a prolific number of volatiles. The main reactions involved are oxidation and degradation of both unsaturated and saturated fatty acids. The primary oxidation products, the monohydroperoxides, decompose via an intermediate alkoxy radical, forming a range of aroma volatiles. Such decompositions are summarized in Figures 2.13 and 2.14 (MacLeod and Ames, 1988), explaining the generation of many aliphatic hydrocarbons, alcohols, aldehydes, ketones, acids, lactones and 2-alkylfurans. Esters are formed from esterification of alcohols and acids. Other cooked beef aroma components created by lipid degradation are several benzenoids, e.g. benzaldehyde, benzoic acid, alkylbenzenes and naphthalene.

Lipid oxidation starts in raw beef and continues during cooking. Even in lean muscle, the intramuscular lipids are a source of a very large number of volatiles, many of which are present in relatively high concentrations (Bailey and Einig, 1989; Buckholz, 1989). In fact, they create quite a nuisance effect, analytically speaking, because they dominate gas chromatograms of aroma isolates from lean beef and hinder the detection of trace components.

The role of lipids in beef flavour has been considerably clarified by the work of Mottram and his colleagues. They showed that the addition of adipose tissue to lean beef does not give a proportional increase in lipid-derived volatiles, indicating that the intramuscular lipids are the major source of volatile components (Mottram *et al.*, 1982). It has also been shown that species-specific flavour precursors are present in lean beef, although the addition of fat induces fat/lean interactions of some kind which enhance species differences. Intramuscular lipids consist of marbling fat (mainly triglycerides) and structural or membrane lipids, which are largely phospholipids, and which contain a relatively higher content of unsaturated fatty acids. Selected removal of the inter- and intramuscular triglycerides from lean beef causes no significant chemical or sensory aroma differences, but removal of both triglycerides and phospholipids generates marked chemical and sensory differences (Mottram and Edwards, 1983). The aroma is less meaty and more roasted, and it contains less lipid oxidation products but higher concentrations of certain heterocyclic compounds, including some alkylpyrazines (Mottram and Edwards, 1983). This implies that, in beef, lipids (or their degradation products) may inhibit the formation of some heterocycles specifically generated from Maillard reactions (Mottram and Edwards, 1983).

This hypothesis was tested using model systems, e.g. of glycine/ribose

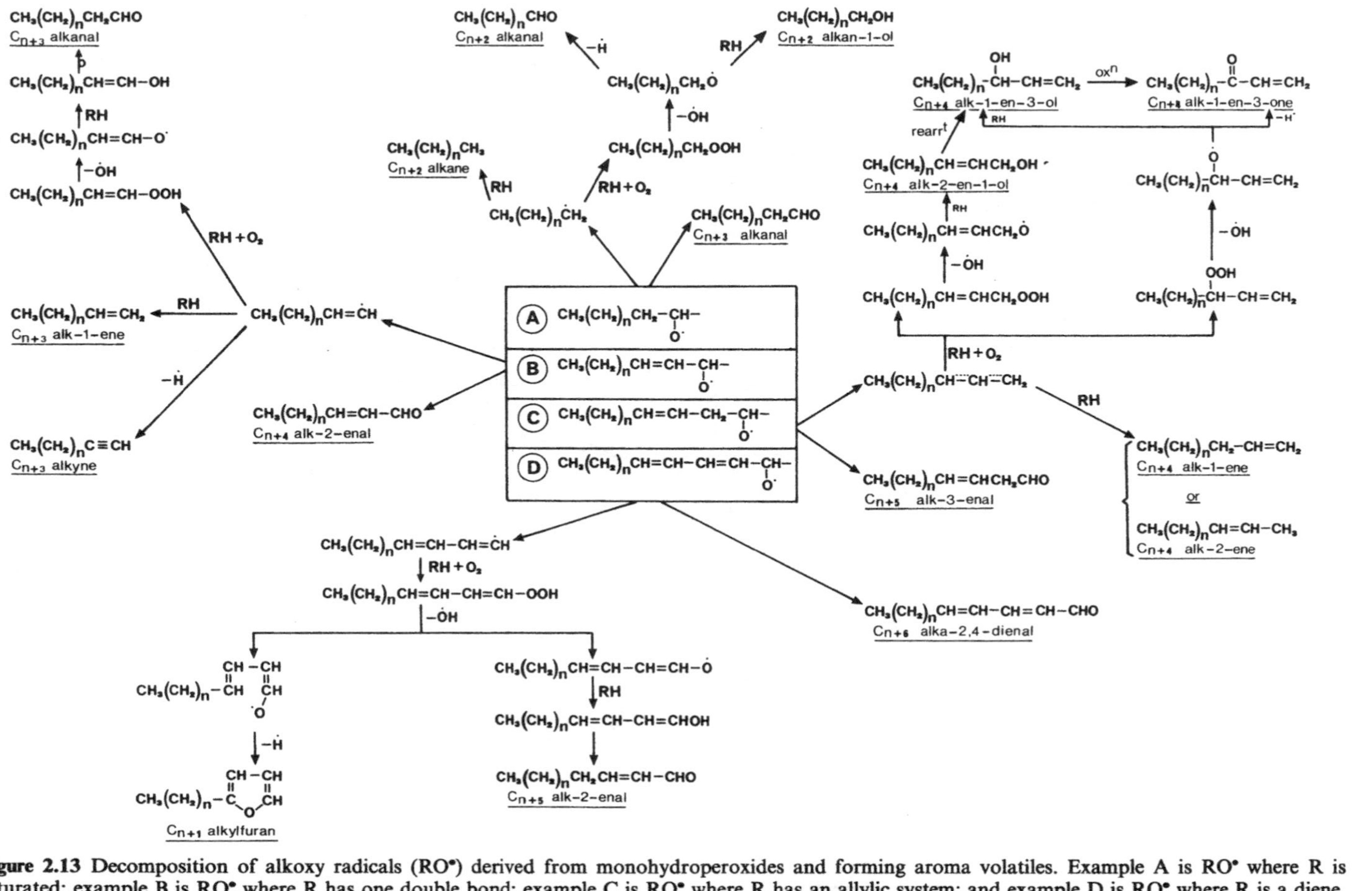

Figure 2.13 Decomposition of alkoxy radicals (RO•) derived from monohydroperoxides and forming aroma volatiles. Example A is RO• where R is saturated; example B is RO• where R has one double bond; example C is RO• where R has an allylic system; and example D is RO• where R is a diene. (Reprinted with permission from MacLeod, G. and Ames, J.M. (1988) *CRC Crit. Revs. Food Sci. and Nutr.*, **27**, 219–400. Copyright CRC Press, Inc., Boca Raton, FL.)

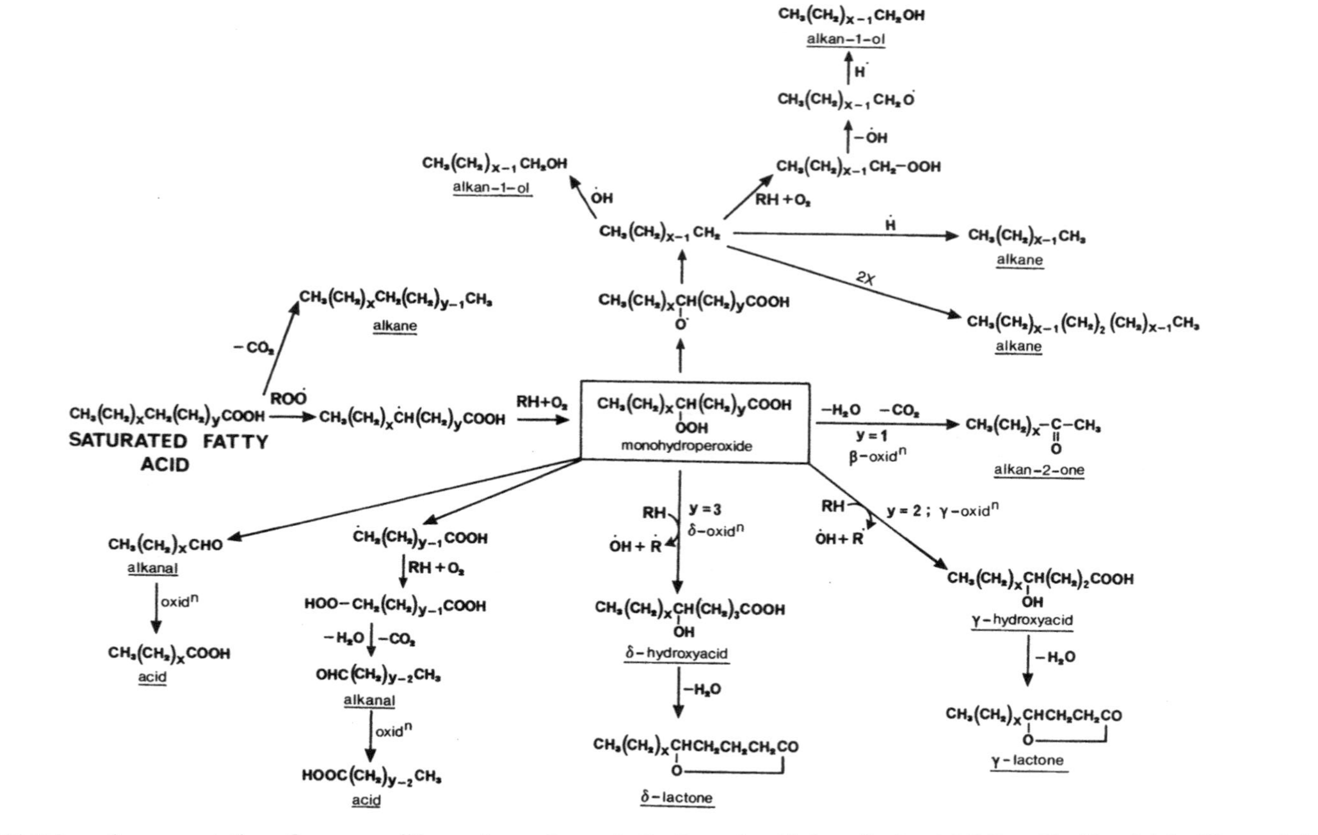

Figure 2.14 Schematic representation of some possible reaction pathways in the thermal oxidation of saturated fatty acids. (Reprinted with permission from MacLeod, G. and Ames, J.M. (1988) *CRC Crit. Revs Food Sci. and Nutr.*, **27**, 219–400. Copyright CRC Press, Inc., Boca Raton, FL.)

and of cysteine/ribose, both in the presence and absence of lecithin (Whitfield *et al.*, 1988; Farmer *et al.*, 1989; Mottram and Salter, 1989; Farmer and Mottram, 1990; Mottram, 1987; Salter *et al.*, 1988). The same overall effect was observed in both cases (i.e. a lower concentration of some heterocyclic compounds generated by the amino acid/ribose reaction. For some compounds in the cysteine system, e.g. 2-methylfuran-3-thiol (7) and thiophene-2-thiol, the decrease was as much as 65% or more (Mottram and Salter, 1989). This is explained by competition by lecithin degradation products for H_2S and NH_3 derived from cysteine, and this was proved to occur by the additional presence in the reaction product mixture of some different heterocycles known to arise from such reactions. The possible interaction of different lipids was also tested. For example, the addition of beef triglyceride had no effect on the aroma of the cysteine/ribose reaction product mixture, but the addition of beef phospholipid caused a significant increase of meaty aroma, a significantly decreased concentration of the normal Maillard reaction heterocycles and a significantly increased concentration of new heterocycles specific to lipid/Maillard interactions (Mottram and Salter, 1989). Examples of such products are 2-pentylpyridine (from deca-2,4-dienal/NH_3 interaction), 2-pentyl-, 2-hexyl-, and 2-hex-1-enyl- thiophenes (from alka-2,4-dienals/H_2S or from 2-alkylfurans/H_2S), heptane-1-thiol and octane-1-thiol (from alkan-1-ols/H_2S) and 4,5-dimethyl-2-pentylthiazole (from hexanal/diacetyl/H_2S/NH_3). The formation of heterocyclic compounds with long-chain alkyl substituents has also been confirmed in model systems of deca-2,4-dienal and dec-2-enal with H_2S (van den Ouweland *et al.*, 1989), of deca-2,4-dienal with cysteine or glutathione (Zhang and Ho, 1989) and from acetol/NH_3/pentanal or hexanal reaction (Chiu *et al.*, 1990).

Therefore, in beef, some lipid is necessary for a full meaty aroma, but the intramuscular tissue phospholipids (which constitute only about 1% of the muscle composition) are sufficient, and the triglycerides are not essential (Mottram and Edwards, 1983). Interactions between water-soluble and phospholipid-derived components occur and could be important.

2.4.5 Selected aroma components of high sensory significance

Grosch and his co-workers have recently developed a screening procedure for highlighting the most important volatile compounds in aroma isolates. In this technique, the volatiles of an extract are 'arranged' in order of their flavour significance according to their 'flavour dilution factors' (FD factors). The FD factor of any compound is proportional to its 'aroma value' which is defined as the ratio of the concentration of the flavour compound to its odour threshold. When the technique was applied to cooked beef, Gasser and Grosch (1988) identified 35 compounds posses-

Table 2.2 Compounds of cooked beef aroma possessing relatively high flavour dilution factors

Flavour dilution factor	Aroma component	Odour quality
512	2-Methylfuran-3-thiol	Meaty, sweet, sulphurous
	Unknown	Roasted
	Methional	Cooked potato
	Non-2(*E*)-enal	Tallowy, fatty
	Deca-2(*E*),4(*E*)-dienal	Fatty, fried potato
	β-Ionone	Violets
	bis(2-methyl-3-furyl) disulphide	Meaty
256	2-Acetyl-1-pyrroline	Roasted, sweet
	Oct-1-en-3-one	Mushroom
	Phenylacetaldehyde	Honey, sweet
128	2-Acetylthiazole	Roasted
	Nona-2(*E*),4(*E*)-dienal	Fatty
64	Octan-2-one	Fruity, musty
	Oct-2(*E*)-enal	Fruity, fatty, tallowy
	Decan-2-one	Musty, fruity
	Unknown	Sulphurous, onion
	Dodecan-2-one	Musty, fruity
32	Hept-2(*E*)-enal	Fatty, tallowy
	Octa-1,5(*Z*)-dien-3-one	Geranium, metallic
	Unknown	Musty, fatty
	Benzothiazole	Pyridine, metallic
16	Hexanal	Green
	Hex-2(*E*)-enal	Green
	Heptan-2-one	Fruity, musty
	Heptanal	Green, fatty, oily
	Dimethyl trisulphide	Cabbage, sulphurous
	Benzylthiol	Sulphurous
	Nona-2(*E*),6(*Z*)-dienal	Cucumber
	Undecan-2-one	Tallowy, fruity
	Tridecan-2-one	Rancid, fruity, tallowy
8	Oct-1-en-3-ol	Mushroom
	Nonan-2-one	Fruity, musty
	Nonanal	Tallowy, green
	Unknown	Tallowy, cardboard
	5-Methylthiophene-2-carboxaldehyde	Mouldy, sulphurous
	Unknown	Sulphurous
	3-Acetyl-2,5-dimethylthiophene	Sulphurous
	A deca-2,4-dienal (not *E*,*E*)	Fatty
4	2-Methyl-3-(methylthio)furan	Sulphurous
	2-Acetylthiophene	Sulphurous, sweet

From Gasser and Grosch (1988).

sing FD factors ⩾4. These are listed, together with the odour qualities apportioned to each, in Table 2.2. Seventeen aroma components (of which 15 were identified) had relatively high FD factors (⩾64), thereby contributing with high aroma values to the flavour of the cooked beef. Only two compounds were described as meaty, i.e. 2-methylfuran-3-thiol (7) and bis(2-methyl-3-furyl)disulphide (9). Both possessed the highest FD factor measured (512), highlighting their considerable odour potency. The odour thresholds of these two compounds were determined as 0.0025–0.01 $ng.l^{-1}$ (air) and 0.0007–0.0028 $ng.l^{-1}$ (air) respectively (Gasser and Grosch, 1990a,b). Gasser and Grosch have also reported a virtually identical study on chicken volatiles (Gasser and Grosch, 1990a) and on commercial meat flavourings (Gasser and Grosch, 1990b). The major differences between beef and chicken were that the sulphur compounds bis(2-methyl-3-furyl)disulphide (9) (meaty) and methional (2) (cooked potato) predominated in beef whereas volatiles from oxidation of unsaturated lipids, in particular deca-2(*E*),4(*E*)-dienal (fatty) and γ-dodecalactone (tallowy, fruity), prevailed in chicken (Gasser and Grosch, 1990a). Interestingly, while the concentration of the important beef aroma compound, bis(2-methyl-3-furyl)disulphide (9), was significantly lower in chicken than in beef, by contrast, its reduction product, 2-methylfuran-3-thiol (7), was present at approximately the same level in both. Gasser and Grosch suggest that the relatively higher level of linoleic acid in chicken captures most of the available gaseous oxygen for peroxidation reactions, thus protecting the thiol against oxidation to the disulphide. This hypothesis links with the previously mentioned findings of Whitfield *et al.* (1988) who showed a very significant decrease in concentration of 2-methylfuran-3-thiol (7) from a cysteine/ribose model system when lecithin was added. They suggested that carbonyl compounds (from the lecithin) were preferentially capturing reactants such as H_2S. Since the combined levels of 2-methylfuran-3-thiol (7) and its disulphide were much lower in chicken than in beef, this difference could well be due to the reactions proposed by Whitfield *et al.* Furthermore, the higher FD factor for methional in beef, compared with chicken volatiles, tends to indicate a partial inhibition of the Strecker degradation of methionine on heating chicken (Gasser and Grosch, 1990a).

A family of furans with sulphur substituents in the 3-position—as exemplified by compounds 7–10 of Figure 2.1 and others also discussed in this paper, but so far unidentified in beef—are likely to be of extreme importance in cooked beef aromas. Many have been synthesized, patented, and described sensorially (Werkhoff *et al.*, 1989, 1990; Evers *et al.*, 1976; Tressl and Silwar, 1981; Gasser and Grosch, 1988). These four compounds have only recently been reported from cooked beef however (Werkhoff *et al.*, 1989, 1990; MacLeod and Ames, 1986; Gasser and Grosch, 1988). Van den Ouweland *et al.* have proposed that an essential structural requirement for meaty aroma is a 5- or 6-membered ring, which is more or less planar and substituted with an enol, thiol and a methyl group adjacent to the thiol, as exemplified by the compounds shown in Scheme 2.1 (van den Ouweland, 1989):

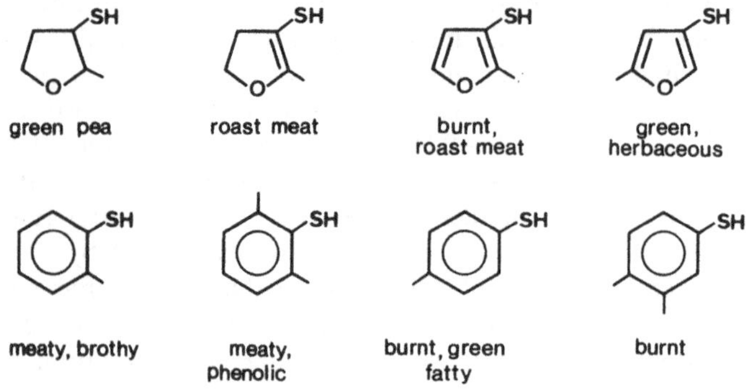

green pea | roast meat | burnt, roast meat | green, herbaceous

meaty, brothy | meaty, phenolic | burnt, green fatty | burnt

Scheme 2.1

A comprehensive investigation into structure/activity correlation in meaty compounds has been reported by Dimoglo *et al.* (1988). They concluded that the generalized molecular fragment shown in Scheme 2.2 accounts for meaty odour:

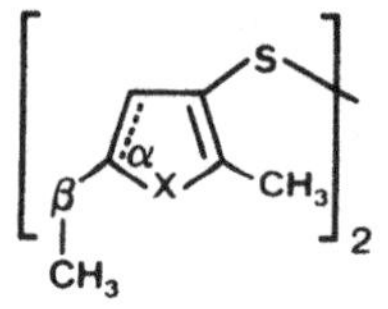

Scheme 2.2

where X = O, S and β = a group coplanar with the α-carbon and the methyl group carbon (e.g. C=O) or an atom different from O, e.g. S. For furan and thiophene derivatives, the methyl group on C_2 plays an important role in meaty odour; additionally, two furan rings and/or ⩾2 sulphur atoms favour meatiness. They also showed that all meaty compounds contain the general structural fragment XH_2 where X = O, N, S and H_2 = two hydrogen atoms belonging, as a rule, to a methyl group. The methyl group must rotate freely. For X = O, N and inter-atomic distance (X–H_1) of 0.262–0.277 nm, compounds are described as 'meaty, meat sauce-like, meat soup odour'; for X = S with X–H_1 distance of 0.278–0.306 nm, compounds possess a 'roast meat' odour (Scheme 2.3).

The extremely low odour thresholds for compounds 9 and 7 have already been mentioned. Related structures are likely to be potent odorants too. It follows therefore that only minute traces of these types of compounds need be present for them to be aroma effective, creating enormous analytical difficulties for their detection.

2-methylfuran-3-thiol
(meat sauce, soup)

2-methylthiophene -3-thiol
(roast meat)

* denotes XH_2 fragments

Scheme 2.3

2.5 Conclusion

Because of these difficulties, many researchers have investigated relevant model systems. Such studies have led to the characterization of important volatiles, mostly sulphur compounds, the presence of which can then be investigated retrospectively in natural cooked beef aromas. The extreme success of this philosophy is now evident, particularly so in the recent literature. Several compounds exhibiting meaty/beefy odours have been fully characterized, both chemically and sensorially, and formation mechanisms from defined precursors have been proposed. A few of these compounds have also been isolated from beef itself, with strong indications that others are inherently present in trace quantities. There is no doubt that the successes of the last decade have evolved from such approaches.

References

Airaudo, C.B., Gayte-Sorbier, A. and Armand, P. (1987). Stability of glutamine and pyroglutamic acid under model system conditions. *J. Food Sci.*, **52**, 1750–1752.

Arctander, S. (1969). *Perfume and Flavor Chemicals*, published by the author, New Jersey.

Bailey, M.E. and Einig, R.G. (1989). Reaction flavors of meat. In *Thermal Generation of Aromas*, eds. T.H. Parliment, R.J. McGorrin and C.-T. Ho. *Amer. Chem. Soc.*, Washington DC, pp. 421–432.

Baltes, W. (1979). Rostaromen. *Deutsch. Lebensm. Rundsch.*, **75**, 2–7.

Bodrero, K.O., Pearson, A.M. and Magee, W.T. (1981). Evaluation of the contribution of flavor volatiles to the aroma of beef by surface response methodology. *J. Food Sci.*, **46**, 26–31.

Brinkman, H.W., Copier, H., de Leuw, J.J.M. and Tjan, S.B. (1972). Components contributing to beef flavor. *J. Agric. Food Chem.*, **20**, 177–181.

Buckholz, L.L. Jr. (1989). Maillard technology as applied to meat and savory flavors. In *Thermal Generation of Aromas*, eds. T.H. Parliment, R.J. McGorrin and C.-T. Ho. *Amer. Chem. Soc.*, Washington DC, pp. 406–420.

Ching, J.C.-Y. (1979). *Volatile Flavor Compounds from Beef and Beef Constituents*. Ph.D. Thesis, University of Missouri.

Chiu, E.-M., Kuo, M.-C., Bruechert, L.J. and Ho, C.-T. (1990) Substitution of pyrazines by aldehydes in model systems. *J. Agric. Food Chem.*, **38**, 58–61.

de Rijke, D., van Dort, J.M. and Boelens, H. (1981). Shigematsu variation of the Maillard reaction. In *Flavour '81*, ed. P. Schreier. de Gruyter, Berlin, pp. 417–431.

Dimoglo, A.S., Gorbachov, M.Y., Bersuker, I.B., Greni, A.I., Vysotskaya, L.E., Stepanova, O.V. and Lukash, E.Y. (1988). Structural and electronic origin of meat odour of organic heteroatomic compounds. *Nahrung*, **32**, 461–473.

Dwivedi, B.K. and Arnold, R.G. (1973). Chemistry of thiamine degradation in food products and model systems. *J. Agric. Food Chem.*, **21**, 54–60.

Evers, W.J., Heinsohn, H.H. Jr., Mayers, B.J. and Sanderson, A. (1976). Furans substituted at the three position with sulfur. In *Phenolic, Sulfur and Nitrogen Compounds in Food Flavors*, eds. G. Charalambous and I. Katz. *Amer. Chem. Soc.*, Washington DC, pp. 184–193.

Farmer, J.J. and Mottram, D.S. (1990). Recent studies on the formation of meat-like aroma compounds. In *Flavour Science and Technology*, eds. Y. Bessiere and A.F. Thomas. Wiley, Chichester, UK., pp. 113–116.

Farmer, L.J., Mottram, D.S. and Whitfield, F.B. (1989). Volatile compounds produced in Maillard reactions involving cysteine, ribose and phospholipid. *J. Sci. Food Agric.*, **49**, 347–368.

Flament, I., Kohler, M. and Aschiero, R. (1976). Sur l'arome de viande de boeuf grille: II. Dihydro-6,7-5H-cyclopenta[*b*]pyrazines, identification et mode de formation. *Helv. Chim. Acta*, **59**, 2308–2313.

Flament, I., Sonnay, P. and Ohloff, G. (1977). Sur l'arome de boeuf grille: III. Pyrrolo[1,2-*a*]pyrazines, identification et synthese. *Helv. Chim. Acta*, **60**, 1872–1883.

Furia, T.W. and Bellanca, N. (1975). *Fenaroli's Handbook of Flavor Ingredients*. 2nd edn, Vol. 1, CRC Press, Ohio.

Galt, A.M. and MacLeod, G. (1983). The application of factor analysis to cooked beef aroma descriptors. *J. Food Sci.*, **48**, 1354–1355.

Gasser, U. and Grosch, W. (1988). Identification of volatile flavour compounds with high aroma values from cooked beef. *Z. Lebensm. Unters. Forsch.*, **186**, 489–494.

Gasser, U. and Grosch, W. (1990a). Primary odorants of chicken broth. *Z. Lebensm. Unters. Forsch.*, **190**, 3–8.

Gasser, U. and Grosch, W. (1990b). Aroma extract dilution analysis of commercial meat flavourings. *Z. Lebensm. Unters. Forsch.*, **190**, 511–515.

Gayte-Sorbier, A., Airaudo, C.B. and Armand, P. (1985). Stability of glutamic acid and monosodium glutamate under model system conditions. *J. Food Sci.*, **50**, 350–360.

Guntert, M., Bruning, J., Emberger, R., Kopsel, M., Kuhn, W., Thielmann, T. and Werkhoff, P. (1990). Identification and formation of some selected sulfur-containing flavor compounds in various meat model systems. *J. Agric. Food Chem.*, **38**, 2027–2041.

Haefeli, R.J. and Glaser, D. (1990). Taste response and thresholds obtained with the primary amino acids in humans. *Lebensm. Wiss. Technol.*, **23**, 523–527.

Hartman, G.J., Carlin, J.T., Scheide, J.D. and Ho, C.-T. (1984). Volatile products formed from the thermal degradation of thiamine at high and low moisture levels. *J. Agric. Food Chem.*, **32**, 1015–1018.

Hirvi, T., Honkanen, E. and Pyysalo, T. (1980). Stability of 2,5-dimethyl-4-hydroxy-3(2H)furanone and 2,5-dimethyl-4-methoxy-3(2H)furanone in aqueous buffer solutions. *Lebensm. Wiss. Technol.*, **13**, 324–325.

Ho, C.-T. and Hartman, G.J. (1982). Formation of oxazolines and oxazoles in Strecker degradation of DL-alanine and L-cysteine with 2,3-butanedione. *J. Agric. Food Chem.*, **30**, 793–794.

IFF Inc. (1979). *Brit. Pat.* 1,543,653.

Jones, N.R. (1969). Meat and fish flavors; significance of ribomononucleotides and other metabolites. *J. Agric. Food Chem.*, **17**, 712–716.

Kato, H. and Nishimura, T. (1987). Taste components and conditioning of beef, pork and chicken. In *Umami: One of the Basic Tastes,* eds. Y. Kawamura and M.R. Kare. Marcel Dekker. New York.

Kato, S., Kurata, T. and Fujimaki, M. (1973). Volatile compounds produced by the reaction of L-cysteine or L-cystine with carbonyl compounds. *Agric. Biol. Chem.*, **37**, 539–544.

Katz, I. (1981). Recent progress in some aspects of meat flavor chemistry. In *Flavor Research: Recent Advances*, eds. R. Teranishi, R.A. Flath and H. Sugisawa. Marcel Dekker, New York, pp. 217–229.

Kawai, T., Irie, M. and Sakaguchi, M. (1985). Degradation of 2,4,6-trialkyltetrahydro-1,3,5-thiadiazines during storage. *J. Agric. Food Chem.*, **33**, 393–397.

Kawamura, Y. (1990). Umami: one of the basic tastes. *Food Technol. Inter. Europe*, 151–155.

Kubota, K., Kobayashi, A. and Yamanishi, T. (1980). Some sulfur-containing compounds in cooked odor concentrate from boiled Antarctic krills. *J. Agric. Food Chem.*, **44**, 2677–2682.

Kuchiba, M., Kaizaki, S., Matoba, T. and Hasegawa, K. (1990). Depressing effect of salts on thermal degradation of inosine 5′-monophosphate and guanosine 5′-monophosphate in aqueous solution. *J. Agric. Food Chem.*, **38**, 593–598.

Kuninaka, A. (1981). Taste and flavour enhancers. In *Flavor Research: Recent Advances*, eds. R. Teranishi and R.A. Flath. Marcel Dekker, New York, pp. 305–353.

Maarse, H. and Visscher, C.A. (1989). *Volatile Compounds in Food--Qualitative and Quantitative Data*. TNO–CIVO, Zeist, The Netherlands.

MacLeod, G. (1986). The scientific and technological basis of meat flavours. In *Developments in Food Flavours*, eds. G.G. Birch and M.G. Lindley. Elsevier Applied Science, London, pp. 191–223.

MacLeod, G. and Ames, J.M. (1986). 2-Methyl-3-(methylthio)furan: a meaty character impact compound identified from cooked beef. *Chem. Ind.*, 175–177.

MacLeod, G. and Ames, J.M. (1988). Soy flavor and its improvement. *CRC Crit. Revs Food Sci. Nutr.*, **27**, 219–400.

MacLeod, G. and Seyyedain-Ardebili, M. (1981). Natural and simulated meat flavors (with particular reference to beef). *CRC Crit. Revs Food Sci. Nutr.*, **14**, 309–437.

Macy, R.L. Jr., Naumann, H.D. and Bailey, M.E. (1970). Water soluble flavor and odor precursors of meat. *J. Food Sci.*, **35**, 81–83.

Mottram, D.S. (1987). The effect of lipid on the formation of volatile heterocyclic compounds in the Maillard reaction. In *Flavour Science and Technology*, eds. M. Martens, G.A. Dalen and H. Russwurm Jr. Wiley, Chichester, UK, pp. 29–34.

Mottram, D.S. and Edwards, R.A. (1983). The role of triglycerides and phospholipids in the aroma of cooked beef. *J. Sci. Food Agric.*, **34**, 517–522.

Mottram, D.S. and Leseigneur, A. (1990). The effect of pH on the formation of aroma volatiles in meat-like Maillard systems. In *Flavour Science and Technology*, eds. Y. Bessiere and A.F. Thomas. Wiley, Chichester, UK, pp. 121–124.

Mottram, D.S. and Salter, L.J. (1989). Flavor formation in meat-related Maillard systems containing phospholipids. In *Thermal Generation of Aromas*, eds. T.H. Parliment, R.J. McGorrin and C.-T. Ho. *Amer. Chem. Soc.*, Washington DC, pp. 442–451.

Mottram, D.S., Edwards, R.A. and Macfie, H.J.H. (1982). A comparison of the flavour volatiles from cooked beef and pork meat systems. *J. Sci. Food Agric.*, **33**, 934–944.

Mulders, E.J. (1973). Volatile components from the non-enzymic browning reaction of the cysteine/cystine–ribose system. *Z. Lebensm. Unters. Forsch.*, **152**, 193–201.

Mussinan, C.J., Wilson, R.A., Katz, I., Hruza, A. and Vock, M.H. (1976). Identification and flavor properties of some 3-oxazolines and 3-thiazolines isolated from cooked beef. In *Phenolic, Sulfur and Nitrogen Compounds in Food Flavors*, eds. G. Charalambous and I. Katz. *Amer. Chem. Soc.*, Washington DC, pp. 133–145.

Nishimura, O., Mihara, S. and Shibamoto, T. (1980). Compounds produced by the reaction of 2-hydroxy-3-methyl-2-cyclopenten-1-one with ammonia and hydrogen sulfide. *J. Agric. Food Chem.*, **28**, 39–43.

Nishimura, T., Rhue, M.R., Okitani, A. and Kato, H. (1988). Components contributing to the improvement of meat taste during storage. *Agric. Biol. Chem.*, **52**, 2323–2330.

Ohloff, G. and Flament, I. (1978). Heterocyclic constituents of meat aroma. *Heterocycles*, **11**, 663–695.

Okumura, J., Yanai, T., Yajima, I. and Hayashi, K. (1990). Volatile products formed from L-cysteine and dihydroxyacetone thermally treated in different solvents. *Agric. Biol. Chem.*, **54**, 1631–1638.

Pittet, A. and Hruza, D. (1974). Comparative study of flavor properties of thiazole derivatives. *J. Agric. Food Chem.*, **22**, 264–269.

Reineccius, G.A. and Liardon, R. (1985). The use of charcoal traps and microwave desorption for the analysis of headspace volatiles above heated thiamine solutions. In *Topics in Flavour Research*, eds. R.G. Berger, S. Nitz and P. Schreier. Eichhorn, Marzling–Hangenhan, pp. 125–136.

Roedel, W. and Kruse, H.P. (1980). Present problems of meat aroma research. *Nahrung*, **24**, 129–139.

Salter, L.J., Mottram, D.S. and Whitfield, F.B. (1988). Volatile compounds produced in Maillard reactions involving glycine, ribose and phospholipid. *J. Sci. Food Agric.*, **46**, 227–242.

Scanlan, R.A., Kayser, S.G., Libbey, L.M. and Morgan, M.E. (1973). Identification of volatile compounds from heated L-cysteine-HCl-D-glucose. *J. Agric. Food Chem.*, **21**, 673–675.

Self, R., Casey, J.C. and Swain, T. (1963). The low boiling volatiles of cooked foods. *Chem. Ind.*, 863–864.

Shaoul, O. and Sporns, P. (1987). Hydrolytic stability at intermediate pHs of the common purine nucleotides in food. *J. Food Sci.*, **52**, 810–812.

Sheldon, S.A., Russell, G.F. and Shibamoto, T. (1986). Photochemical and thermal activation of model Maillard reaction systems. In *Amino–Carbonyl Reactions in Food and Biological Systems*, eds. M. Fujimaki, M. Namiki and H. Kato. Elsevier Science, New York, pp. 145–154.

Shibamoto, T. (1980). Heterocyclic compounds found in cooked meats. *J. Agric. Food Chem.*, **28**, 237–243.

Shu, C.-K. and Ho, C.-T. (1988). Effect of pH on the volatile formation from the reaction between cysteine and 2,5-dimethyl-4-hydroxy-3(2H)furanone. *J. Agric. Food Chem.*, **36**, 801–803.

Shu, C.-K. and Ho, C.-T. (1989). Parameter effects on the thermal reaction of cystine and 2,5-dimethyl-4-hydroxy-3(2H)furanone. In *Thermal Generation of Aromas*, eds. T.H. Parliment, R.J. McGorrin and C.-T. Ho. *Amer. Chem. Soc.*, Washington DC, pp. 229–241.

Shu, C.-K., Hagedorn, M.L., Mookherjee, B.D. and Ho, C.-T. (1985a). Volatile components of the thermal degradation of cystine in water. *J. Agric. Food Chem.*, **33**, 438–442.

Shu, C.-K., Hagedorn, M.L., Mookherjee, B.D. and Ho, C.-T. (1985b). pH effect on the volatile components in the thermal degradation of cysteine. *J. Agric. Food Chem.*, **33**, 442–446.

Shu, C.-K., Hagedorn, M.L., Mookherjee, B.D. and Ho, C.-T. (1985c). Volatile components of the thermal degradation of 2,5-dimethyl-4-hydroxy-3(2H)furanone. *J. Agric. Food Chem.*, **33**, 446–448.

Shu, C.-K., Hagedorn, M.L., Mookherjee, B.D. and Ho, C.-T. (1985d). Two novel 2-hydroxy-3(2H)thiophenones from the reaction between cystine and 2,5-dimethyl-4-hydroxy-3(2H)furanone. *J. Agric. Food Chem.*, **33**, 638–641.

Shu, C.-K., Hagedorn, M.L. and Ho, C.-T. (1986). Two novel thiophenes identified from the reaction between cysteine and 2,5-dimethyl-4-hydroxy-3(2H)furanone. *J. Agric. Food Chem.*, **34**, 344–346.

Takken, H.J., van der Linde, L.M., de Valois, P.J., van Dort, H.M. and Boelens, M. (1976). Reaction products of 2,3-dicarbonyl compounds, aldehydes, hydrogen sulfide and ammonia. In *Phenolic, Sulfur and Nitrogen Compounds in Food Flavors*, eds. G. Charalambous and I. Katz. *Amer. Chem. Soc.*, Washington DC, pp. 114–121.

Tonsbeek, C.H.T., Plancken, A.J. and von der Weerdhof, T. (1968). Components contributing to beef flavor: isolation of 4-hydroxy-5-methyl-3(2H)furanone and its 2,5-dimethyl homolog from beef broth. *J. Agric. Food Chem.*, **16**, 1016–1021.

Tonsbeek, C.H.T., Koenders, E.B., van der Zijden, A.S.M. and Losekoot, J.A. (1969). Components contributing to beef flavor: natural precursors of 4-hydroxy-5-methyl-3(2H)furanone in beef broth. *J. Agric. Food Chem.*, **17**, 397–400.

Tressl, R. and Silwar, R. (1981). Investigation of sulfur-containing components in roasted coffee. *J. Agric. Food Chem.*, **29**, 1078–1082.

Tressl, R., Grunewald, H.-G., Silwar, R. and Bahri, D. (1979). Chemical formation of flavor substances. In *Progress in Flavour Research*, eds. D.G. Land and H.E. Nursten. *Applied Science*, London, pp. 197–213.

Tressl, R., Helak, B., Grunewald, H.-G. and Silwar, R. (1983). Formation of flavour components from proline, hydroxyproline and sulphur-containing amino acids. In *Colloque International sur les Aromes Alimentaires*, eds. J. Adda and H. Richard. *Tech. Doc. Lav.*, Paris, pp. 207–230.

Tressl, R., Helak, B., Martin, N. and Kersten, E. (1989). Formation of amino acid specific Maillard products and their contribution to thermally generated aromas. In *Thermal Generation of Aromas*, eds. T.H. Parliment, R.J. McGorrin and C.-T. Ho. *Amer. Chem. Soc.*, Washington DC, pp. 156–171.

van den Ouweland, G.A.M. and Peer, H.G. (1975). Components contributing to beef flavor. Volatile compounds produced by the reaction of 4-hydroxy-5-methyl-3(2H)furanone and its thio analog with hydrogen sulfide. *J. Agric. Food Chem.*, **23**, 501–505.

van den Ouweland, G.A.M., Olsman, H. and Peer, H.G. (1978). Challenges in meat flavor research. In *Agricultural and Food Chemistry: Past, Present and Future*, ed. R. Teranishi. AVI, Westport, pp. 292–314.

van den Ouweland, G.A.M., Demole, E.P. and Enggist, P. (1989). Process meat flavor development and the Maillard reaction. In *Thermal Generation of Aromas*, eds. T.H. Parliment, R.J. McGorrin and C.-T. Ho. *Amer. Chem. Soc.*, Washington DC, pp. 433–441.

van der Linde, L.M., van Dort, J.M., de Valois, P., Boelens, H. and de Rijke, D. (1979). Volatile components from thermally degraded thiamine. In *Progress in Flavour Research*, eds. D.G. Land and H.E. Nursten. *Applied Science*, London, pp. 219–224.

Vernin, G. (1979). Heterocycles in food aromas. I. Structure and organoleptic properties. *Parfum Cosmetiq. Aromes*, **29**, 77–87.

Vernin, G. (1982). *The Chemistry of Heterocyclic Flavouring and Aroma Compounds*. Ellis Horwood, Chichester, UK.

Werkhoff, P., Emberger, R., Guntert, M. and Kopsel, M. (1989). Isolation and characterisation of volatile sulfur-containing meat flavor components in model systems. In *Thermal Generation of Aromas*, eds. T.H. Parliment, R.J. McGorrin and C.-T. Ho. *Amer. Chem. Soc.*, Washington DC, pp. 460–478.

Werkhoff, P., Bruning, J., Emberger, R., Guntert, M., Kopsel, M., Kuhn, W. and Surburg, H. (1990). Isolation and characterisation of volatile sulfur-containing meat flavor components in model systems. *J. Agric. Food Chem.*, **38**, 777–791.

Whitfield, F.B., Mottram, D.S., Brock, S., Puckey, D.J. and Salter, L.J. (1988). Effect of phospholipid on the formation of volatile heterocyclic compounds in heated aqueous solutions of amino acids and ribose. *J. Sci. Food Agric.*, **42**, 261–272.

Wilson, R.A., Vock, M.H., Katz, I. and Shuster, E.J. (1974). *Brit. Pat.* 1,364,747.

Yayalan, V. (1990). In search of alternative mechanisms for the Maillard reaction. *Trends in Food Sci. and Technol.*, **1**, 20–22.

Zhang, Y. and Ho, C.-T. (1989). Volatile compounds formed from thermal interaction of 2,4-decadienal with cysteine and glutathione. *J. Agric. Food Chem.*, **37**, 1016–1020.

Zhang, Y., Chien, M. and Ho, C.-T. (1988). Comparison of the volatile compounds obtained from thermal degradation of cysteine and glutathione in water. *J. Agric. Food Chem.*, **36**, 992–996.

3 The flavour of pork

C.-T. HO, Y.-C. OH and M. BAE-LEE

3.1 Introduction

Due to its excellent nutritional value and unique sensory properties, meat has always contributed an important part of our diet. There has been much research aimed at understanding the chemistry associated with meat flavour.

However, most publications on meat flavour have been devoted to the flavour of beef. The flavour of pork has not received much attention. According to a recent review by Mottram (1991), there have been at least 70 publications over the past 30 years reporting on volatiles found in beef, whereas pork volatiles have been reported in only 15 publications.

This chapter reviews the few recent publications on pork flavour.

3.2 Role of lipid degradation products in pork flavour

Lipid-derived volatile compounds dominate in the flavour profile of pork cooked at temperatures below 100°C. Table 3.1 lists some of the volatile compounds generated from the degradation of lipid in the volatiles of cooked pork.

During cooking, thermal and oxidative degradation of triglycerides from depot fat and tissue phospholipids occurs simultaneously. In the absence of oxygen, lipids are thermally degraded through dehydration, decarboxylation, hydrolysis, dehydrogenation and carbon–carbon cleavage. Through hydrolysis, free fatty acids are released. Other products formed during thermal degradation include saturated and unsaturated hydrocarbons, β-keto acids, methylketones, lactones and esters (Nawar, 1969).

Volatile compounds in cooked pork are also formed by autoxidation. The accepted mechanism of lipid oxidation involves hydroperoxide formation, followed by production of fragments containing various functional groups. Autoxidation is a free radical chain reaction and is initiated by extraction of a hydrogen atom from the lipid molecule. For lipids with two or more double bonds in a non-conjugated system, extraction of the hydrogen atoms from the methylene groups between the double bonds is stabilized by delocalization of the free radical over five carbons. Reaction of oxygen with the lipid free radical creates peroxy radicals, and the

Table 3.1 Volatile lipid oxidation products identified in cooked pork

Compounds	Sample A (g/kg)	Sample B (relative abundance)	Sample C (ng/100 g)
Aldehydes			
Hexanal	12.66	L	245
Heptanal			221
3-Methylhexanal	0.65		
Octanal			52
Nonanal		L	209
Dodecanal	[tr]		
Tridecanal	0.25		
Tetradecanal	0.40		
Hexadecanal	0.65		
Octadecanal	1.19		
2-Hexenal	[tr]		
2-Heptenal	0.34		
(*E*)-2-octenal	0.99		
2-Nonenal	0.39		
(*Z*)-4-decenal		S	
2-Undecenal	0.39		
2-Dodecenal	0.43		
(*E*)-2-tridecenal		S	
(*Z*)-2-tridecenal		L	
(*E*)-2-tetradecenal		S	
(*Z*)-2-tetradecanal		S	
17-Octadecenal	[tr]		
16-Octadecenal	8.34		
15-Octadecenal	0.70		
9-Octadecenal	0.81		
2,4-Nonadienal	[tr]		
(*E,E*)-2,4-decadienal	0.69	S	
(*E,Z*)-2,4-decadienal	0.41	S	
2,4-Undecadienal	[tr]	S	
Ketones			
2-Heptanone	0.20		
2-Tetradecanone		S	
2-Hexadecanone		S	
2,3-Octanedione	0.88		
Alcohols			
1-Pentanol			6
1-Hexanol			47
1-Heptanol		M	30
1-Octanol			35
(*E*)-2-heptenol		S	
1-Octen-3-ol		L	1003
1-Nonanol		S	
1-Nonen-3-ol	0.75		
Alkylfurans			
2-Pentylfuran		M	30

Sample A: Ground pork (250–450 g) was placed in a 2 l beaker. Distilled water was added to obtain a meat-to-water ratio of 4:1 (w/w), and the contents were heated at 85°C until the meat slurry reached a constant temperature of 73°C. The slurry was held at that temperature for 10 min (Ramarathnam *et al.*, 1991).
Sample B: 2 kg of pork and 500 ml of water were heated to boiling then refluxed for 7 h (Chou and Wu, 1983).
Sample C: 100 g of minced meat was cooked in Synthene bags for 25 min in a boiling water bath (Mottram, 1985).
L = large; S = small; M = medium; tr = trace.

reaction is then propagated by formation and decomposition of hydroperoxides.

Hydroperoxides formed during lipid oxidation are typically cleaved to give a lipid oxy-radical and a hydroxy radical.

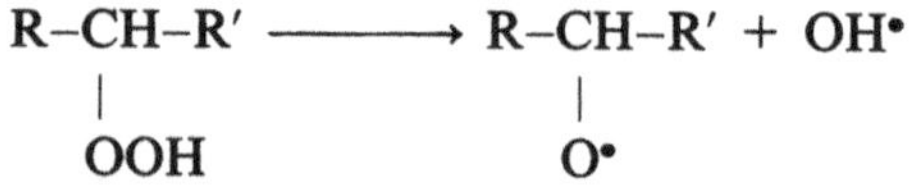

$$\underset{\displaystyle\text{OOH}}{\text{R–}\overset{}{\underset{|}{\text{CH}}}\text{–R}'} \longrightarrow \underset{\displaystyle\text{O}^{\bullet}}{\text{R–}\underset{|}{\text{CH}}\text{–R}'} + \text{OH}^{\bullet}$$

Frankel (1982) illustrated the possible reactions of the resulting lipid oxy-radical as shown below.

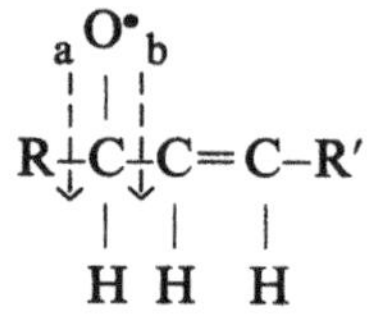

$$\begin{array}{c} {}_{a}\text{O}^{\bullet}{}_{b} \\ \text{R}\downarrow\text{C}\downarrow\text{C}=\text{C}-\text{R}' \\ \text{H}\ \ \text{H}\ \ \ \text{H} \end{array}$$

Of the volatiles produced by these reactions, aldehydes are the most significant flavour compounds. Aldehydes can be produced by scission of the lipid molecules on either side of the radical. The products formed by these scission reactions depend on the fatty acids present, the hydroperoxide isomers formed, and the stability of the decomposition products. Temperature, time and degree of autoxidation are variables which affect thermal oxidation (Frankel, 1982).

As shown in Table 3.1, aldehydes are the major components identified in the volatiles of cooked pork. Octanal, nonanal and 2-undecenal are oxidation products of oleic acid, and hexanal, 2-nonenal and 2,4-decadienal are major volatile oxidation products of linoleic acid. Oleic acid and linoleic acid are the two most abundant unsaturated fatty acids of pork (Schliemann *et al.*, 1987). The unusually high concentration of 1-octen-3-ol reported by Mottram (1985) and Chou and Wu (1983) in cooked pork may be derived from the 12-hydroperoxide of arachidonic acid. Figure 3.1 shows the possible mechanism for its formation. 1-Octen-3-ol has been known to be the character-impact compound of cooked mushroom, however, the precursor of 1-octen-3-ol in mushroom was identified to be linoleic acid with the unusual 10-hydroperoxide as the intermediate (Whitefield and Last, 1991).

2-Pentylfuran identified in cooked pork is a well-known autoxidation product of linoleic acid and has been known as one of the compounds responsible for the reversion flavour of soybean oil (Ho *et al.*, 1978). Figure 3.2 shows the probable mechanism for its formation. The conjugated diene radical generated from the cleavage of the 9-hydroxy radical of linoleic acid may react with oxygen to produce vinyl hydroperoxide. The vinyl hydroperoxide will then undergo cyclization via the alkoxy radical to yield 2-pentylfuran (Frankel, 1982).

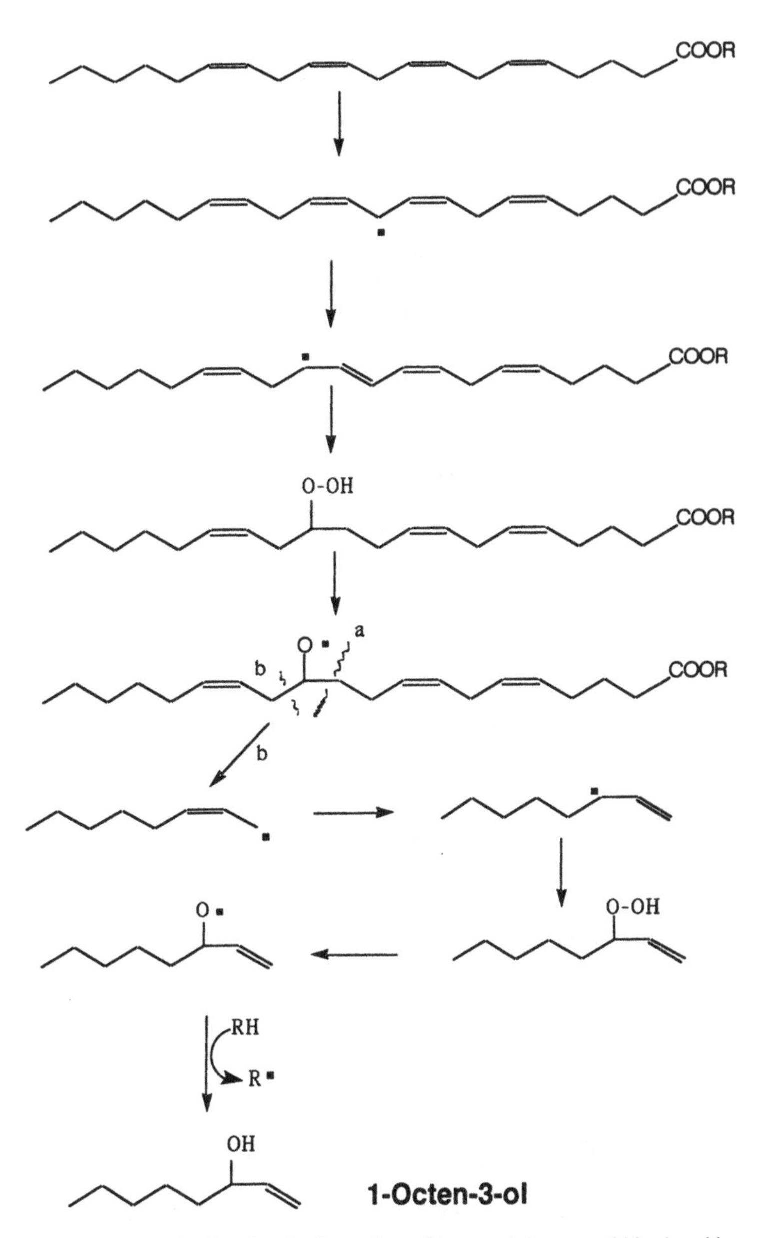

Figure 3.1 Mechanism for the formation of 1-octenol from arachidonic acid.

The concentration of lean and fat tissues in the flavour of cooked meats has been the subject of a number of studies. Hornstein and Crowe (1960, 1963) found that aqueous extracts of beef, pork and lamb had similar aromas when heated, but fats yielded the species-characteristic aromas.

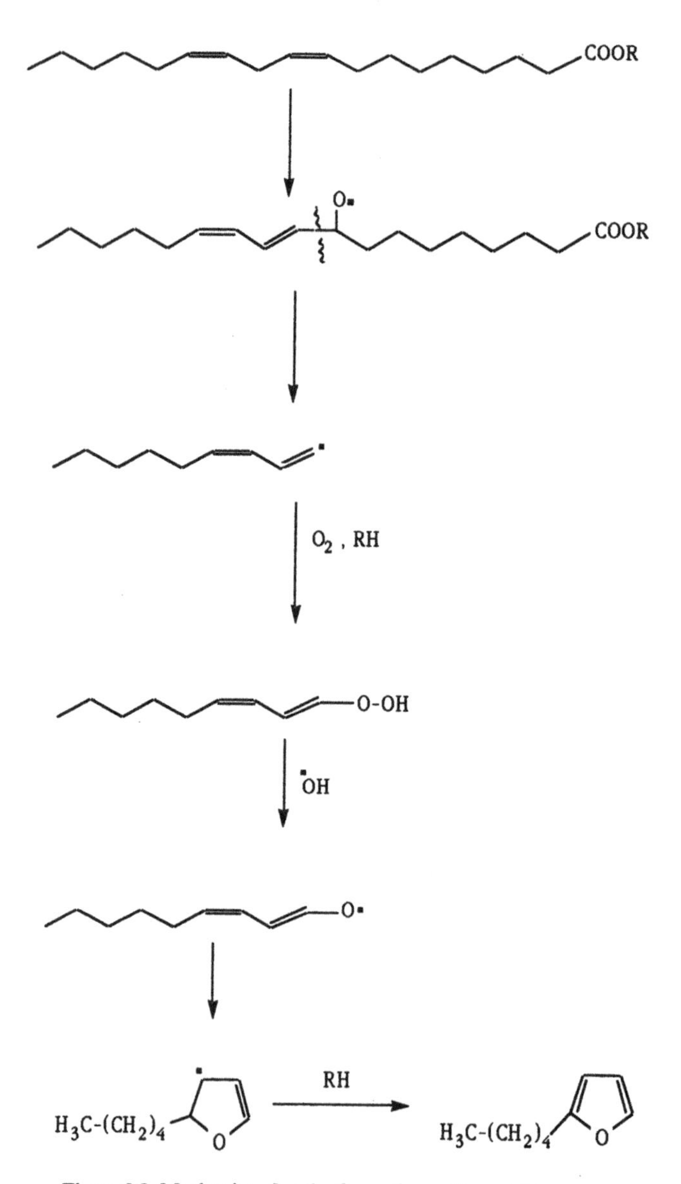

Figure 3.2 Mechanism for the formation of 2-pentylfuran.

However, fat tissue contains all the normal components of cells, including proteins, amino acids, salts and sugars, as well as lipids. Although species-characteristic aromas have been found on heating fatty tissues, lipids alone are not responsible (Mottram *et al.*, 1982).

In one study, Mottram (1979) used triangle taste-tests to differentiate minced pork and beef meat-cakes cooked with and without added fat. The addition of 10% subcutaneous fat (beef and pork) enabled the panel to distinguish the lean meats more easily, no matter which fat was used.

Mottram *et al.* (1982) also reported a comparison of the flavour volatiles from cooked beef and pork meat systems. The addition of minced subcutaneous adipose tissue to the meat-cakes increased the lipid content by up to nine-fold. The addition of pork fat to either beef or pork lean resulted in a substantial increase in hexanal but only small changes in most other volatiles. The general lack of relationship between volatiles and adipose fat level suggested that the triglycerides of the adipose tissues may not be the major source of the volatiles and that the intramuscular triglycerides and phospholipids may be more important.

3.3 Heterocyclic constituents of pork aroma

The large numbers of heterocyclic compounds reported in the aroma volatiles are associated with roasted, grilled or pressure-cooked meat rather than boiled meat where the temperature does not exceed 100°C (Mottram, 1985). In pork liver pressure-cooked at 163°C, over 70% of the total volatiles were furans and pyrazines (Mussinan and Walradt, 1974), while in boiled pork, the volatiles were dominated by aliphatic aldehydes and alcohols with only very small quantities of heterocyclic compounds. In an excellent report by Mottram (1985), the effects of cooking conditions on the formation of volatile heterocyclic compounds in pork were compared. Table 3.2 shows the volatile heterocyclic compounds in well-done grilled, roasted and boiled pork. Well-done grilled pork contained 66 heterocyclic compounds including pyrazines, thiazoles, thiophenes, furans and pyrroles. Pork cooked by less severe roasting or boiling contained considerably fewer heterocyclic compounds.

The alkylpyrazines accounted for almost 80% of the total headspace volatiles of well-done grilled pork (Mottram, 1985). Alkylpyrazines generally produce a roasted nut-like sensory impression (Ohloff and Flament, 1978). The contribution of these pyrazines to the flavour of cooked pork will depend upon their odour threshold values as well as their concentrations and odour characters. Methyl- and dimethylpyrazines have relatively high odour thresholds but much lower values have been observed for pyrazines with increasing substitutions (Guadangi *et al.*, 1972); for example 2-ethyl-3,6-dimethylpyrazine has a threshold of 0.4 parts in 10^9 parts water. It has been well-accepted that alkylpyrazines are formed from a reaction of the degraded nitrogenous substances, NH_3, RNH_2 from proteins, peptides, amino acids, and phospholipids, and the α-dicarbonyl compounds in food (Shibamoto, 1980).

Table 3.2 Major volatile heterocyclic components of pork cooked under different conditions

Compounds	Grilled (well-done)	Roast (ng/100 g)	Boiled
Pyrazines			
Methylpyrazine	91	3	1
2,5-Dimethylpyrazine	2448	4	tr
2,6-Dimethylpyrazine	1913	6	3
2,3-Dimethylpyrazine	220	2	–
2-Ethyl-6-methylpyrazine	545	–	–
2-Ethyl-5-methylpyrazine	302	–	–
2-Ethyl-3-methylpyrazine	28	–	–
Trimethylpyrazine	1952	–	–
2-Ethyl-5,6-dimethylpyrazine	30	–	–
2-Ethyl-3,6-dimethylpyrazine	1155	–	–
2,5-Diethylpyrazine	10	–	–
2-Ethyl-3,5-dimethylpyrazine	370	–	–
2-Methyl-5-propylpyrazine	10	–	–
Tetramethylpyrazine	70	–	–
5,6-Diethyl-2-methylpyrazine	45	–	–
3,5-Diethyl-2-methylpyrazine	80	–	–
Dimethylpropylpyrazine	2	–	–
3,6-Diethyl-2-methylpyrazine	100	–	–
Butyldimethylpyrazine	10	–	–
Triethylpyrazine	10	–	–
Acetylpyrazine	7	–	–
Pentylmethylpyrazine	15	–	–
5-Methyl-6,7-dihydro-5*H*-cyclopentapyrazine	5	–	–
Pentyldimethylpyrazine	40	–	–
2,5-Dimethyl-6,7-dihydro-5*H*-cyclopentapyrazine	11	–	–
Acetylmethylpyrazine	5	–	–
3,5-Dimethyl-6,7-dihydro-5*H*-cyclopentapyrazine	2	–	–
Pyridines			
2-Methylpyridine	17	–	–
2,6-Dimethylpyridine	8	–	–
2-Ethylpyridine	10	–	–
4-Methylpyridine	8	–	–
2,5-Dimethylpyridine	10	–	–
2,4-Dimethylpyridine	5	–	–
2,3-Dimethylpyridine	8	–	–
4-Ethylpyridine	40	–	–
2-Pentylpyridine	80	–	–
2-Acetylpyridine	5	–	–
Thiazoles			
4-Methylthiazole	1	–	–
2,4-Dimethylthiazole	3	–	–
2,5-Dimethylthiazole	3	–	–
2-Ethyl-4-methylthiazole	tr	–	–
4-Ethyl-2-methylthiazole	tr	–	–
4,5-Dimethylthiazole	10	–	–
Trimethylthiazole	8	–	–
5-Ethyl-2-methylthiazole	1	–	–
4-Ethyl-2,5-dimethylthiazole	1	–	–
4-Ethyl-5-methylthiazole	1	–	–
5-Ethyl-4-methylthiazole	5	–	–
5-Ethyl-2,4-dimethylthiazole	5	–	–

Table 3.2 Continued

Compounds	Grilled (well-done)	Roast (ng/100 g)	Boiled
2,5-Diethyl-4-methylthiazole	tr	–	–
4-Methyl-5-vinylthiazole	tr	–	–
2-Acetylthiazole	15	9	54
Benzothiazole	tr	tr	2
Thiophenes			
2-Ethylthiophene	tr	–	–
2-Methyl-5-ethylthiophene	tr	–	–
A thiophene (MW 126)	tr	–	–
Dihydro-2-methyl-3(2*H*)-thiophenone	15	–	–
2-Acetylthiophene	2	–	–
Furans			
2-Pentylfuran	100	824	30
Dihydro-2-methyl-3(2*H*)-furanone	24	–	–
2-Furanmethanethiol	10	–	–
2-Acetylfuran	73	–	–
Oxazole			
2,4,5-Trimethyloxazole	2	–	–
Pyrroles			
Pyrrole	3	–	–
1-(2-Furfuryl)pyrrole	2	–	–

tr = trace

Bicyclic pyrazines such as 5-methyl-, 2,5-dimethyl- and 3,5-dimethyl-6,7-dihydro-5*H*-cyclopentapyrazine were identified in well-done grilled pork. These cyclopentapyrazines were formed from the cyclotenes, 1,2-dicarbonyl compounds and ammonia (Figure 3.3). Two pentyl-substituted and one butyl-substituted pyrazines were also observed in grilled pork. The formation of higher carbon number-substituted pyrazines is hypothesized to be due to the intervention of lipid-derived aldehydes.

Like alkylpyrazines, alkylthiazole identified in pork arises from Maillard-type reaction between amino acids and sugars or other carbonyl compounds. From Table 3.2, it is demonstrated that the requirement for the formation of these compounds is severe heating. However, acetylthiazole was present in large amounts of boiled pork suggesting that a different mechanism is involved in its production. Four alkylthiazoles, namely 4-methylthiazole, 4,5-dimethylthiazole, 4-methyl-5-ethylthiazole and 4-methyl-5-vinylthiazole, identified in pork may also be thermal degradation products of thiamine (Figure 3.4) (Ho *et al.*, 1989).

Several thiophenes were identified in grilled pork. Thiophenes are responsible for the mild sulphurous odour of cooked meat (Shibamoto, 1980). The sulphur atom in a thiophene ring may come either from amino acid (cysteine or cystine) or from thiamine. Other minor heterocyclic com-

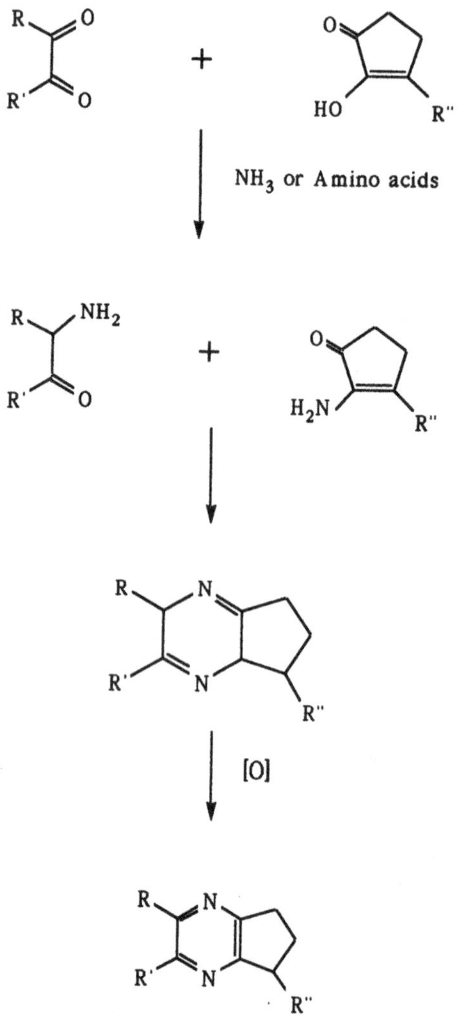

Figure 3.3 Mechanism for the formation of cyclopentapyrazines.

pounds reported in grilled pork, such as furans, pyridine and pyrroles are also well-known Maillard reaction products.

The primary meaty aroma compounds in beef and chicken have been identified as 2-methyl-3-furanthiol and bis-(2-methyl-3-furyl)disulphide (Gasser and Grosch, 1988, 1990). The most probable precursor for these two compounds is thiamine (Hartman *et al.*, 1984; Reineccius and

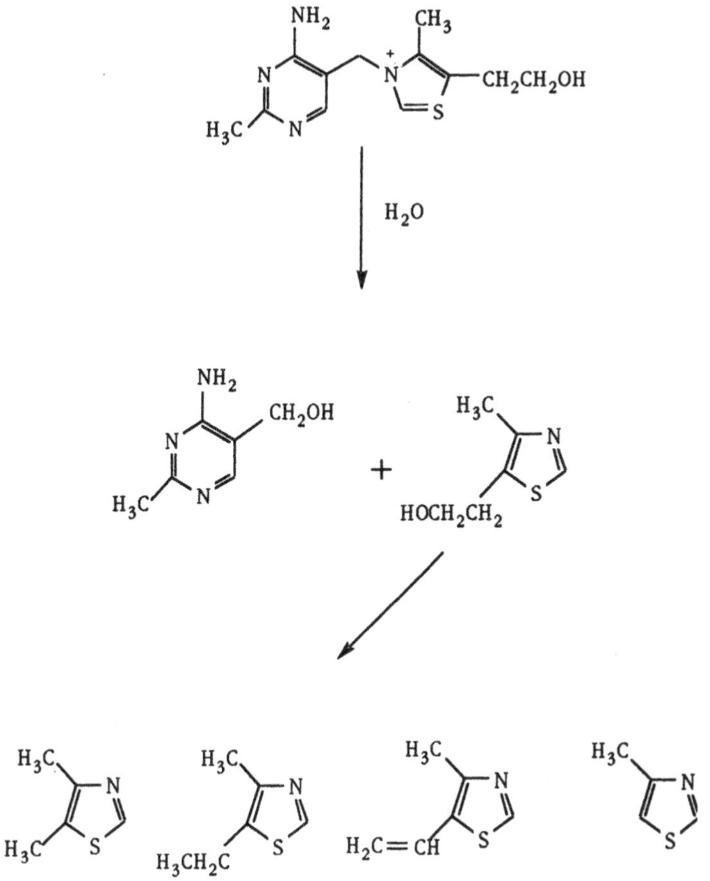

Figure 3.4 Mechanism for the formation of thiazoles from thiamine.

Liardon, 1985). Since the content of thiamine in pork (0.66 mg/100 g) is much higher than in beef (0.08 mg/100 g) (Belitz and Grosch, 1987), the presence of 2-methyl-3-furanthiol and bis-(2-methyl-3-furyl)disulphide in cooked pork flavour is expected.

3.4 Polysulphides in roasted pork

Two polysulphides, 3-methyl-2,4,5-trithiahexane and 4,6-dimethyl-2,3,5,7-tetrathiaoctane were reported in the head-space aroma components of roasted pork (Dubs and Joho, 1978; Dubs and Stussi, 1978). It is postulated that they form from the reaction of acetaldehyde with methanethiol, hydrogen sulphide and dimethyldisulphide as shown in Figure 3.5.

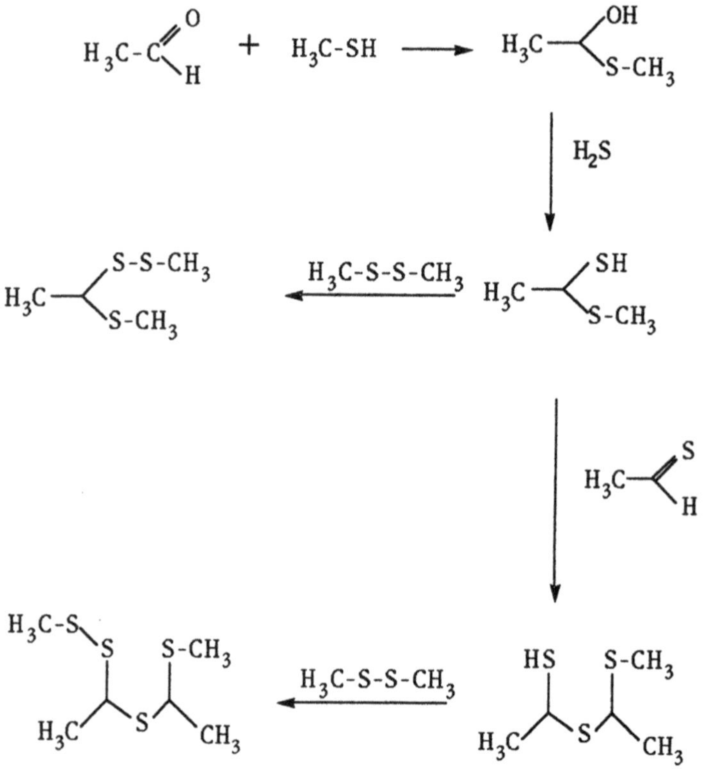

Figure 3.5 Mechanism for the formation of polysulphides in roasted pork.

These nonheterocyclic polysulphides may play an important role in the flavour of pork. With very low odour threshold values of polysulphides, relatively small amounts could have a significant effect. They may contribute directly to the meaty characteristics, or may provide an overall sulphurous note that is a part of meat aroma (Mottram, 1991).

3.5 Effect of ingredients on the flavour of pork

The effect of nitrite and nitrate on the flavour of pork is well-known and is the topic of discussion in another chapter of this book. A recent quantitative comparison of the flavour of cured and uncured pork indicated that the concentration of carbonyl compounds was higher in uncured pork, while they were either present in reduced amounts or not detectable in the cured meat. For instance, the major carbonyl compound, hexanal, was found to be present in the cured meat at a concentration of 12.66 mg/kg,

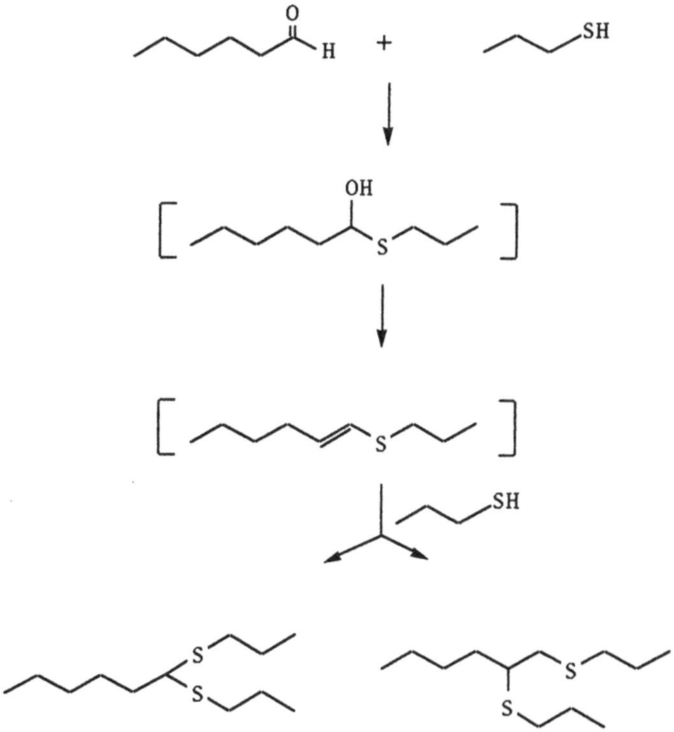

Figure 3.6 Mechanism for the formation of sulphides from the reaction of hexanal with propanethiol.

while only 0.03 mg/kg was present in the cured product, a reduction of 99.8% (Ramarathnam *et al.*, 1991).

The effect of soy sauce on the flavour development in cooked pork showed a similar phenomenon. Many unsaturated aldehydes including 2,4-decadienal, 2-undecenal, 2-dodecenal and 2-tridecenal identified in the volatiles of cooked pork were not found in the volatiles of pork stewed with soy sauce (Chou and Wu, 1983). It is possible that the very reactive carbonyl compounds such as aldehydes can react with nitrite, nitrate or reactive substances in soy sauce during curing or cooking of pork.

The reactivity of aldehydes toward thiols, a class of components in onions was observed. When hexanal was incubated with propanethiol, 1,1-bis(propylthio)-hexane and 1,2-bis(propylthio)-hexane were formed (Kuo and Ho, 1991). Figure 3.6 shows the mechanism for their formation. It is, therefore, expected that the addition of onions to the pork during heating will definitely modify the flavour profile of the cooked pork.

The effect of ingredients on the volatile components generated other than carbonyl compounds has also been observed. In the studies of the

Table 3.3 Effect of soy sauce on the formation of trithiolanes and thialdine in stewed pork

Compound	Relative concentration (%)			
	A	B	C	D
anti-3,5-Dimethyl-1,2,4-trithiolane	0.7	1.8	4.3	5.3
syn-3,5-Dimethyl-1,2,4-trithiolane	1.3	2.2	3.3	5.5
Thialdine	42.9	44.9	41.4	27.2

Cooking conditions: (A), pork cooked with 2.5% soy sauce and 27.5% water; (B), pork cooked with 5.0% soy sauce and 25.0% water; (C), pork cooked with 7.5% soy sauce and 22.5% water; (D), pork cooked with 10.0% soy sauce and 20.0% water.

flavour of soy sauce-stewed pork by Chou and Wu (1983), several alkyl-substituted trithiolanes such as *syn*- and *anti*-3-methyl-5-ethyl-1,2,4-trithiolanes, *syn*- and *anti*-3-methyl-5-propyl-1,2,4-trithiolanes, *syn*- and *anti*-3-methyl-5-butyl-1,2,4-trithiolanes and *syn*- and *anti*-3-methyl-5-isobutyl-1,2,4-trithiolanes were identified in the headspace volatile components of stewed pork. Neither soy sauce nor cooked pork contained these trithiolanes. Apparently, they were generated when pork was stewed with soy sauce.

The amount of soy sauce used in the preparation of stewed pork also showed a significant effect on the quantity of some volatile compounds generated in stewed pork. Table 3.3 shows the relative percentage of 3,5-dimethyl-1,2,4-trithiolanes and 2,4,6-trimethylperhydro-1,3,5-dithiazine (thialdine) in stewed pork samples prepared with different amounts of soy sauce. The amount of *syn*- and *anti*- 3,5-dimethyl-1,2,4-trithiolanes increased with increasing amounts of soy sauce. Soy sauce may provide the acetaldehyde which is the necessary intermediate for the formation of 3,5-dimethyl-1,2,4-trithiolanes. There is no simple explanation for the lesser amount of thialdine generated in the stewed pork samples cooked with a higher concentration of soy sauce.

References

Belitz, H.-D. and Grosch, W. (1987). *Food Chemistry*, Springer-Verlag, Berlin.

Chou, C.C. and Wu, C.M. (1983). The volatile compounds of stewed pork. Research Report 285, Food Industry Research and Development Institute, Hsinchu, Taiwan.

Dubs, P. and Joho, M. (1987). Investigation of the head-space of roasted meat. III. Synthesis of 4,6-dimethyl-2,3,5,7-tetrathiaoctane. *Helv. Chim. Acta.*, **61**, 2809–2812.

Dubs, P. and Stussi, R. (1978). Investigation of the head-space of roasted meat. II. Synthesis of substituted 2,4,5-trithia-hexane. *Helv. Chim. Acta.*, **61**, 2351–2359.

Frankel, E.N. (1982). Volatile lipid oxidation products. *Prog. Lipid Res.*, **22**, 1–33.

Gasser, U. and Grosch, W. (1988). Identification of volatile flavor compounds with high aroma values from cooked beef. *Z. Lebensm. Unters. Forsch.*, **186**, 489–494.

Gasser, U. and Grosch, W. (1990). Primary odorants of chicken broth. *Z. Lebensm. Unters. Forsch.*, **190**, 3–8.

Guadagni, D.G., Buttery, R.G. and Turnbaugh, J.G. (1972). Odour thresholds and similarity ratings of some potato chip components. *J. Sci. Food Agric.*, **23**, 1435–1444.
Hartman, G.J., Carlin, J.T., Scheide, J.D. and Ho, C.-T. (1984). Volatile products formed from the thermal degradation of thiamine at high and low moisture levels. *J. Agric. Food Chem.*, **32**, 1015–1018.
Ho, C.-T., Smagula, M.S. and Chang, S.S. (1978). The synthesis of 2-(1-pentenyl)furan and its relationship to the reversion flavor of soybean oil. *J. Amer. Oil Chem. Soc.*, **55**, 233–237.
Ho, C.-T., Bruechert, L.J., Kuo, M.C. and Izzo, M.T. (1989). Formation of volatile compounds from extruded corn-based model systems. In *Thermal Generation of Aroma*, eds. T.H. Parliment, R.J. McGorrin and C.-T. Ho. *ACS Symp. Ser. 409*, American Chemical Society, Washington, DC., pp. 504–511.
Hornstein, I. and Crowe, P.F. (1960). Flavor studies on beef and pork. *J. Agric. Food Chem.*, **8**, 494–498.
Hornstein, I. and Crowe, P.F. (1963). Meat flavor: lamb. *J. Agric. Food Chem.*, **11**, 147–149.
Kuo, M.C. and Ho, C.-T. (1991). Unpublished results.
Mottram, D.S. (1979). The flavor of cooked meats. *Agric. Res. Counc. Meat Res. Inst. (Bristol) Biennial Rep. 1977–79*, 87–88.
Mottram, D.S. (1991). Meat. In *Volatile Compounds in Foods and Beverages*, ed. H. Maarse, Marcel Dekker, New York, pp. 107–177.
Mottram, D.S. (1985). The effect of cooking conditions on the formation of volatile heterocyclic compounds in pork. *J. Sci. Food Agric.*, **36**, 377–382.
Mottram, D.S., Edwards, R.A. and MacFie, H.J.H. (1982). A comparison of the flavor volatiles from cooked beef and pork meat systems. *J. Sci. Food Agric.*, **33**, 934–944.
Mussinan, C.J. and Walradt, J.P. (1974). Volatile constituents of pressure-cooked pork liver. *J. Agric. Food Chem.*, **22**, 827–831.
Nawar, W.W. (1969). Thermal degradation of lipids: a review. *J. Agric. Food Chem.*, **17**, 18–29.
Ohloff, G. and Flament, I. (1978). Heterocyclic constituents of meat aroma. *Heterocycles*, **11**, 663–695.
Ramarathnam, N., Rubin, L.J. and Diosady, L.L. (1991). Studies of meat flavor. 1. Qualitative and quantitative differences in uncured and cured pork. *J. Agric. Food Chem.*, **39**, 344–350.
Reineccius, G.A. and Liardon, R. (1985). The use of charcoal traps and microwave desorption for the analysis of headspace volatiles above heated thiamine solutions. In *Topics in Flavor Research.*, eds. R.G. Berger, S. Nitz and P. Schreier. Eichhorn, Marzling–Hangenhan, pp. 125–136.
Schliemann, J., Wolm, G., Schrodter, R. and Ruttloff, H. (1987). Chicken flavor – formation, composition and production. Part 1. Flavor precursors. *Nahrung.*, **31**, 47–56.
Shibamoto, T. (1980). Heterocyclic compounds found in cooked meats. *J. Agric. Food Chem.*, **27**, 237–243.
Whitefield, F.B. and Last, J.H. (1991). Vegetables. In *Volatile Compounds in Foods and Beverages*, ed. H. Maarse, Marcel Dekker, New York, pp. 203–281.

4 The flavour of poultry meat

H. SHI and C.-T. HO

4.1 Introduction

Poultry meat has become a significant constituent of the North American diet. Numerous reviews on poultry flavour have been published (Wilson and Katz, 1972; Steverink, 1981; Ramaswamy and Richards, 1982) over the past few years. Thus far, more than 450 components have been characterized in cooked poultry meat (Mottram, 1991). This chapter contains a review of some of the chemical reactions which are responsible for the formation of volatile compounds significant to the aroma of cooked poultry meat.

4.2 Primary odorants of chicken broth

The most important study on chicken flavour in recent years is probably the one published by Gasser and Grosch (1990). Using aroma extract dilution analysis, they identified 16 primary odour compounds in chicken broth. Fourteen of these compounds were structurally identified as 2-methyl-3-furanthiol, 2-furfurylthiol, methional, 2,4,5-trimethylthiazole, nonanal, 2-*trans*-nonenal, 2-formyl-5-methylthiophene, *p*-cresol, 2-*trans*-4-*trans*-nonadienal, 2-*trans*-4-*trans*-decadienal, 2-undecenal, β-ionone, γ-decalactone and γ-dodecalactone. When the primary odorants of chicken were compared with those resulting from the aroma extract dilution analysis of broth from beef (Table 4.1), the major differences were that 2-*trans*-4-*trans*-decadienal (fatty) and γ-dodecalactone (tallowy, fruity) prevailed in the chicken broth, whereas the sulphur compounds, bis(2-methyl-3-furyl)disulphide (meat-like aroma) and methional (cooked potato-like), predominated in broth prepared from beef.

4.3 Sulphur-containing compounds in chicken flavours

2-Methyl-3-furanthiol identified in Gasser and Grosch (1990) as the most important flavour compound contributing to the meaty flavour of chicken broth has been recognized as a character impact compound in the aroma of cooked beef (Gasser and Grosch, 1988) and canned tuna fish (Withy-

Table 4.1 Comparison of flavour dilution factors of odorants appearing in broths from chicken and beef

Compounds	Flavour dilution factor		Odour description
	Chicken	Beef	
2-Methyl-3-furanthiol	1024	512	Meat-like, sweet
bis(2-Methyl-3-furyl)disulphide	<16	2048	Meat-like
2-Furfurylthiol	512	512	Roasty
2,5-Dimethyl-3-furanthiol	256	<16	Meaty
3-Mercapto-2-pentanone	128	32	Sulphurous
Methional	128	512	Cooked potato
2,4,5-Trimethylthiazole	128	<16	Earthy
2-Formyl-5-methylthiophene	64	64	Sulphurous
Phenylacetaldehyde	16	64	Honey-like
2-*trans*-4-*trans*-Decadienal	2048	<16	Fatty
2-*trans*-4-*cis*-Decadienal	128	<16	Fatty, tallowy
2-Undecenal	256	<16	Tallowy, sweet
γ-Dodecalactone	512	<16	Tallowy, fruity
γ-Decalactone	64	<16	Peach-like
Nonanal	64	<16	Tallowy, green
2-*trans*-Nonenal	64	<16	Tallowy, fatty
2-*trans*-4-*trans*-Nonadienal	64	<16	Fatty
β-Ionone	64	<16	Violet-like
p-Cresol	64	<16	Phenolic

combe and Mussinan, 1988). 2-Methyl-3-furanthiol and its oxidative dimer, bis-(2-methyl-3-furyl)disulphide, possessing characteristic meat flavour notes have been found by Evers *et al.* (1976) among the volatile products from heating thiamine hydrochloride with cysteine hydrochloride and hydrolysed vegetable protein. These two compounds were also found in volatile products of thiamine degradation (van der Linde *et al.*, 1979; Hartman *et al.*, 1984; Reineccius and Liardon, 1985) and in heated yeast extract (Ames and MacLeod, 1985). Thiamine has, therefore, been recognized as the precursor for the formation of the meaty aroma compounds, 2-methyl-3-furanthiol and bis-(2-methyl-3-furyl)disulphide. However, thiamine is not the sole source of 2-methyl-3-furanthiol. It was found that when ribose or inosine 5′-monophosphate reacted with cysteine of glutathione it formed significant quantities of 2-methyl-3-furanthiol (Farmer and Mottram, 1990; Grosch *et al.*, 1990; Zhang and Ho, 1989a). The mechanisms for the formation of 2-methyl-3-furanthiol and bis-(2-methyl-3-furyl)disulphide are shown in Figure 4.1.

The formation of 2-methyl-3-furanthiol from either ribose or IMP requires the participation of sulphur-containing amino acids, cysteine or cystine, or peptide, glutathione. Cysteine, cystine and glutathione will liberate hydrogen sulphide which is a major reactant for meaty aroma generation. Rapid evolution of hydrogen sulphide from glutathione occurs

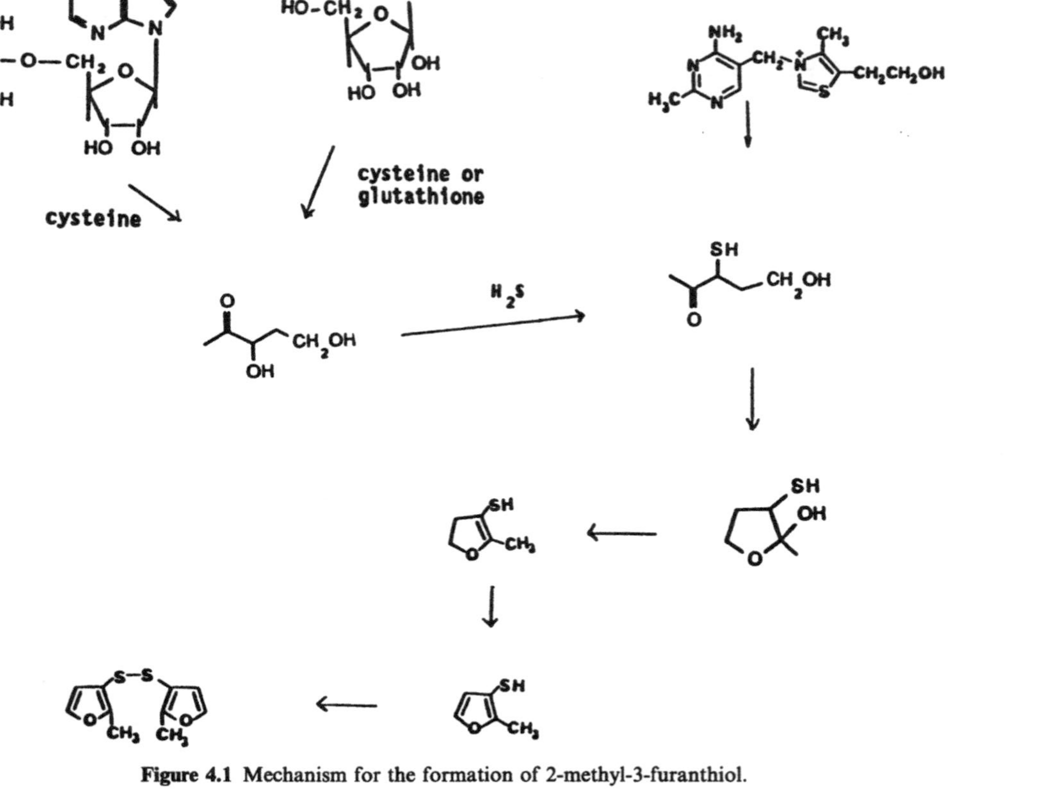

Figure 4.1 Mechanism for the formation of 2-methyl-3-furanthiol.

during the early stages of cooking (Ohloff *et al.*, 1985), but cysteine is the main hydrogen sulphide precursor on prolonged heating (Mecchi *et al.*, 1964).

Another compound structurally related to 2-methyl-3-furanthiol and identified as a primary odorant in chicken broth was 2-furfurylthiol (Gasser and Grosch, 1990). This compound possesses a threshold of 5 ppt, with aroma qualities such as roasted and sulphury, and has been recognized as the most important compound in roasted coffee (Tressl, 1989). It has been thermally generated from the reaction of furfural and cysteine (Tressl and Silwar, 1981).

The thermal degradation of cysteine and cystine in aqueous solution has been studied by Shu *et al.* (1985a,b). Two diastereomers of 3,5-dimethyl-1,2,4-trithiolane were identified as the major degradation products when either cysteine or cystine was heated at 160°C. However, when cysteine was degraded at 180°C, approximately frying temperature, both 3,5-dimethyl-1,2,4-trithiolanes and 2,4,6-trimethylperhydro-1,3,5-dithiazines were identified as major products (Zhang *et al.*, 1988). Degradation of glutathione at 180°C, on the other hand, generated only the 3,5-dimethyl-1,2,4-trithiolanes as major products (Zhang *et al.*, 1988). Although, 3,5-dimethyl-1,2,4-trithiolanes have not been identified in the recent study on the flavour of chicken broth (Gasser and Grosch, 1990), they have been previously reported to be present in cooked chicken (Horvat, 1976). Both 3,5-dimethyl-1,2,4-trithiolanes and 2,4,6-trimethylperhydro-1,3,5-dithiazines (thialdine) were identified as major volatile components of fried chicken (Tang *et al.*, 1983).

According to the patent literature (Wilson *et al.*, 1974), thialdine is useful as an ingredient of chicken flavour. It is interesting that the organoleptic quality of thialdine and other dithiazines was found to be affected by pH (Kawai, 1991). At pH 8.1 and 3.5, thialdine had a medium roasted shrimp flavour. When the pH approached the equivalence point at 2.5, a sweet odour was detected. At a pH of 1.6, thialdine had the weak flavour of edible mushrooms (Shiitake).

Although the mechanisms for the formation of dithiazines have been proposed by Boelens *et al.* (1974) and Shu *et al.* (1985), the complete understanding of the mechanism was achieved only recently by the extensive experimental study of Kawai (1991). The most plausible mechanism for the formation of dithiazine proposed by Kawai is shown in Figure 4.2. This mechanism revealed some significant characteristics, observed by Kawai (1991) as follows: (1) the reactions occur under basic conditions; (2) aldehydes react with ammonia prior to the participation of hydrogen sulphide; (3) elimination of ammonia and addition of hydrogen sulphide occurs stepwise in the mechanism; and (4) the mechanism for the formation of dithiazines is independent of the mechanism for the formation of either trithiolanes or dialkyl polysulphides.

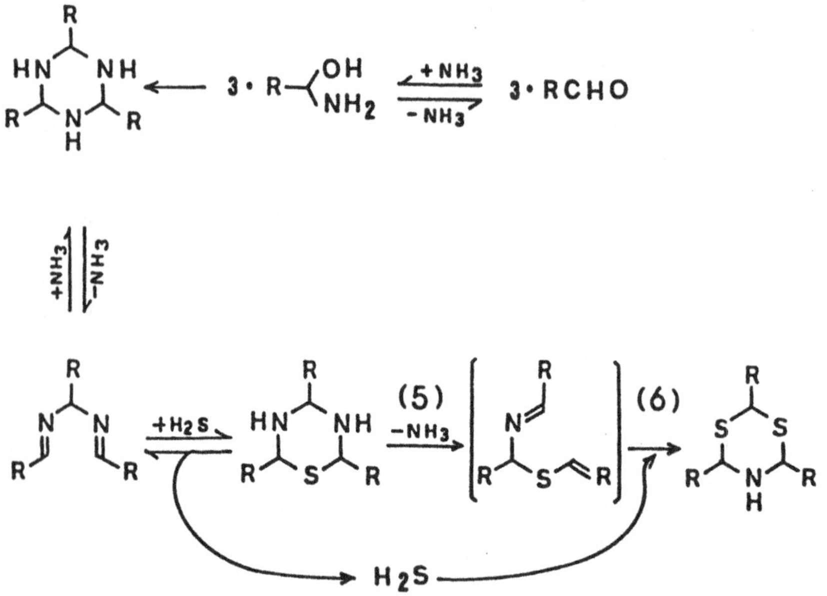

Figure 4.2 Possible mechanism for the formation of dithiazines (Kawai, 1991).

Interestingly, the higher homologue of 3,5-dimethyl-1,2,4-trithiolane, 3,5-diisobutyl-1,2,4-trithiolane was identified in the volatiles isolated from fried chicken flavour (Hartman *et al.*, 1984). This trithiolane possesses roasted, roasted-nut, crisp bacon-like and pork rind-like aromas and flavours, and has been produced in a model system containing isovaleraldehyde, ammonia and hydrogen sulphide (Shu *et al.*, 1981). Figure 4.3 shows the mechanism of the formation of 3,5-diisobutyl-1,2,4-trithiolane as proposed by Shu *et al.* (1981). Isovaleraldehyde arises via Strecker degradation of leucine. Ammonia and hydrogen sulphide may arise via thermal degradation of amino acids and cysteine or cystine. In addition to 3,5-dimethyl-1,2,4-trithiolane and 3,5-diisobutyl-1,2,4-trithiolane, two long-chain alkyl-substituted trithiolanes, namely, 3-methyl-5-butyl-1,2,4-trithiolane and 3-methyl-5-pentyl-1,2,4-trithiolane, were also reported to be present in fried chicken flavour (Hwang *et al.*, 1986). The identification of 3-methyl-5-butyl-1,2,4-trithiolane and 3-methyl-5-pentyl-1,2,4-trithiolane in food flavour suggests that pentanal and hexanal were involved in the formation of these compounds. Pentanal and hexanal are major thermal and oxidative decomposition products of lipids.

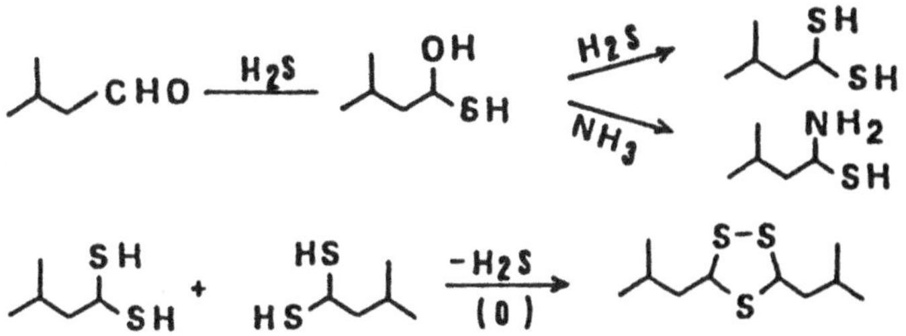

Figure 4.3 Possible mechanism for the formation of 3,5-diisobutyl-1,2,4-trithiolane (Shu *et al.*, 1985c).

4.4 Lipid oxidation products in chicken flavour

The major difference between the flavour of chicken broth and that of beef broth is the abundance of 2,4-decadienal and γ-dodecalactone in the chicken broth (Table 4.1). Both are well-known lipid oxidation products. The role of lipid-derived carbonyl compounds in poultry flavour has been extensively reviewed by Ramaswamy and Richards (1982).

A total of 193 compounds has been reported by Noleau and Toulemonde (1986) in the flavour of roasted chicken. Forty-one of them are lipid-derived aldehydes. When the aroma components of cooked chicken and cooked papain hydrolysates of chicken meat were qualitatively and quantitatively analysed, 23 of the 66 compounds reported were lipid-derived aldehydes (Schroll *et al.*, 1988). Table 4.2 lists the quantitative data for selected aldehydes identified in these two studies. The most abundant aldehydes identified in chicken flavour are hexanal and 2,4-decadienal. In view of the much lower odour threshold of 2,4-decadienal (0.00007 mg/kg) compared to hexanal (0.0045 mg/kg) (van Gemert and Nettenbreijer, 1977), the 2,4-decadienal should be the more important odorant for chicken flavour. Hexanal and 2,4-decadienal are the primary oxidation products of linoleic acid. The autoxidation of linoleic acid generates 9- and 13-hydroperoxides of linoleic acid. Cleavage of 13-hydroperoxide will lead to hexanal, and the breakdown of 9-hydroperoxide will lead to 2,4-decadienal (Ho *et al.*, 1989). Subsequent retro-aldolization of 2,4-decadienal will produce 2-octenal and hexanal (Josephson and Lindsay, 1987). Figure 4.4 shows their formation mechanism. 2,4-Decadienal is known to be one of the most important flavour contributors to deep-fat-fried foods (Ho *et al.*, 1987). As shown in Table 4.2, the enzymic hydrolysis of chicken with papain increased the concentration of 2,4-decadienal, as the aroma of cooked meat improved.

In addition to 2,4-decadienal, another compound more polar than 2,4-decadienal having a mass spectrum as shown in Figure 4.5 was found in

Table 4.2 Selected aldehydes identified in chicken flavour

Aldehyde	Concentration (mg per kg)		
	Roasted chicken[a]	Cooked chicken meat[b]	Cooked chicken meat[b] (papain treated)
Butanal	0.133		
Pentanal	0.319		
Hexanal	1.804	25.6	17.2
Heptanal	0.212	2.1	1.5
Octanal	0.422	2.3	1.2
Nonanal	0.467	1.7	1.3
Decanal	0.052	0.3	0.3
Undecanal	0.058		
Dodecanal	0.022		
Tridecanal	0.151		
Tetradecanal	0.125	0.2	0.7
Pentadecanal	0.383		
Hexadecanal	19.788	1.4	9.8
Heptadecanal	0.276	0.1	0.1
Octadecanal	2.664		
trans-2-Butenal	tr		
cis-2-Pentenal	tr		
trans-2-Pentenal	0.085	1.1	0.2
cis-2-Hexenal	tr		
trans-2-Hexenal	0.060	0.3	0.4
trans-2-Heptenal	0.104	1.2	1.5
cis-2-Octenal	0.004		
trans-2-Octenal	0.195	3.7	2.0
trans-2-Nonenal	0.084		
cis-2-Decenal	0.003		
trans-2-Decenal	0.139	1.0	1.2
cis-2-Undecenal	0.002		
trans-2-Undecenal	0.139	0.4	1.1
trans-Dodecenal	0.002	0.3	0.1
trans,cis-2,4-Nonadienal	tr		
trans,trans-2,4-Nonadienal	tr	0.3	0.5
trans,cis-2,4-Decadienal	0.051	1.0	2.7
trans,trans-2,4-Decadienal	0.137	5.2	13.7
trans,trans-2,4-Undecadienal	0.001	0.2	0.2

[a]Noleau and Toulemonde (1986)
[b]Schroll *et al.* (1988)
tr = trace

the flavour of fried chicken (Tang and Ho, 1991). The same compound was also reported in the volatile components of roasted chicken fat as an unidentified component (Noleau and Toulemonde, 1987). It was recently structurally identified by Schieberle and Grosch (1991) as *trans*-4,5-epoxy-*trans*-2-decenal. It was characterized as one of the most potent odorants of the crumb flavour of wheat bread. *trans*-4,5-Epoxy-*trans*-2-decenal having a low odour threshold of approximately 1.5 pg/l (air) (Schieberle and Grosch, 1991) is probably an important flavour contributor of fried

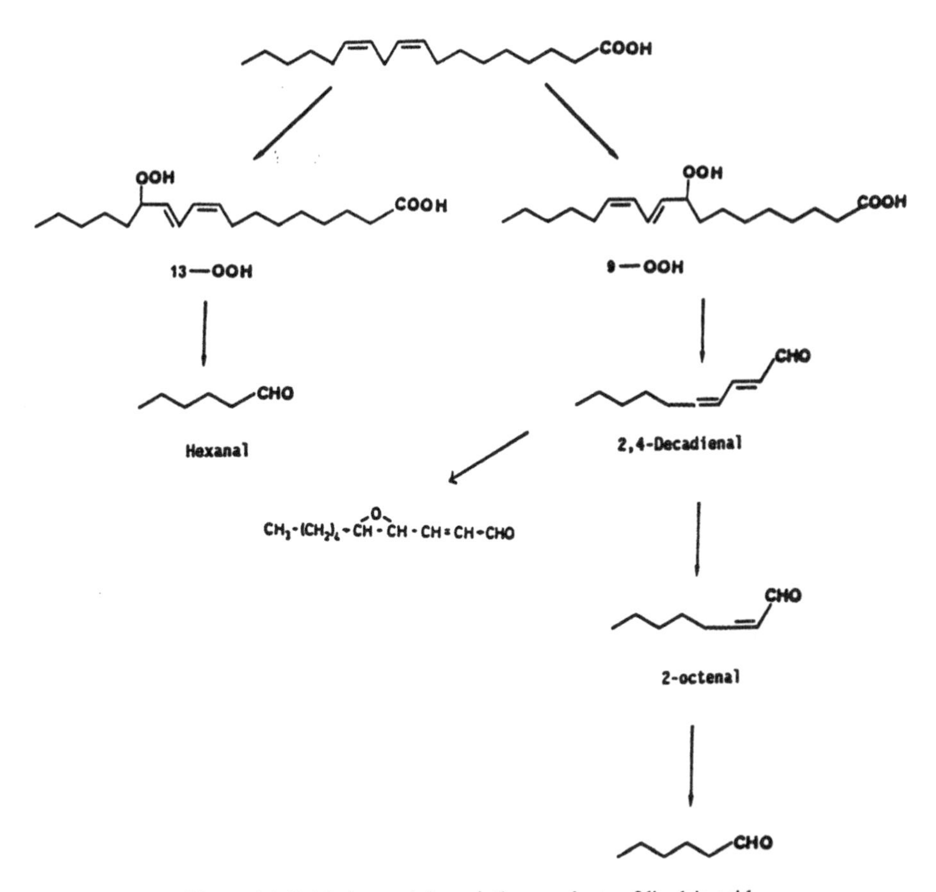

Figure 4.4 Oxidation and degradation products of linoleic acid.

chicken. *trans*-4,5-Epoxy-*trans*-2-decenal, an obvious oxidation product of 2,4-decadienal, was reported to be one of the major products when pure 2,4-decadienal was incubated at 80°C for 3 weeks. In the presence of dipropyl disulphide, a possible antioxidant, the formation of *trans*-4,5-epoxy-*trans*-2-decenal was completely inhibited (Kuo, 1991).

Meal from poultry byproducts is a relatively inexpensive protein source and is used in pet foods. The major volatile flavour compounds present in the meal have been reported to be hexanal, 3-octen-2-one, 1-pentanol, pentanal, heptanal, octanal, 1-heptanol, 1-octanol and 1-octen-3-ol. All of these compounds are well-known lipid oxidation products (Greenberg, 1981).

Oxidative deterioration of lipids is the major cause of the off-flavour development in poultry meat (Lillard, 1987). The rancidity of fats in mechanically deboned chicken meat has been reported to present a sub-

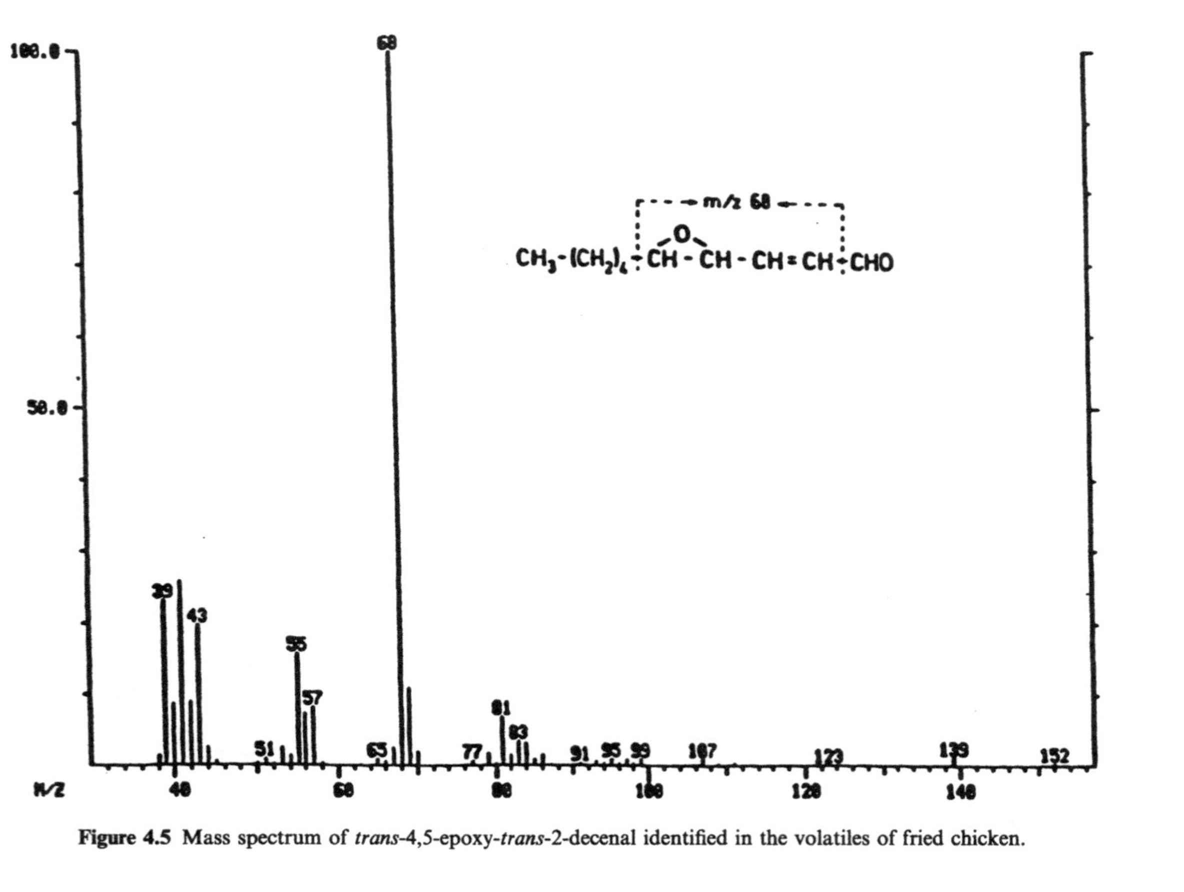

Figure 4.5 Mass spectrum of *trans*-4,5-epoxy-*trans*-2-decenal identified in the volatiles of fried chicken.

Table 4.3 Fatty acid composition of the phospholipid and triglyceride fractions of chicken meat

Type of chicken meat	Lipid type	Fatty acid composition (%)		
		Saturated	Monounsaturated	Polyunsaturated
Breast	Triglyceride	33.8	42.7	25.6
	Phospholipid	35.3	21.1	41.4
Leg	Triglyceride	33.0	42.3	24.8
	Phospholipid	39.1	16.4	43.5

From Pikul *et al.* (1984)

stantial off-flavour problem (Dimick *et al.*, 1972) and the haem pigments as the catalysts for lipid oxidation have been reported to be predominantly responsible for the off-flavour formation (Lee *et al.*, 1975). Phospholipids have also been shown to play an important role in lipid oxidation (Katz *et al.*, 1986). Phospholipid concentration in cooked chicken and turkey skin was determined by Dimick and MacNeil (1970), who reported that the fractions which were high in phospholipids were highly unstable and produced high concentrations of carbonyl compounds. Table 4.3 lists the fatty acid composition of the phospholipid and triglyceride fractions of chicken meat (Lillard, 1987). It is clear that the higher content of polyunsaturated fatty acids in the phospholipid fraction makes it highly susceptible to oxidation.

4.5 Heterocyclic compounds in chicken flavour

4.5.1 Pyrazines

Alkylpyrazines have been recognized as important trace flavour components of a large number of cooked, roasted, toasted and deep-fat fried foods (Maga, 1982). As a rule, alkylpyrazines are described as being roasted, nut-like or toasted in flavour character and are highly desirable in foods. Formation pathways for alkylpyrazines have been proposed by numerous researchers (Shibamoto and Bernhard, 1977; Flament, 1981).

Table 4.4 lists pyrazines identified in the flavours of fried chicken and roasted chicken. No pyrazine was identified in the volatiles of chicken broth. This indicates that high temperature and low moisture favour the generation of pyrazines. The formation of alkyl pyrazines, such as 2-butylpyrazine and 2-methyl-3-butylpyrazine, is hypothesized to be due to the intervention of aldehydes. The presence of aldehydes, originating from lipid degradation, could facilitate addition to the metastable dihydropyrazine compound resulting in longer chain substituted pyrazines (Huang *et*

Table 4.4 Pyrazines identified in chicken flavour

Compound	Fried chicken[a]	Roasted chicken[b]
Pyrazine	+	
2-Methylpyrazine	+	+
2,3-Dimethylpyrazine	+	+
2,5-Dimethylpyrazine	+	
2,6-Dimethylpyrazine	+	+
Trimethylpyrazine	+	+
2-Isopropylpyrazine	+	
2-Methyl-3-ethylpyrazine	+	+
2-Methyl-6(5)-ethylpyrazine	+	+
2-Butylpyrazine	+	
2,3-Dimethyl-5-ethylpyrazine	+	+
2,5-Dimethyl-3-ethylpyrazine	+	
2,6-Dimethyl-3-ethylpyrazine		+
2,6-Diethylpyrazine	+	+
2-Methyl-5,6-diethylpyrazine	+	
2-Methyl-3,5-diethylpyrazine	+	
2-Methyl-3-butylpyrazine	+	
2-Methyl-5-vinylpyrazine	+	
2-Methyl-6-vinylpyrazine	+	
2-Isopropenylpyrazine	+	
6,7-Dihydro-5*H*-cyclopentapyrazine	+	
2-Methyl-6,7-dihydro-5*H*-cyclopentapyrazine	+	

[a]Tang *et al.* (1983)
[b]Noleau and Toulemonde (1986)

al., 1987). Figure 4.6 shows the mechanism for the formation of long-chain alkyl-substituted pyrazines.

4.5.2 Pyridines

The occurrence of pyridines in food has been reviewed (Vernin and Vernin, 1982). 2-Alkylpyridines were proposed to form from the corresponding unsaturated n-aldehydes with ammonia upon heat treatment (Buttery *et al.*, 1977; Ohnishi and Shibamoto, 1984). Table 4.5 lists pyridines identified in the volatiles of fried and roasted chicken.

2-Pentylpyridine was identified in fried chicken flavour. This compound has a strong fatty and tallow-like odour and was the major product in the volatiles generated from the thermal interaction of valine and linoleate (Henderson and Nawar, 1981). Recently, the mechanism for the formation of 2-pentylpyridine in the reaction of 2,4-decadienal with either cysteine or glutathione (γ–glu–cys–gly) in aqueous solution at high temperature (180°C) was studied (Zhang and Ho, 1989). The quantities of five major volatile components generated for these two systems are listed in Table 4.6. A greater amount of 2-pentylpyridine and a lesser amount of hexanal were observed in the system of glutathione suggesting that 2,4-decadienal

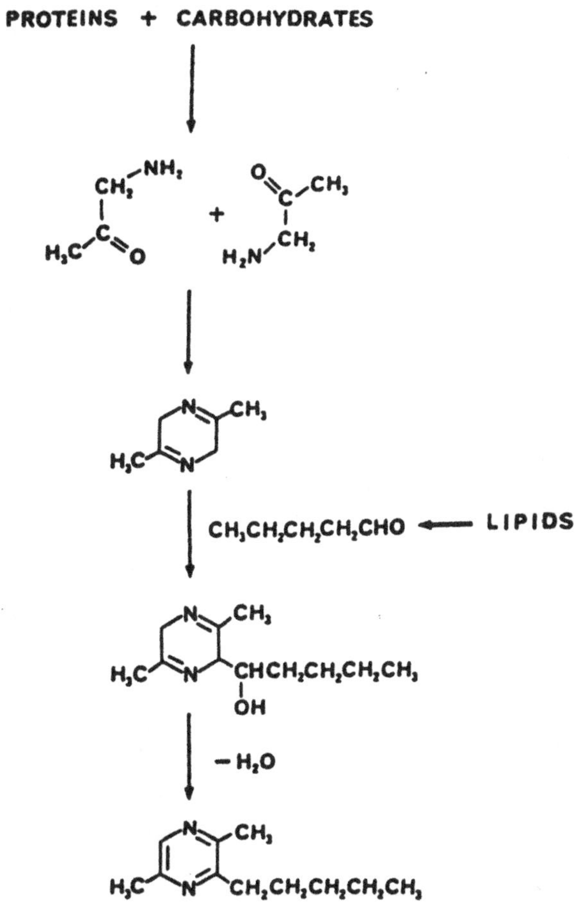

PROTEINS + CARBOHYDRATES

$CH_3CH_2CH_2CH_2CHO$ ← LIPIDS

$-H_2O$

Figure 4.6 Mechanism for the formation of pyrazines.

was involved in the formation of a Schiff base with the amino group in glutathione directly and thus left less free 2,4-decadienal to go to autoxidation or retro-aldolization. According to the pathway proposed by Henderson and Nawar (1981), 2-pentylpyridine, which accounted for almost one half of the total volatiles identified in the reaction of glutathione and 2,4-decadienal, can be formed via the Schiff base intermediate from the reaction of 2,4-decadienal with ammonia (Henderson and Nawar, 1981). However, it is known that the formation of other compounds, dithiazine and thiadiazine, requires the presence of free ammonia (Boelens *et al.*, 1974). The absence of dithiazine or thiadiazine formed in the 2,4-decadienal/glutathione indicates that no free ammonia is available;

Table 4.5 Pyridines identified in chicken flavours

Compound	Fried chicken[a]	Roasted chicken[b]
Pyridine	+	+
2-Methylpyridine	+	+
3-Ethylpyridine	+	+
4-Ethylpyridine	+	
2-Methyl-5-ethylpyridine	+	+
2-Ethyl-3-methylpyridine	+	
2-Butylpyridine	+	
2-Pentylpyridine	+	
2-Isobutyl-3,5-dipropylpyridine	+	

[a]Tang *et al.* (1983)
[b]Noleau and Toulemonde (1986)

Table 4.6 Quantitation of some major flavour components in model systems of 2,4-decadienal/cysteine and 2,4-decadienal/glutathione

Component	System	
	A	B
	(mg/mol)	
Total volatiles	2627.3	2511.0
Hexanal	278.6	45.4
3,5-Dimethyl-1,2,4-trithiolane	140.9	368.5
2,4,6-Trimethylperhydro-1,3,5-thiadiazine	828.5	–
2,4,6-Trimethylperhydro-1,3,5-dithiazine	284.2	–
2-Pentylpyridine	501.5	1219.0
Sum	2033.7	1632.9
% of total volatiles	77.4	65.0

System A is 2,4-decadienal with cysteine; system B is 2,4-decadienal with glutathione.

this suggests that free ammonia may not be necessary for the formation of 2-pentylpyridine. It is possible that the amino group for amino acids or peptides, condenses directly with the aldehydic group of 2,4-decadienal and is then followed by an electrocyclic reaction and aromatization to form 2-pentylpyridine (Figure 4.7).

2-Isobutyl-3,5-diisopropylpyridine was identified in fried chicken and has a roasted cocoa-like aroma (Hartman *et al.*, 1984a). Figure 4.8 shows the mechanism for the formation of this compound as proposed by Shu *et al.* (1985c). It involves the reaction of aldehyde and ammonia at high temperatures and is known as the Chichibabin condensation.

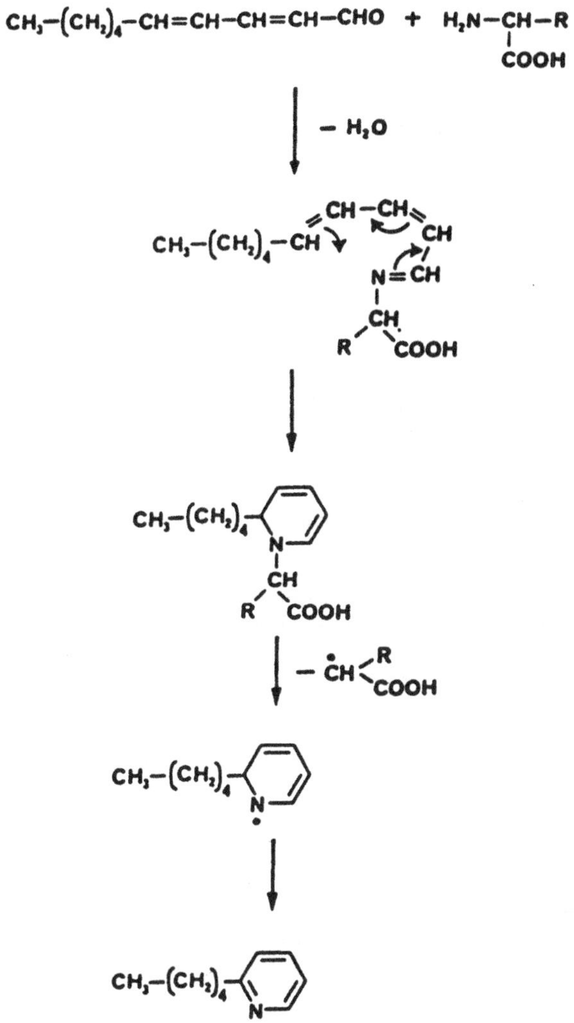

$CH_3-(CH_2)_4-CH=CH-CH=CH-CHO + H_2N-CH(COOH)-R$

$\downarrow -H_2O$

$CH_3-(CH_2)_4-CH=CH-CH=CH-CH=N-CH(R)COOH$

$\downarrow$

$CH_3-(CH_2)_4$–(dihydropyridine ring)–N–CH(R)COOH

$\downarrow - \dot{C}H(R)COOH$

$CH_3-(CH_2)_4$–(pyridinyl radical, N•)

$\downarrow$

$CH_3-(CH_2)_4$–(pyridine)

Figure 4.7 Mechanism for the formation of 2-pentylpyridine.

4.5.3 Pyrroles

Pyrrole may have been the first individual heterocyclic compound to be isolated from foods. Table 4.7 lists some pyrroles identified in chicken flavours. Some pyrrole derivatives have a pleasant flavour. For example, 2-acetylpyrrole has a caramel-like flavour. Pyrroles have not received as much attention as flavour compounds as have other heterocyclic compounds such as pyrazines and thiazoles (Shibamoto, 1989).

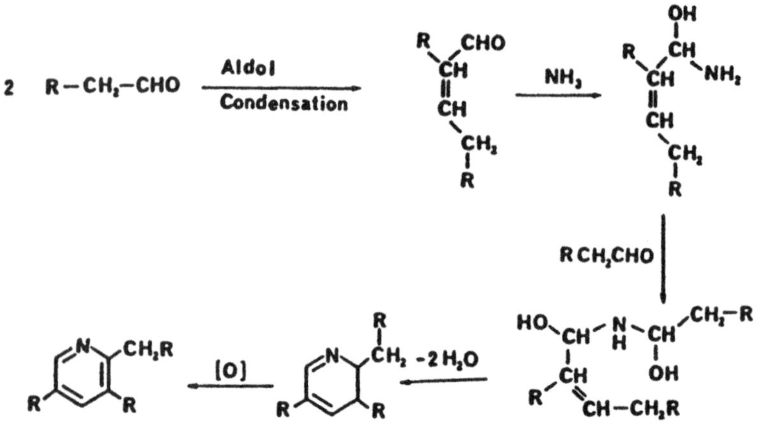

Figure 4.8 Mechanism for the formation of 2-isobutyl-3,5-diisopropylpyridine.

Table 4.7 Pyrroles identified in chicken flavours

Compound	Fried chicken[a]	Roasted chicken[b]
Pyrrole	+	+
N-methylpyrrole		+
2-Methylpyrrole	+	
2-Ethylpyrrole		+
N-acetylpyrrole		+
2-Acetylpyrrole	+	
2-Isobutylpyrrole	+	
N-isobutylpyrrole	+	
N-(2-butanoyl)pyrrole	+	
N-furfurylpyrrole		+

[a]Tang *et al.* (1983)
[b]Noleau and Toulemonde (1986)

4.5.4 *Thiazoles*

Thiazoles are a class of compounds possessing a five-membered ring with sulphur and nitrogen in the 1 and 3 positions, respectively. The potential for thiazole derivatives as flavorants is evident from the work of Stoll *et al.* (1967) who found the strong nut-like odour of a cocoa extract to be due to a trace amount of 4-methyl-5-vinylthiazole. Since then, numerous thiazoles have been identified in food flavours.

The exact origin of thiazoles remains unclear. They can form through the thermal degradation of cystine or cysteine (Shu *et al.*, 1985a,b), or by

Table 4.8 Thiazoles identified in chicken flavours

Compound	Fried chicken[a]	Roasted chicken[b]
Thiazole	+	+
2-Methylthiazole	+	+
2,4,5-Trimethylthiazole	+	
2-Methyl-4-ethylthiazole	+	
2-Methyl-5-ethylthiazole		+
2,4-Dimethyl-5-ethylthiazole	+	
2-Isopropyl-4,5-dimethylthiazole	+	
2,5-Dimethyl-4-butylthiazole	+	
2-Isopropyl-4-ethyl-5-methylthiazole	+	
2-Butyl-4,5-dimethylthiazole	+	
2-Butyl-4-methyl-5-ethylthiazole	+	
2-Pentyl-4,5-dimethylthiazole	+	
2-Hexyl-4,5-dimethylthiazole	+	
2-Heptyl-4,5-dimethylthiazole	+	
2-Heptyl-4-ethyl-5-methylthiazole	+	
2-Octyl-4,5-dimethylthiazole	+	

[a]Tang *et al.* (1983)
[b]Noleau and Toulemonde (1986)

the interaction of sulphur-containing amino acids and carbonyl compounds (Zhang and Ho, 1989b; Hartman and Ho, 1984). Thiazoles have been identified as volatile components of thermally degraded thiamine (Hartman *et al.*, 1984b).

Table 4.8 lists alkylthiazoles identified in fried chicken flavour and roasted chicken flavour. 2-Pentyl-4-methyl-5-ethylthiazole has a strong paprika pepper flavour and 2-heptyl-4,5-dimethylthiazole has a strong spicy flavour. 2-Octyl-4,5-dimethylthiazole has a sweet fatty aroma (Ho and Jin, 1984). They are probably important contributors to the flavour of fried foods. These and other thiazoles that have been identified have long-chain alkyl substitutions on the thiazole ring. The involvement of frying fat or fat decomposition products in the formation of these compounds has been suggested.

4.6 Conclusion

Remarkable progress has been made in poultry flavour research in recent years. The use of aroma extract dilution analysis to study the flavour of chicken broth is undoubtedly the most outstanding achievement. We now know that the chemical components responsible for the meaty note of chicken meat are no different from those of beef broth. It is the fatty aroma compounds that make the flavour of chicken broth significantly different from that of beef broth. This is very helpful to flavorists interested in compounding savoury flavours of various types. The use of aroma extract dilution analysis or other similar methods such as Charm analysis (Acree *et al.*, 1984) to study other poultry meat products, such as fried chicken or roasted chicken, is highly desirable.

The volatiles resulting from the Maillard reaction and lipid oxidation are obviously the major sources of flavour compounds identified in poultry meat. We begin to see the incorporation of lipid-derived aldehydes into Maillard reaction products, such as long-chain, alkyl-substituted pyrazines, thiazoles and trithiolanes. Other aspects of the interaction between the Maillard reaction and lipid oxidation such as the antioxidative effect of the Maillard reaction products need to be explored further.

A recent study by Kawai (1991) on the effect of pH on the formation and sensory quality of perhydrodithiazines is very interesting. How pH influences the flavour of chicken broth remains to be determined.

References

Acree, T.E., Bernard, J. and Cunningham, D.G. (1984). The analysis of odor-active volatiles in gas chromatographic effluents. In *Analysis of Volatiles. Method and Application*, ed. P. Schreier. Walter de Gruyter, Berlin, pp. 251–267.

Ames, J.M. and MacLeod, G. (1985). Volatile components of a yeast extract composition. *J. Food Sci.*, **50**, 125–131.

Boelens, M., van der Linde, L.M., de Valois, P.J., van Dort, H.M. and Takken, H.J. (1976). Organic sulfur compounds from fatty aldehydes, hydrogen sulfide, thiols and ammonia as flavor constituents. *J. Agric. Food Chem.*, **22**, 1071–1076.

Buttery, R.G., Ling, L.C., Teranishi, R. and Mon, T.R. (1977). Roasted lamb fat: basic volatile components. *J. Agric. Food Chem.*, **25**, 1227–1230.

Dimick, P.S. and MacNeil, J.H. (1970). Poultry product quality. 2. Storage time–temperature effects of carbonyl composition of cooked turkey and chicken skin fractions. *J. Food Sci.*, **35**, 186–190.

Dimick, P.S., MacNeil, J.H. and Grunden, L.E. (1972). Poultry product quality. Carbonyl composition and organoleptic evaluation of mechanically deboned poultry meat. *J. Food Sci.*, **37**, 544–546.

Evers, W.J., Heinsohn, H.H. Jr., Mayers, B.J. and Sanderson, A. (1976). Furans substituted at the three position with sulfur. In *Phenolic, Sulfur and Nitrogen Compounds in Food Flavors*, eds. G. Charalambous and I. Katz. American Chemical Society, Washington, DC., pp. 184–193.

Farmer, L.J. and Mottram, D.S. (1990). Recent studies on the formation of meat-like aroma compounds. In *Flavor Science and Technology*, eds. Y. Bessiere and A.F. Thomas. John Wiley & Sons, Chichester, UK, pp. 113–116.

Flament, I. (1981). Some recent aspects of the chemistry of naturally occurring pyrazines. In *The Quality of Food and Beverages (Vol. 1)*, eds. G. Charalambous and G. Inglett. Academic Press, New York, pp. 42–48.

Gasser, U. and Grosch, W. (1988). Identification of volatile flavor compounds with high aroma values from cooked beef. *Z. Lebensm. Unters. Forsch.*, **186**, 489–494.

Gasser, U. and Grosch, W. (1990). Primary odorants of chicken broth. *Z. Lebensm. Unters. Forsch.*, **190**, 3–8.

Greenberg, M.J. (1981). Characterization of poultry byproduct meal flavor volatiles. *J. Agric. Food Chem.*, **29**, 831–834.

Grosch, W., Sen, A., Guth, H. and Zeiler-Hilgart, G. (1990). Quantitation of aroma compounds using a stable isotope diluted assay. In *Flavor Science and Technology*, eds. Y. Bessiere and A.F. Thomas. John Wiley & Sons, Chichester, UK, pp. 191–194.

Hartman, G.J. and Ho, C.-T. (1984). Volatile products of the reaction of sulfur-containing amino acids with 2,3-butanedione. *Lebensm.-Wiss. u. -Technol.*, **17**, 171–174.
Hartman, G.J., Carlin, J.T., Hwang, S.S., Bao, Y., Tang, J. and Ho, C.-T. (1984a). Identification of 3,5-diisobutyl-1,2,4-trithiolane and 2-isobutyl-3,5-diisopropylpyridine in fried chicken flavor. *J. Food Sci.*, **49**, 1398 and 1400.
Hartman, G.J., Carlin, J.T., Scheide, J.D. and Ho, C.-T. (1984b). Volatile products formed from the thermal degradation of thiamine at high and low moisture levels. *J. Agric. Food Chem.*, **32**, 1015–1018.
Henderson, S.K. and Nawar, W.W. (1981). Thermal interaction of linoleic acid and its esters with valine. *J. Amer. Oil Chem. Soc.*, **58**, 632–635.
Ho, C.-T. and Jin, Q.Z. (1984). Aroma properties of some alkylthiazoles. *Perfumer & Flavorist*, **9**(6), 15–18.
Ho, C.-T., Carlin, J.T., Huang, T.C., Hwang, L.S. and Hau, L.B. (1987). Flavor development in deep-fat fried foods. In *Flavor Science and Technology*, eds. M. Martens, G.A. Dalen and H. Russwurm, Jr. John Wiley & Sons, Chichester, UK, pp. 35–42.
Ho, C.-T., Zhang, Y., Shi, H. and Tang, J. (1989). Flavor chemistry of Chinese foods. *Food Rev. Int.*, **5**, 253–287.
Horvat, R.J. (1976). Identification of some volatile compounds in cooked chicken. *J. Agric. Food Chem.*, **24**, 953–958.
Huang, T.-C., Bruechert, L.J., Hartman, T.G., Rosen, R.T. and Ho, C.-T. (1982). Effects of lipids and carbohydrates on thermal generation of volatiles from commercial zein. *J. Agric. Food Chem.*, **35**, 985–990.
Hwang, S.S., Carlin, J.T., Bao, Y., Hartman, G.J. and Ho, C.-T. (1986). Characterization of volatile compounds generated from the reaction of aldehyde with ammonium sulfide. *J. Agric. Food Chem.*, **34**, 538–542.
Josephson, D.B. and Lindsay, R.C. (1987). Retro-aldol degradation of unsaturated aldehydes: role in the formation of *c*4-heptenal from *t*2, *c*6-nonadienal in fish, oyster and other flavors. *J. Amer. Oil Chem. Soc.*, **64**, 132–138.
Katz, M.A., Dugan, L.R., Jr. and Dawson, L.E. (1966). Fatty acids in neutral lipids and phospholipids from chicken tissues. *J. Food Sci.*, **31**, 717–720.
Kawai, T. (1991). *Thiadiazines and Dithiazines in Volatiles from Heated Seafoods, and Mechanisms of their Formation*, 2nd ed., Shiono Koryo Kaisha, Osaka.
Kuo, M.C. (1991). Formation of Volatile Polysulfides from Welsh Onions and Scallions. Ph.D. dissertation, Rutgers University, New Brunswick, NJ.
Lee, Y.B., Hargus, G.L., Kirkpatrick, J.A., Berner, D.L. and Forsythe, R.H. (1975). Mechanism of lipid oxidation in mechanically deboned chicken meat. *J. Food Sci.*, **40**, 964–967.
Lillard, D.A. (1987). Oxidative deterioration in meat, poultry, and fish. In *Warmed-Over Flavor of Meat*, eds. A.J. St. Angelo and M.E. Bailey. Academic Press, Orlando, pp. 41–67.
Maga, J.A. (1982). Pyrazines in foods. *CRC Crit. Rev. Food Sci. Nutr.*, **16**, 1–115.
Mecchi, E.P., Pippen, E.L. and Lineweaver, H. (1964). Origin of hydrogen sulfide in heated chicken muscle. *J. Food Sci.*, **29**, 393–399.
Mottram, D.S. (1991). Meat. In *Volatiile Compounds in Foods and Beverages*, ed. H. Maarse. Marcel Dekker, New York, pp. 107–177.
Noleau, I. and Toulemonde, B. (1986). Quantitative study of roasted chicken flavor. *Lebensm.-Wiss. u. -Technol.*, **19**, 122–125.
Noleau, I. and Toulemonde, B. (1987). Volatile components of roasted chicken fat. *Lebensm.-Wiss. u. -Technol.*, **20**, 37–41.
Ohloff, G., Flament, I. and Pickenhagen, W. (1985). Flavor chemistry. *Food Rev. Int.*, **1**, 99–148.
Ohnishi, S. and Shibamoto, T. (1984). Volatile compounds from heated beef fat and beef fat with glycine. *J. Agric. Food Chem.*, **32**, 987–992.
Pikul, J., Leszcaynski, D.E. and Kummerow, F.I. (1984). *J. Food Sci.*, **49**, 704–709.
Ramaswamy, H.S. and Richards, J.F. (1982). Flavor of Poultry Meat--A Review. *Can. Inst. Food Sci. Technol. J.*, **15**, 7–18.
Reineccius, G.A. and Liardon, R. (1985). The use of charcoal traps and microwave desorption for the analysis of headspace volatiles above heated thiamine solutions. In *Topics in*

Flavor Research, eds. R.G. Berger, S. Nitz and P. Schreier. Eichhorn, Marzling–Hangenhan, pp. 125–136.

Schieberle, P. and Grosch, W. (1991). Potent odorants of the wheat bread crumb. *Z. Lebensm. Unters. Forsch.*, **192**, 130–135.

Schroll, W., Nitz, S. and Drawert, F. (1988). Determination of aroma compounds in cooked chicken meat and hydrolysate. *Z. Lebensm. Unters. Forsch.*, **187**, 558–560.

Shibamoto, T. (1989). Volatile flavor chemicals formed by the Maillard reaction. In *Thermal Generation of Aromas*, eds. T.H. Parliment, R.J. McGorrin and C.-T. Ho. ACS Symp. Ser. 409, American Chemical Society, Washington, DC., pp. 134–142.

Shibamoto, T. and Bernhard, R.A. (1977). Investigation of pyrazine formation pathways in sugar–ammonia model systems. *J. Agric. Food Chem.*, **25**, 609–614.

Shu, C.K., Mookherjee, B.D. and Vock, M.H. (1981). Flavoring with a mixture of 2,5-dimethyl-3-acetylfuran and 3,5-diisobutyl-1,2,4-trithiolane. US Patent 4,263,331.

Shu, C.K., Hagedorn, M.L., Mookherjee, B.D. and Ho, C.-T. (1985a). pH effect on the volatile components in the thermal degradation of cysteine. *J. Agric. Food Chem.*, **33**, 442–446.

Shu, C.K., Hagedorn, M.L., Mookherjee, B.D. and Ho, C.-T. (1985b). Volatile components of the thermal degradation of cystine in water. *J. Agric. Food Chem.*, **33**, 438–442.

Shu, C.K., Mookherjee, B.D., Bondarovich, H.A. and Hagedorn, M.L. (1985c). Characterization of bacon odor and other flavor components from the reaction of isovaleraldehyde and ammonium sulfide. *J. Agric. Food Chem.*, **33**, 130–132.

Steverink, A.T.G. (1981). Review of the literature on the chemical constituents of chicken meat flavor and suggestions for future research. In *Quality of Poultry Meat, Proceedings of the 5th European Symposium*, eds. R.W.A.W. Mulder, C.W. Scheele and C.H. Veerkamp, pp. 356–371.

Stoll, M., Dietrich, P., Sundte, E. and Winter, M. (1967). Recherches sur les aromes. 15. Sur l'arome du cacao. *Helv. Chim. Acta.*, **50**, 2065–2067.

Tang, J. and Ho, C.-T. (1991). Unpublished results.

Tang, J., Jin, Q.Z., Shen, G.-H., Ho, C.-T. and Chang, S.S. (1983). Isolation and identification of volatile compounds from fried chicken. *J. Agric. Food Chem.*, **31**, 1287–1292.

Tressl, R. and Silwar, R. (1981). Investigation of sulfur-containing components in roasted coffee. *J. Agric. Food Chem.*, **29**, 1078–1082.

Tressl, R. (1989). Formation of flavor components in roasted coffee. In *Thermal Generation of Aromas*, eds. T.H. Parliment, R.J. McGorrin and C.-T. Ho. ACS Symp. Ser. 409, American Chemical Society, Washington, DC., pp. 285–301.

van der Linde, L.M., van Dort, J.M., de Valois, P., Boelens, H. and de Rijke, D. (1979). Volatile compounds from thermally degraded thiamine. In *Progress in Flavour Research*, eds. D.G. Land and H.E. Nursten, Applied Science, London, pp. 219–224.

Van Germert, L.J. and Nettenbrijer, A.H. (1977). *Compilation of Odour Threshold Values in Air and Water*. National Institute for Water Supply, Voorburg, The Netherlands.

Vernin, G. and Vernin, G. (1982). Heterocyclic aroma compounds in foods: occurrence and organoleptic properties. In *The Chemistry of Heterocyclic Flavouring and Aroma Compounds*; ed. G. Vernin. John Wiley & Sons, New York, pp. 72–150.

Wilson, R.A. and Katz, I. (1972). Review of literature on chicken flavor and report of isolation of several new chicken flavor components from aqueous cooked chicken broth. *J. Agric. Food Chem.*, **20**, 741–747.

Wilson, R.A., Vock, M.H., Katz, I. and Shuster, S.J. (1974). British Patent 1,364,747.

Withycombe, D.A. and Mussinan, C.J. (1988). Identification of 2-methyl-3-furanthiol in the steam distillate from canned tuna fish. *J. Food Sci.*, **53**, 658–659.

Zhang, Y. and Ho, C.-T. (1989a). Formation of meat-like aroma compounds from thermal reaction of inosine 5-monophosphate with cysteine and glutathione. *J. Agric. Food Chem.*, **39**, 1145–1148.

Zhang, Y. and Ho, C.-T. (1989b). Volatile compounds formed from thermal interaction of 2,4-decadienal with cysteine and glutathione. *J. Agric. Food Chem.*, **37**, 1016–1020.

Zhang, Y., Chien, M. and Ho, C.-T. (1988). Comparison of the volatile compounds obtained from thermal degradation of cysteine and glutathione in water. *J. Agric. Food Chem.*, **36**, 992–996.

5 Sheepmeat odour and flavour

O.A. YOUNG, D.H. REID, M.E. SMITH and
T.J. BRAGGINS

5.1 Introduction

Sheepmeat (the flesh of *Ovis aries*) is eaten by millions of people all over the world and is probably eaten in every country to some extent. There are no religious or cultural taboos on eating sheepmeat, which contrasts sharply with the taboos that apply to beef (Hindu) and pork (Moslem, Jewish). Nevertheless, many people avoid sheepmeat because they object to its odour (especially during cooking) and/or its flavour. The Chinese even have a special word for the disagreeable cooking odour of sheepmeat, 'soo', meaning sweaty, sour (Wong, 1975). Even in those Western countries that have a greater acceptance of sheepmeat, many dislike it, particularly the meat from mature animals with its apparently stronger odour and flavour. Also, the relatively high melting point of sheep fat contributes to a 'waxy' mouthfeel that is unacceptable to many. On a cool plate the fat tends to harden rapidly, which contrasts with the more oily character of, say, pork fat.

These and other factors conspire to limit the consumption of sheepmeat in countries where consumers are affluent and have a wide choice of meats. The United States and Canada are good examples. In 1987, their (bone-in) annual consumptions were a mere 0.7 and 0.9 kg/person respectively (Smith and Young, 1991). At the other end of the scale, Mongolians, poor by Western standards, consume about 60 kg/person/year.

Sheepmeat consumption is also affected by historical agricultural practices that have led to current dietary traditions. As a domestic animal, the sheep is suited to arid climates such as are found on either side of the equator. Many oil-rich Middle Eastern countries, for example Kuwait (35 kg/person/year), Saudi Arabia 21 kg, Libya 18 kg and Iran 9 kg have a history of sheepmeat consumption and remain large per caput consumers. They import sheepmeat, indicating a specific demand for it. Evidently their populations like mutton odour and flavour, simply disregard it, or use spices to modify it. All three factors may be important.

This review examines the chemicals thought responsible for and the factors affecting sheepmeat's characteristic odour and flavour in an effort to identify ways in which that odour and flavour can be modified. The objective is clear, to sell more sheepmeat to people who currently disdain it because they do not like the way it smells and tastes.

5.2 Assessment of sheepmeat odour and flavour by sensory panels and chemical analysis

Sensory panels assessing sheepmeat odour and flavour must be asked appropriate questions. Crouse (1983) noted that in 24 sensory studies of sheepmeat odour or flavour, 16 used a hedonic scale. Hedonic scales are useful in studies seeking a greater market share for sheepmeat, but an intensity scale is essential for studying 'muttonness' in a scientific way. With hedonic scales the responses of those who like a strong odour and flavour can cancel out responses of those who dislike it. In both types of assessment the use of ethnic panels should be considered, although this can be expensive. In New Zealand, Japanese tourists have been used as assessors.

Methods of chemical analysis of sheepmeat cooking volatiles are the same as those for other species. Most current methods rely on capillary or packed-column gas chromatography using various types of detectors, as the means of separating and identifying the components of complex mixtures.

For chromatography, the volatiles must first be trapped in quantities large enough to satisfy the sensitivity requirements of the detector used. Two advanced trapping methods are head-space sampling, and adsorption on porous polymers, both followed by cryofocussing the volatiles at the start of the column (Suzuki and Bailey, 1985; MacLeod and Ames, 1986; St. Angelo *et al.*, 1987). To cryofocus the volatiles from porous polymers such as Tenax™ requires an initial desorption step, accomplished by heating at temperatures up to 260°C. Cooking volatiles are usually generated at lower temperatures than this, so there is a risk that artefacts will be produced. A way to avoid this is to elute with a solvent. However, solvent methods suffer from the risk of introducing impurities or losing some of the volatiles when evaporating excess solvent.

Another approach is to elute with supercritical fluids. Supercritical carbon dioxide is an excellent eluant (Raymer and Pellizzari, 1987). A preliminary study by one of us (TJB), using supercritical carbon dioxide to elute mutton volatiles from Tenax TA™, showed that additional compounds can be recovered compared to thermal desorption.

Once cryofocussed at the start of the chromatography column, the volatiles are separated by applying a temperature gradient. Often the column effluent is split so that each effluent component or group of them can be simultaneously measured and smelt. But for identifying a characteristic species odour, descriptions of individual components may be of limited value, since a species odour is likely to be composed of a mixture of volatiles.

5.3 The tissue source of mutton odour and flavour

Even before sheepmeat is cooked, specific odours are evident. Workers handling sheep carcasses develop a distinctive odour on their hands as they become smeared with fats and presumably other tissue components. Although there is no proof that the odour is fat-derived, subcutaneous fat is a likely source. The chemistry of this effect has never been examined, and the relationship of it to cooking odour is unknown.

The tissue source of cooked mutton odours, whether from the fat, the lean or both, has received much attention. In a classic paper, Hornstein and Crowe (1963) proposed that lean meat provides the basic meaty flavour common to beef, pork and sheepmeat, whereas the fat is responsible for the species flavour. There is evidence to support this model in sheepmeat. Wasserman and Talley (1968) found that lamb fat was distinctive to the point that the addition of beef fat to veal did not increase panel recognition of veal as beef, whereas addition of lamb fat to veal significantly increased the false identification of veal as lamb. Pearson *et al.* (1973) made water extracts of lean beef and lamb. Panelists smelling the heated extracts could not differentiate between the two species. When the species' fat was added to the respective extract, the panel noticed a difference between the two samples, but still could not pick lamb from beef. However, samples of pure boiled lamb fat were distinguished from equivalent beef fat samples. Wenham (1974) trimmed ewe mutton and beef to nearly zero visible fat, then back-blended mutton and beef fat in various ratios. Although the assessment scale was hedonic, when the added mutton fat content reached 20%, the patties became unacceptable, unlike those with 20% beef fat. Wenham also showed that blends of lean mutton and lean beef could be distinguished (but not correctly identified). Wenham argued that any difference between beef and mutton lean was due to the content of non-visible fat. In this experiment it was higher for mutton lean (10.7%) than beef lean (6.4%). Using a panel selected on ability to distinguish lamb, beef and pork, Brennand and Lindsay (1982) clearly showed that fatty tissues were the most significant source of mutton flavours, but were not so strongly distinguishing for beef and pork flavours.

Echoing Wenham's (1974) comment about non-visible fat, MacLeod and Seyyedain-Ardebili (1981) concluded in a major review that if fat is an important contributor to species flavour, the intramuscular (non-visible) fat is sufficient to generate species flavour. Moody (1983) also concluded there is sufficient fat in lean meat to allow the development on cooking of normal (species) flavour. Lean tissue lipids contain a relatively high proportion of unsaturated fatty acids, which are highly susceptible to oxidation. A number of authors have implicated fat oxidation products as contributors to species odour and flavour (see later). However, lipid oxidation products from depot fats will also contribute. Depot fat tissue has cell membranes that contain the susceptible fatty acids, and storage triglycerides of domestic animals contain high proportions of unsaturated fatty acids that are capable of being oxidized.

Mutton tallow has a distinct sheepy odour, which can develop in even steam-deodorized samples (Hoffmann and Meijboom 1968; Brown, 1989). Hoffmann and Meijboom attributed tallow odour to fatty acid oxidation products, several of which they identified. Although tallow odour and mutton cooking odour are probably different, the capacity of mutton tallow to produce odorous compounds with a 'sheep note' clearly implicates tissue fat as a major odour source.

A New Zealand company, Tenon Developments Ltd., has extracted more than 96% of the fat from raw mutton, using supercritical carbon dioxide in a pilot-scale extraction process. The company has told us that most of the mutton odour appears to be extracted with the fat. Results of earlier experiments in our laboratory (unpublished) with liquid carbon dioxide (not supercritical) support Tenon's observation. All in all, the evidence indicates that sheepmeat odour and flavour come more from fatty than from lean tissue, although lean tissues undoubtedly make a contribution.

5.4 Chemical components involved in sheepmeat odour and flavour

Most of the chemical analyses directed at sheepmeat odour and flavour have centred on trapping volatiles released during cooking. This is a sensible approach as consumers sense odour first, and what they smell before the meal will certainly colour their appreciation of the food. But odour and flavour of foods are inextricably linked (Meilgaard *et al.*, 1987), so what determines odour will to some extent determine flavour.

A given cooking volatile from any species is seldom unique and there is general agreement that species-characteristic odour will be represented by a blend of components, each having its own detection threshold and each present at different concentrations. Nonetheless, several schools of thought have emerged, focusing on certain classes of compounds as being the most dominant contributors to mutton odour. Such a simplified approach is understandable considering the complexity of gas chromatographs of volatiles.

Because of the undoubted contribution of fat to cooking odours, several groups have proposed that oxidation products of fat are important contributors to mutton odour. The oxidation products are mainly alkanes, aldehydes, ketones, alcohols and lactones. Hornstein and Crowe (1963) and Jacobson and Koehler (1963) identified carbonyls as important contributors to mutton odour. More recent studies have extended their observations. Caporaso *et al.* (1977) rendered mutton fat (at 50°C) and judged it to have the characteristic odour even after this minimal heating. The rendered fat was then heated to simulate oven temperatures and steam extracted. The residual fat had no odour. Instead, odour compounds were present in the extract, specifically the neutral fraction rather than the acidic or basic fractions. Subsequent gas chromatographic–mass spectrometric analysis coupled with sensory evaluation allowed the group to shortlist ten aldehydes, three ketones and one lactone as significant contributors to mutton odour. Work by Dimick *et al.* (1966) and Watanabe and Sato (1968) implicated fatty lactones in mutton odour, although only uncooked fat was examined in these two studies.

Cross and Ziegler (1965) evaluated volatiles from cured and uncured cooked pork, beef, and chicken, and found that the volatiles of cured and uncured chicken and beef, after having been stripped of aldehydes and ketones by passage through a 2,4-dinitrophenylhydrazine solution, possessed an odour very similar to that of cured ham. They suggested that the flavour of cured meat is the basic meat flavour, derived from precursors other than fats, and that the different cooking odours of the various types of cooked meat depend on the types of carbonyl compounds derived from fat oxidation.

Expanding on this hypothesis, Rubin and Shahidi (1988) proposed that meat flavours can be rationally divided into two groups that are relevant here: the fundamental meat flavour obtained by cooking, or more generally, by preventing oxidation; and the flavour of uncured cooked meat, with species differentiation largely due to different types and concentrations of carbonyls from lipid oxidation.

No systematic study has been published that conclusively confirms Cross and Ziegler's hypothesis. If true, it could provide the means of minimizing mutton odour for markets where this was important.

Other studies suggest that the contribution of carbonyls to mutton flavour is not significant. Hornstein and Crowe (1963) found that lamb fat developed a mutton aroma when heated either in air or in nitrogen (where lipid oxidation would presumably be limited by a lack of oxygen), but the aromas of beef and pork fats were quite different following heating in a vacuum compared with samples heated in air (Hornstein and Crowe, 1960). Locker and Moore (1977) found that 'bacon' made from mutton has a pronounced mutton flavour in the fat, and Berry *et al.* (1989) found that chopped-and-formed bacon containing 12% lean mutton had a significantly less desirable flavour than commercial bacon or chopped-and-formed bacon containing pork lean. These findings suggest that the undesirable flavour of both fat and lean mutton persists despite minimizing oxidation by, for example, curing. Lipid oxidation and mutton odour is assessed later in this review.

A quite different involvement of fats was described by Wong *et al.* (1975a). They showed that the volatile fatty acids from mutton fat contained branched-chain or unsaturated acids with eight to ten carbon atoms that contributed to the characteristic odour of cooked mutton. Two acids, 4-methyloctanoic (especially) and 4-methylnonanoic, were identified as being important. In a related study this group showed that lacing a bland mutton sample with 4-methyloctanoic acid significantly ($p < 0.01$) increased panel scores for degree of mutton flavour (Wong *et al.*, 1975b). The acid was particularly concentrated in the fat of sheep that panelists described as very muttony. It was absent in beef. 'Goatiness' was also attributed to this acid. But the authors concluded that although the branched chain acids contributed to mutton flavour, other factors were important too. Using the same branched-chain acids as Wong *et al.*, Brennand and Lindsay (1982) arrived at the same conclusion.

Branched-chain fatty acids are featured in much of the work on diet (see later). Miller *et al.* (1986) showed that concentrations of branched chain fatty acids were low in intramuscular fats (triglycerides and phospholipids) as opposed to subcutaneous fats. Since subcutaneous fats are strongly implicated in mutton odour, this difference in concentration tends to support the hypothesis of Wong *et al.* On the other hand Crouse *et al.* (1982) could find no significant correlation between 'lamb flavour' intensity and one of Wong's indicator branched-chain fatty acids, namely 4-methylnonanoic acid. Crouse *et al.* even concluded that fatty acids had little effect on variation in lamb flavour intensity. More recently, Purchas *et al.* (1986) found that there were no individual fatty acids, or groups of them, which showed consistent relationships with sheepmeat flavour characteristics. In spite of these objections to Wong's theory, there is no doubt that branched-chain fatty acids dominate the fatty acid component in the volatiles of cooking sheepmeat, but the same is not true of beef or pork (Baines and Mlotkiewicz, 1984).

Other factors noted by Wong *et al.* such as neutral volatiles from cooking mutton, were examined by Nixon *et al.* (1979). As expected, many of the compounds identified had been previously described by Caporaso *et al.* (1977), but Nixon *et al.* observed that the formation of aldehydes and ketones comprising the 'muttony' shortlist of Caporaso *et al.* depended on the cooking method. But irrespective of method, the volatiles always smelt 'sheepy' to quote Nixon *et al.* Moreover, Dwivedi (1975) had shown that the 'muttony' volatiles also occurred in beef volatiles. Although the work of Caporaso *et al.* is flawed on these grounds, Nixon *et al.* could propose no alternative theory. Both works stand as worthwhile libraries of mutton volatiles.

Arguing that common experience shows that lean mutton can also yield the mutton odour, Wong and Mabrouk (1979) assessed mutton odour in the aqueous fraction of a chloroform–methanol extract of lean mutton. The amino acid fraction was the chief source of the odour, but the specific components responsible were, unfortunately, not identified.

Another school of thought is that basic components of volatiles significantly contribute to mutton odour (Buttery *et al.*, 1977). Of the several pyrazines and pyridines identified, 2-ethyl-3,6-dimethylpyrazine and 2-pentylpyridine were considered likely candidates as specific contributors to mutton odour. These workers proposed that 2-pentylpyridine could be formed from ammonia and a fat oxidation product, deca-2,4-dienal. The former would be formed from the breakdown of amino acids, polymerized or otherwise, present in sheep fat. Ammonia is a well-known volatile of cooking meats and is particularly common in sheepmeat volatiles (Baines and Mlotkiewicz, 1984). This is perhaps due to the fact that the sugar content of sheepmeat is lower than that of pork and beef (Macy *et al.*, 1964a). During cooking, the amino acids of sheepmeat are less likely to react with sugars (to form Maillard products) and more likely to be thermally degraded to ammonia.

Pyrazines and pyridines are two examples of the several classes of heterocyclic volatiles produced from any cooking meat. They include furan derivatives and a variety of *N*- and *S*-heterocyclics. Apart from Buttery *et al.* (1977), no authors have claimed that a given heterocyclic is significantly responsible for mutton odour, but heterocyclics undoubtedly contribute to the odour. The odour thresholds of many heterocyclic compounds are extremely low (Mussinan *et al.*, 1976). The thiamine degradation products 2-methyl-3-(methyldithio)furan and bis-(2-methyl-3-furyl) disulphide can be sensed in water at 1 part in 10^8 and 2 parts in 10^{14}, respectively (MacLeod, 1986).

Sulphur compounds have had special attention in sheepmeat odour and flavour. Wool, a fibre unique to sheep, is rich in cystine. Cramer (1983) proposed this may have resulted in unique mechanisms for sulphur storage, notably in fatty tissue (see below); unidentified sulphur stores could supply compounds or precursors that would make the cooking odour of lamb different from that of other species.

When meat is cooked, sulphur-containing volatiles are generated primarily as a result of degradation of sulphur amino acids. The most dominant sulphur compound in meat volatiles is H_2S (Nixon *et al.*, 1979). It has its own smell and can act as a precursor for other odorous compounds (see later). Kunsman and Riley (1975) found that on a fresh weight basis, the depot fat tissues of beef and lamb gave off much larger amounts of H_2S than the lean tissues, and that lamb samples produced considerably more H_2S than beef samples (Figure 5.1, bottom curves). This result was at the root of Cramer's proposal regarding sulphur storage in sheep.

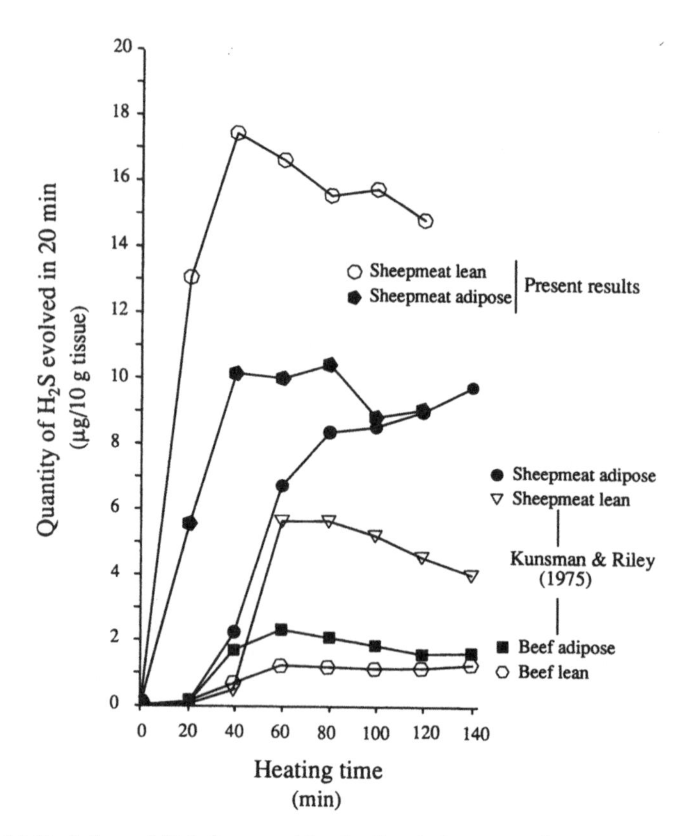

Figure 5.1 Evolution of H_2S from cooking beef and sheepmeat tissue. In contrast to the results of Kunsman and Riley (1975) (bottom four lines), the present results show that more H_2S is evolved from the lean tissue than from adipose on cooking.

The finding that fat released more H_2S than lean on cooking is surprising considering that the protein (and hence amino acid) content of lean tissue is much higher than that of fat. The relative rates of H_2S evolution from cooking sheepmeat have been reassessed in the our laboratory, using a cooking procedure adapted from Pepper and Pearson (1969). Regardless of sex or age of the sheep, more H_2S evolved from the lean than from the fat when expressed on a fresh weight basis (Figure 5.1, top curves). This result is clearly contrary to that of Kunsman and Riley, although they used a different cooking method. Our results in Figure 5.1 cast some doubt on Cramer's proposed sulphur stores in sheep fat. Furthermore, Ockerman *et al.* (1982) showed that haired sheep, which are indigenous to tropical regions, were no more or less muttony than woolled sheep. Nonetheless, the observation made by Kunsman and Riley (1975), that lamb evolved more H_2S than beef, may well be valid (Figure 5.1).

The literature often states that (lean) lamb has a high cysteine/cystine content (see for example, Baines and Mlotkiewicz, 1984) and that this is the basic cause of enhanced H_2S generation. However, amino acid composition data (Paul and Southgate, 1978; Chrystall and West, 1991) for beef and lamb show no significant difference in methionine and half cystine contents. Clearly though, these data do not show what proportion of each amino acid is free or polymerized in proteins. Differences in distribution could affect cooking odour, as free amino acids are probably more labile. The only published clue to the compositions of free amino acids is that provided by Macy *et al.* (1964b), who found that the free amino acid contents of lean beef, pork and lamb were qualitatively similar.

One sulphur precursor that has been implicated in enhanced H_2S formation in sheepmeat is glutathione. Glutathione gives off H_2S more rapidly than do sulphur amino acids polymerized in proteins (Cramer, 1983), although the latter contain much more sulphur. Macy *et al.* (1964b) noted that glutathione was present in water extracts of lean lamb but not in those of lean beef and pork. However, as glutathione is an essential part of the free radical scavenging system common to mammalian cells (Munday and Winterbourn, 1989), it is unlikely to be absent from fresh beef and pork.

Schutte (1974) described how thiamine degradation products could play a part in meat odour. Lean lamb contains about 0.14 mg of thiamine per 100 g, double that in beef, 0.07 mg/100 g. Pork is higher still at 0.89 mg/100 g (Paul and Southgate, 1978). Although thiamine concentrations are insignificant compared to the total methionine and cysteine concentrations for beef and lamb—around 500 and 250 mg/100 g, respectively—there is general agreement that thiamine degradation products play a very significant part in determining meatiness. But because the concentration of thiamine in lamb is not outstanding, it seems unlikely that primary degradation products of thiamine are responsible for mutton odour.

Collectively, the data suggest that the sulphur contents of lean beef and sheepmeat are not fundamentally different. But what differences there are in the distribution of sulphur might be sufficient to alter the path of sulphur compound degradation, particularly when the concentrations of other reactants, precursors or derived, vary from species to species. Reviews by Mabrouk (1976), Baines and Mlotkiewicz (1984) and Cramer (1983) discuss the H_2S- and methanethiol-dependent reactions that lead to cyclic and aliphatic sulphur compounds. (Methanethiol is a Strecker degradation product of methionine.) As an example of how interaction might occur, Baines and Mlotkiewicz (1984) proposed that sulphur compounds, simple and complex, can be retained by fatty tissue. Their presence there would allow them to interact with lipid oxidation products, to produce a host of complex sulphur compounds. Many sulphur-containing volatiles have exceptionally low aroma thresholds and as a result may be effective contributors to mutton odour and flavour at very low concentrations, as they are in beef (MacLeod, 1986; Werkhoff *et al.*, 1990). Species differences in concentrations of odour precursors, such as lipid oxidation products, could lead to the formation of sulphur compounds unique to cooked sheepmeat volatiles that have not been isolated from other species. Nixon *et al.* (1979) identified a number of sulphur compounds unique to sheepmeat volatiles, although there was no evidence that these are singularly important in sheepmeat odour.

There is clearly no consensus as to the cause of the characteristic sheepmeat odour. The following sections review the factors that are known to affect sheepmeat odour and flavour. Where possible the information will be discussed with respect to the classes of compounds noted above.

5.5 Factors affecting sheepmeat odour and flavour

5.5.1 Pre-slaughter factors

Age. An odour difference between lambs and cull ewes (usually three or more years old) can be noticed during carcass dressing. This is not documented in the scientific literature, but is well known to those who have worked in a sheep slaughterhouse. This simple observation implies that odorous compounds accumulate with age. Trapping and identifying these volatiles has never been attempted. The raw and cooked odours might be related; a study of the raw-meat volatiles would not be complicated by the host of compounds present in volatiles released during cooking but not specific to mutton odour.

It is not clear whether the odour of cooking sheepmeat is so distinct that it is present at any age, or whether mutton has a significantly different odour and flavour from lamb. People who dislike sheepmeats generally claim the former, but others apparently have different sensory perceptions of lamb and mutton.

It is generally believed that as animals grow older their meat becomes more strongly flavoured. There is some published evidence for sheepmeat that supports this view (Paul *et al.*, 1964; Batcher *et al.*, 1969; Misock *et al.*, 1976), but others have reported the reverse (Weller *et al.*, 1962). In all these studies panelists were not questioned specifically on mutton odour or flavour, although some participants in the studies carried out by Weller *et al.* commented on 'muttony' flavours. Panelists taking part in the studies carried out by Misock *et al.* agreed that a 'very strong aroma' in cooked fat was undesirable. In these studies any change in disagreeable odour or flavour with age was minimal. However, it is perhaps significant that these four studies examined animals no older than 16 months. A pronounced mutton odour and flavour might not be obvious until animals are much older.

Accepting that mutton odour probably increases in intensity with age, no one apparently has used this as a means of studying the chemicals responsible. With identical or genetically similar sheep it should be possible to identify compounds that are unique to the volatiles from older sheep, or are present at changed concentrations. Modern computer-assisted analysis should make detailed comparisons of the volatiles relatively easy.

Of the many biochemical changes that sheep must undergo as they age, four are here noted with respect to mutton odour. The lean meat of older sheep contains more haem iron, present as myoglobin. New Zealand meat retailers call carcasses from older sheep 'red sheep', reflecting this colour change. During cooking, haem iron and iron released from haem can act as catalysts of lipid oxidation (Rhee, 1988; Johns *et al.*, 1989) with its multitude of downstream effects.

Sheep become fatter as they age. The fat tends to accumulate in subcutaneous depots rather than intermuscularly or around organs such as kidneys (Broad and Davies, 1980; Butler-Hogg, 1984). Since fats are strongly implicated in mutton odour (see earlier), it seems reasonable that the increased fattiness alone may be important in odour. This may be especially true when meat is roasted as a whole cut. Precursors such as ammonia and H_2S will have more opportunity to interact with lipid oxidation products in a thicker subcutaneous fat layer. The importance of the interaction of lean and fat in mutton odour was noted by Cramer *et al.* (1967). As appealing as this theory might be, Field *et al.* concluded in a 1983 review that plane of nutrition, which results in different amount of fat cover at the same age, has no significant effect on sheepmeat flavour. On the other hand, in the diet experiments that support the conclusion of Field *et al.* (see later), mutton odour intensity of whole roasted meats was not assessed.

As sheep age, the fatty acid composition of their depot fats changes. In general the fats become more saturated (Barnicoat and Shorland, 1952; Spillane and L'Estrange, 1977; Miller *et al.*, 1986). Although this may relate to animal fatness rather than age (Bensadoun and Reid, 1965), older sheep being generally fatter, the fact remains that a changed fatty composition has the potential to change odour by way of volatile free fatty acids and fatty acid oxidation products. Also, age effects have been observed in the content of branched-chain fatty acids, although the results of two studies are conflicting (Bensadoun and Reid, 1965; Spillane and L'Estrange, 1977), the former showing an increase with age and the latter a decrease. The hypothesis of Wong *et al.* (1975a,b) requires that the concentration of branched-chain fatty acids, specifically around C9 and C10, must increase with age if it is accepted that older animals have more mutton odour than younger. Unfortunately, the two conflicting studies just mentioned addressed mainly long-chain branched fatty acids. Another age-related comparison can be made between the studies of Miller *et al.* (1986) and Busboom *et al.* (1981). Miller *et al.* noted that regardless of energy level in the diet, the subcutaneous fat of old cull ewes contained a lower concentration of 4-methyl branched acids than did fat from lambs used in studies carried out by Busboom *et al.* An age-dependent study that simultaneously compared panel assessment of mutton odour and branched chain fatty acid content would be a useful test of the hypothesis put forward by Wong *et al.*

The fourth age-related change concerns pH. Johnson and Vickery (1964) showed that the higher the pH of meat, the more H_2S is produced on heating; a pH increase from 5.6 to 6.6 would increase the evolution of H_2S by about 60%. The dark-cutting condition in beef is due to a low glycogen content in muscles at the time of death, resulting in a rigor pH greater than 6.0. Not much is known of the condition in sheep, but it certainly exists. Petersen (1983) found that about 7% of randomly chosen lambs had rigor pH values greater than 6. Shorthose (1989) has some evidence that the occurrence of high rigor pH is much greater in older sheep. This may be related to nutritional stress. In view of the role of H_2S in odour, a high rigor pH might be linked to increases in mutton odour. However, experimental evidence suggests otherwise. Bramblett *et al.* (1963) found that the aroma and flavour of lamb treated with adrenalin before slaughter to reduce muscle glycogen were unchanged compared to controls. Curiously though, pH was not measured in this experiment which was designed to test palatability as affected by stress-induced pH changes. Recent work by Graafhuis *et al.* (1991) shows that mean pH differences (5.84 versus 6.38) have no effect on odour or flavour.

Diet. No other factor is so amenable to experimentation as diet. Many studies have been performed that compare the organoleptic effects of diets. Many of these have been done in the United States, a relatively minor sheep producer. The motive behind these studies was to seek ways of enhancing the acceptability of sheepmeat because consumption per person is very low, less than 1 kg per year. It is likely that no diet applied to date has been able to change mutton odour/flavour to the point that the species is unrecognizable, although this may reflect a failure to transfer experimental know-how to production and marketing.

Considering the size of the Australian and New Zealand sheep industries, surprisingly little work has been done on the relationship between diet and odour/flavour, and no findings have been applied commercially. This may be due in part to the following reasons. Historically, their captive markets in Europe required cheap meat, which the two countries could easily supply provided the feed regime was forage-based, largely grass and clover. Further, the sheep being a dual-purpose animal (meat, wool) meant that at times the meat could be regarded almost as a by-product of the wool industry. There has been little incentive to change diet, whatever the organoleptic effect.

The major feeding options for sheep are grasses, legumes, and cereals. A complicating effect in feeding trials is the energy content of the feeds: highest in cereals, next highest in legumes and lowest in grasses. The conclusion of Field *et al.* (1983), that fat cover has no significant effect on sheepmeat flavour, was supported by experiments where the diet composition did not change when feeding level was varied. Paul *et al.* (1964) found no difference in lean meat flavour between fatty and less fatty cuts, and Kirton *et al.* (1981) found that palatability differences between pasture-fed animals that gained, maintained, or lost weight were insignificant. In other studies (for example, Jacobs *et al.*, 1973; Crouse *et al.*, 1978), changes in fatness were achieved by varying the diet composition. Even so, the conclusions were similar. In assessing mutton odour and flavour, the conclusion of Field *et al.* is important, since most experiments discussed below are complicated by variations in metabolizable energy and carcass fatness.

New Zealand pastures for sheep predominantly comprise a legume (white clover) and perennial ryegrass. Cramer *et al.* (1967) studied the two feeds in isolation and found that meat from sheep fed on clover had a significantly more intense flavour and odour than meat from grass-fed sheep. The word 'muttonness' was not used in sensory panel work. Several groups of Australian and New Zealand workers (Czochanska *et al.*, 1970; Shorland *et al.*, 1970; Nicol and Jagusch, 1971; Park *et al.*, 1972, 1975; Purchas *et al.*, 1986) generally agreed with the sensory results of Cramer *et al.*, although only Park *et al.* (1972) clearly asked panelists about mutton odour and flavour intensity. In that study, average mutton aroma and flavour values were slightly higher for lucerne-fed (another legume) animals, but the result was not statistically significant. 'Foreign flavour' and 'foreign aroma' averages were significantly higher for lucerne-fed than for grass-fed animals. It seems that the clover-dependent increase in intensity noted in the studies carried out by Cramer *et al.* might have been due to a foreign note. Adding to the confusion, Nicol and Jagusch (1971) obtained conflicting results from their trained and untrained panels.

Nixon (1981) carried out a repeat experiment that compared the effect of five legumes with that of ryegrass. Panelists, selected on their ability to discriminate mutton flavour, were asked to judge samples on 'muttonness' and 'off-flavours'. The latter were defined as not contributing to muttonness. There was no difference in muttonness between any of the samples. Off-flavour was significantly higher in the grass-fed samples than in any of the legume-fed. In experiments that compared grass-fed with tropical legumes-fed, Park and Minson (1972) found that the meat acceptability was usually not significantly different.

Researchers in the United States have studied a wide range of forage-based feeds. Few differences in acceptability were found in the two major studies, Crouse *et al.* (1978) and Kemp *et al.* (1981). Although muttonness was not assessed, if any sample had been found to be different in a panelist's perception of muttonness, we feel sure the authors would have reported it.

On balance it appears that differences in odour and flavour are affected by grass versus legume diets—and this may be seasonally related (Park and Minson, 1972; Park *et al.*, 1975)—but these differences do not relate to muttonness. Suzuki and Bailey (1985) reported that clover-fed lambs had higher concentrations of two particular ketones and certain diterpenoids than corn-fed lambs; these volatiles may be the cause of legume flavour.

Other feeds also affect odour and flavour. Spurway (1972) noted that some panelists described the cooking odour of rape-fed sheepmeat as reminiscent of boiled cabbage. But his work does not suggest that grazing on oats, rape, or vetch (a legume), fundamentally alters sheepmeat odour and flavour from conventional pasture grazing.

Another feeding option for sheep is a high-energy grain diet. Many authors have reported that sheep fats become soft and oily on such a diet (see for example, Field *et al.*, 1978; Hansen and Czochanska, 1978; Busboom *et al.*, 1981; Miller *et al.*, 1986). This is because increased propionate concentrations in the rumen translate to increased concentrations of odd-chain and branched-chain fatty acids in storage fats (Garton *et al.*, 1972). This can also happen in sheep fed a legume diet (Cramer *et al.*, 1967). Propionate can replace acetate as a primer in fatty acid synthesis, yielding odd-chain fatty acids (Horning *et al.*, 1961), and, by way of methylmalonate as an analogue of malonate, branched-chain acids. Barley-fed sheep used in studies carried out by Wong *et al.* (1975b) were significantly more muttony than sheep fed the control diets, and this increase, they claimed, was caused by the higher concentrations of C9 and C10 branched-chain fatty acids in the fat of barley-fed sheepmeat. By contrast, Locker (1980) noted that 'bland' was the most common descriptor for barley-fed lamb in the same panel study that decided that pasture-fed sheepmeat had a 'stronger flavour'. Muttonness was not assessed. (In the same experiment a diet of maize silage resulted in a 'porky' character.)

It is clear that meat from grain-fed sheep does have a different odour and flavour profile, but whether the character is capable of reducing or masking muttonness to the point of economic advantage is not known. In this respect Miller *et al.* (1986) believed that lean grass-fed old ewes could be fed corn supplements for short periods to lower the fat melting point and increase flavour desirability. On the other hand, Field *et al.* (1983) noted that off-flavours in lamb were common when sheep had soft, oily fat arising from a high-energy diet.

In all the feeding experiments discussed so far, the fatty acids entered the sheeps' bloodstream after hydrogenation in the rumen. Australian workers (Ford *et al.*, 1975) fed ewes and lambs unsaturated sunflower oil, protected from hydrogenation by suitable encapsulation. The treatment dramatically increased the linoleic acid content of depot fats over that of control animals fed combinations of oats and a legume. Sensory panel work revealed that the aroma and flavour intensities were reduced in the high linoleic acid group and that different aromas and flavours were present. Muttonness was not addressed. Panelists found the meat flavour unacceptable, but in a population accustomed to eating pasture-fed sheepmeat this is perhaps not surprising. Some panelists preferred the high-linoleic acid lamb. In a later study by this group (Park *et al.*, 1978a), mature sheep were assessed for mutton odour and flavour after a grass diet followed by a protected oil supplement. The Australian panel judged that the protected oil supplement reduced mutton odour and flavour. Other work by this group identified some of the diet-specific volatiles derived from high linoleic acid fat that contributed to the altered flavour and reduced muttonness. *Trans,trans*-deca-2,4-dienal concentration was an order of magnitude higher than in feed-lot or grazed animals (Park *et al.*, 1978b). An unsaturated γ-dodecalactone was held responsible for a sweet or fruity odour. Whether these and other linoleate-derived volatiles reduced or simply masked mutton odour is not known.

In New Zealand, Purchas *et al.* (1979) confirmed the results given by the sensory panel used in studies carried out by Park *et al.* The panel found that meat from sheep fed a protected oil supplement was less acceptable. Sheep fed protected oil were also included in the work of Wong *et al.* (1975b) who evaluated the mutton-like nature of branched-chain fatty acids. The muttonness score of these sheep was the lowest (and significantly so) of the several diet treatments, which included barley, and grass plus legume.

Breed and sex. Cramer *et al.* (1970a) found that a fine-woolled breed, Rambouillet, produced meat that was more muttony ($p < 0.01$) than two coarse-woolled breeds, Columbia and Hampshire. Later work, which extended the study to the very fine-woolled Merino breed, failed to confirm this (Cramer *et al.*, 1970b). The experiment did show, however, that compared with the other breeds, Merinos had an oily subcutaneous fat, which was reflected in the breed's fatty acid composition. Variations in fatty acid composition are frequently implicated in odour and flavour differences (see earlier). Experiments by others assessing the effects of breed or sire breed (see for example, Fox *et al.*, 1962, 1964; Dransfield *et al.*, 1979; Mendenhall and Ercanbrack, 1979; Crouse *et al.*, 1981, 1983) generally yielded no significant differences.

In Australia and New Zealand it is common belief that meat from

Merinos is more muttony than that from other breeds. The origin of this belief might be age-related. Merino is primarily a wool breed. So long as a Merino is adequately producing wool it is likely to be held on-farm rather than slaughtered. As a result, Merinos at slaughter are likely to be older and therefore more muttony if age and muttonness are related. A similar argument might apply to the species as a whole; as a meat animal the sheep is a victim of its success as a fibre-bearing animal.

Since ram lambs grow faster than wethers (castrates) (Kirton *et al.*, 1982; Dransfield *et al.*, 1990), there are economic advantages in growing rams. Considerable research has been directed at the effect of animal sex on flavour differences. Much of the published work shows insignificant differences between ram, wether and ewe meat. This is true even of old animals (Kirton *et al.*, 1983). However, several studies have found ram meat to be significantly less acceptable than ewe meat (Cramer *et al.*, 1970a) or wether meat (Kemp *et al.*, 1972; Misock *et al.*, 1976; Crouse *et al.*, 1981; Field *et al.*, 1984). Misock *et al.* in particular noted that meat from rams had a stronger and more undesirable odour than meat from wethers. Busboom *et al.* (1981) analysed the fat from rams and wethers used in studies carried out by Crouse *et al.* and found that the fat of heavy rams in particular had more branched-chain fatty acids than that from all wethers and the fat was softer. Vimini *et al.* (1984) obtained similar results. There is some indication (Misock *et al.*, 1976; Crouse *et al.*, 1981) that the heavier the ram, the more intense the flavour, again possibly implicating fatty acids in flavour (and odour) differences.

New Zealand is one of the few major sheep producers where the production of entirely males is encouraged (Butler-Hogg and Brown, 1986). There have been no adverse market reports regarding the palatability of New Zealand's sheepmeats. In summary, there are breed and sex effects on odour and flavour, but they seem to be subtle and sometimes contradictory.

5.5.2 Post-slaughter factors

Culinary practice. Sheepmeat is prepared for the table by methods that vary from culture to culture. In Australia and New Zealand, sheepmeat is commonly dry-roasted as a whole piece. The roasting volatiles are distinctive, especially those from older animals. Often a large bunch of rosemary is included in the covered roasting pan. Why this is done is not known, although it is easy to think of explanations. One is that the antioxidant volatiles of rosemary permeate the subcutaneous fat and inhibit or modify the oxidation reactions that take place during cooking. Translated into everyday language this explanation would mean that sheepmeat cooking odour—specifically, the roasting odour—would be reduced. Another explanation is that rosemary volatiles mask mutton odour in some way. Some cooks often place slivers of garlic in the meat so that garlic volatiles permeate the tissues during cooking. Although garlic components are antioxidants and may act as such with sheepmeat, the role of garlic in sheepmeat cuisine may go beyond the antioxidant effect (see later). A third common practice is to apply a mint sauce to the sliced meat on the plate. Mint sauce is a sweet and sour preparation made of vinegar, sugar and crushed mint leaves. Mint is another excellent antioxidant but it is not obvious how the use of an antioxidant might improve the odour and flavour of lamb when added at such a late stage. Moreover, the fact that a fluid has been poured over the meat may be important, as might be the sweetness and the acidity of the sauce.

An interesting feature of retail meat sales in New Zealand is that quantities of minced (ground) beef outsell those of minced sheepmeat by approximately a factor of ten. Price is not the reason. Mincing probably distributes oxygen through the meat, which might lead, in the case of lamb, to undesirable odours during cooking. Alternatively, because the surface area capable of releasing odours is greater in minced meats, the mince will inevitably release more volatiles than will whole-tissue meat.

Specific herbs play an important role in enhancing the enjoyment of sheepmeat, a theme recently explored by Smith and Young (1991). They examined the herbs and spices used in different sheepmeat cuisines, to point the way to innovative sheepmeat products. If unpleasant sheepmeat odours can be inhibited, masked or otherwise changed for the better by the use of unusual herbs and spices, then producers of sheepmeat products should exploit them. For instance, the use of cinnamon, a tropical laurel, is common in Middle Eastern but not in Western sheepmeat cuisines, and presents an opportunity for adding spices.

As expected, cuisines tend to use condiments that historically grew locally (Smith and Young, 1991). Most if not all the latter are antioxidants (Gordon, 1987), but as is discussed later, minimizing oxidation does not necessarily reduce mutton odour and flavour. Garlic is the condiment most widely used in sheepmeats. Along with onions, this member of the lily family is noted for its complex sulphur chemistry. Alliin (*S*-allyl-L-cysteine sulphoxide) comprises about 0.25% of a garlic bulb (Block, 1985). It breaks down to a number of sulphur compounds that could mask volatile sulphur compounds released from cooking sheepmeat. Alternatively, garlic volatiles could react with meat volatiles to produce agreeable odours. If the sulphur compounds from garlic were specifically involved in making mutton more acceptable, this would support Cramer's (1983) theory that sulphur compounds are intimately involved in mutton odour. At a practical level, Klettner *et al.* (1989) found that high levels of mutton can be used in processed meat products for the German market if the appropriate seasoning, particularly with garlic, is used.

Sugars. As discussed earlier, the reduced sugar content in meat alters the pattern of amino acid breakdown during cooking, which in turn will affect the pattern of volatiles produced. Apparently unaware that sheepmeat has a lower sugar content than beef or pork, Hudson and Loxley (1983) studied the effect of pentose sugars on mutton odour and flavour. (Pentose sugars were chosen because a patent (US 836,694) claimed that beef flavours could be formed by heating amino acids with them.) An Australian sensory panel was chosen on the basis of being able to distinguish species flavour. Xylose-treated mutton (2%) had a significantly different odour and flavour from untreated mutton and untreated beef. The scores indicated that the treated mutton was no more similar to mutton than to beef. Panelists preferred the xylose-mutton to lamb, mutton or beef. Unfortunately the panelists were not questioned on intensity of mutton odour or flavour.

Various methods claimed to reduce mutton odour or flavour. The patent literature contains several methods claimed to reduce mutton odour. In light of Asians' common dislike of mutton, Japanese patents are of particular interest. Soaking mutton in a 0.1% solution of either malate, fumarate or succinate has been claimed to be effective in eliminating mutton odour (Japan 80 11,302). Cooking mutton or goat meat in the presence of maltol, ethylmaltol or isomaltol removes the offensive odours, according to Japan patent 80 88,677. Mutton treated with 0.05% to 5% of asparagine, glutamine, alanine, or glycine is claimed to lose its mutton odour on cooking (Japan application 70 91,341). Two of the examples in this application also employed the sugars xylose and glucose. Low molecular weight dextrins remove the cooking odour of mutton and fish, according to another patent (Japan 80 77,875).

Mechanisms for the above treatments or for pentose sugars (Hudson and Loxley, 1983), were not proposed by the authors. There are at least two possibilities. First, the additives are low molecular weight compounds that, as precursors, are likely to substantially alter the pattern of volatiles formation. If true, this would mean that subtle differences between species in the profile of low molecular weight precursors are important in specific odour generation. Second, the reaction products of amino acids and sugars heated together (Maillard reaction products) are very good antioxidants (Bailey *et al.*, 1987). Maltol is a good example (Sato *et al.*, 1973), and xylose is particularly effective as a precursor sugar of these antioxidants (Lingnert and Ericksson, 1981). If the Maillard reaction products derived in the patented treatments reduce mutton odour by reducing oxidation, this would support Rubin and Shahidi's (1988) theory that lipid oxidation products are at the root of species differences regarding odour and flavour. Work reported in the next section suggests that this is not the case.

Other patents (New Zealand 159,518; France 2,597,726) call for the use of certain Chinese herbs and 'Herbs de Provence', respectively, also noted by Smith and Young (1991). Again, the herbs may act as antioxidants and/or mask mutton odour. Hayama (Japan 71 30,780) claimed that boiling mutton in water containing ginger, followed by coating the meat with soya, sucrose, rice wine and egg, followed by drying was effective in enhancing the sensory properties. This claim is not doubted. Other combinations of complex additives are similarly likely to be effective, simply by swamping any mutton odour or flavour present.

A recent patent (Australia 596,801) describes a procedure, involving acetic acid, salt and phosphates, and exposure to ultraviolet light, that is claimed to reduce mutton odour. The mechanism is unstated. It is possible that reduced muttonness might stem from reduced H_2S evolution due to a more acidic environment (Johnson and Vickery, 1964). In this respect a patent (Japan 61 96,970) calls for browning mutton and then cooking in a 1% acetic acid solution as a means of reducing mutton odour.

World patent 87 03454 claims that by passing steam (12.12 KPa (0.12 atm) 45–60°C) through comminuted mutton, the mutton odour of the cooked product approaches that of lamb. The colour data presented for this patent show that the treated raw meat is at risk of turning grey (due to myoglobin denaturation). Interestingly, the patent also extends to the simultaneous use of antioxidants and pentose sugars, methods discussed above.

The Captech® process is an advanced controlled-atmosphere packaging system that maintains the microbiological quality of chilled meat by carefully controlling the packaging atmosphere to be extremely low in oxygen and to consist essentially of pure carbon dioxide. Gill (1988) noted that Captech® lamb had a 'strong ovine odour' when the packages were opened. He proposed that the carbon dioxide stripped the meat of its characteristic odour, leaving a reduced species flavour after cooking. Carbon dioxide is commonly used in supercritical fluid extraction technology to remove, among other things, odours and flavours from foods, although in the Captech® process the supercritical temperature and pressure is not reached.

Curing and anoxia. In contrast to beef and pork, it is uncommon to cure sheepmeats. This may have its origins in the relative hardiness of the species. Slaughter followed by curing and drying of meat presented an alternative to the expensive practice of housing animals over winter. With a hardy species there may have been no need to slaughter and cure. If the species odour and flavour is dependent on fat oxidation, which can be minimized by curing (Rubin and Shahidi, 1988), then one would expect few differences between the flavours of cured meats. If true, this could have implications in marketing sheepmeat products.

Table 5.1 Number of correct responses/number of panelists in triangle tests involving lean meats of four species

Species	Control	Nitrite	Salt + Nitrite
Mutton versus beef	6/9	7/11	8/12
Mutton versus pork	12/12***	11/12**	10/12*
Mutton versus chicken	11/11***	11/11***	11/12**
Beef versus pork	11/12**	12/12***	11/12**
Beef versus chicken	12/12***	12/12***	12/12***
Pork versus chicken	10/10**	12/12***	12/12***

Levels of significance: * $p < 0.05$; ** $p < 0.01$; *** $p < 0.001$.

To test the curing/flavour hypothesis, panelists were asked to differentiate between the flavours of chicken, pork, beef and mutton. Meats were trimmed of all visible fat (but retained intramuscular fat), minced, and treated with either nothing, or 200 ppm $NaNO_2$ or 200 ppm $NaNO_2$ plus 1% NaCl, before cooking and hot presentation. Tasters experienced in meat flavour research were asked to pick the different sample in triangle tests (Table 5.1). Panelists knew the four species involved, but did not know which species were presented in each test. Tests were conducted blind.

The inability of panelists to differentiate between uncured mutton and beef was unexpected, but similar difficulties were encountered by Pearson *et al.* (1973), as noted earlier. Nitrite curing did not alter the flavour perceptions of mutton and beef. The five other tests indicate that curing did not reduce species flavour differences. Because curing significantly reduces the formation of lipid oxidation products, the results indicate that these products are not the major determinants of flavour differences between the lean meats of different species. The results of the mutton versus pork test confirm the findings of Berry *et al.* (1989), noted earlier. The results also suggest that curing, by itself, would be of no practical value in disguising the flavour of sheepmeat served hot.

Oxidation can presumably also be reduced by processing meat totally in the absence of oxygen. An experiment was performed where (initially) pre-rigor lean meat was stored in the presence or total absence of oxygen for several days in the cold. It was also cooked in the controlled atmospheres before immediate presentation to an experienced panel. Panelists scored intensities on scales of 0 to 100 (Tables 5.2 and 5.3). Neither 'mutton odour' intensity nor 'other odour' intensity was affected, neither were flavours (Table 5.2).

However, mutton fat, similarly treated, did show significant differences. In this experiment (Table 5.3) the effect of nitrite on fat was also examined. Fatty tissue, trimmed from carcasses immediately after slaughter, was blended with nothing or 200 ppm $NaNO_2$. Half of the minced fat from

Table 5.2 Mean odour and flavour intensity scores for lean mutton samples, prepared, stored and cooked in the absence or presence of oxygen. Values in the same column bearing different superscripts are significantly different at $p < 0.01$. There were no significant differences

Treatment	Mutton odour	Other odour
Anoxic	33.3^{a}	17.7^{b}
Aerobic	33.5^{a}	19.6^{b}
	Mutton flavour	Other flavour
Anoxic	30.5^{c}	7.20^{d}
Aerobic	31.5^{c}	7.03^{d}

Table 5.3 Mean odour intensity scores for mutton fat samples that were prepared, stored and cooked in the absence or presence of oxygen and/or nitrite. Values in the same column bearing different superscripts are significantly different at $p < 0.01$

Treatment	Mutton odour	Other odour
Cured anoxic	26.9^{b}	29.1^{y}
Control anoxic	38.2^{a}	16.9^{z}
Cured aerobic	26.9^{b}	35.3^{x}
Control aerobic	27.3^{b}	33.5^{xy}

each treatment was immediately packed anoxically and the remaining tissue packed aerobically. After a week in the cold the samples were cooked before immediate presentation to the panel. Mutton odour was more evident in the control (uncured) anoxic treatment than in the others. Nitrite negated this effect, but did not eliminate mutton odour. It restored the score to that of the aerobic treatments. The two aerobic treatments scored highest for 'other odour', probably due to lipid oxidation products (Rubin and Shahidi, 1988). The fact that the control anoxic treatment scored highest for 'mutton odour' and lowest for 'other odour' indicates that lipid oxidation products do not significantly contribute to mutton odour. We are currently studying differences in the spectra of volatiles from these four treatments as one approach to the study of mutton odour and flavour.

The well-known reaction of nitrite, or more accurately nitric oxide, with haem is thought responsible for the pronounced antioxidative effect of nitrite. Nitrite also reacts directly with unsaturated bonds in fats (reviewed by Cassens *et al.*, 1979), and as a result possibly contributes to 'cured' flavour by altering the pattern of fat degradation on cooking. However, curing did not reduce mutton odour from fatty tissue (Table 5.3), confirming an observation by Locker and Moore (1977).

The results in Table 5.3 present a difficulty. Herbs and spices, well-known antioxidants, are effective in improving or otherwise enhancing sheepmeat odour and flavour. Similarly, Maillard reaction products, also good antioxidants that are claimed to reduce muttonness, are prominent in the patent literature. Armed with this information and Rubin and Shahidi's (1988) hypothesis on oxidation, one would logically expect the anoxic treatment of mutton fat to result in reduced mutton odour and flavour. That this was not the case underscores our current lack of knowledge.

5.6 Concluding remarks

Two problems face researchers who straddle the field between pure research and the marketplace. The first is to identify the chemical components responsible for mutton odour and flavour, and the second is to modify the organoleptic properties to make sheepmeat more attractive to consumers who currently dislike it. By applying meat product technologies developed for other species to sheepmeat and by appropriate spicing, more appealing sheepmeat products could almost certainly be made and marketed. But success might be short-lived without knowing the answer to the first problem—what causes the species' characteristic odour and flavour?

We suggest that identifying the components responsible is best approached by a comparative technique where cooked sheepmeats are assessed by a sensory panel to gauge muttonness, and at the same time, chromatographs of volatiles from the samples are compared for differences. The present review has highlighted the pre-slaughter factors of age and diet, and the post-slaughter factors of culinary practice, addition of sugars, anoxia and curing as being capable of yielding some changes in muttonness. By comparing data from experiments involving these variables it should be possible to solve this longstanding problem.

Acknowledgments

This work was funded by the Meat Research and Development Council of New Zealand.

References

Bailey, M.E., Shin-Lee, S.Y., Dupuy, H.P., St. Angelo, A.J. and Vercellotti, J.R. (1987). In *Warmed-over Flavor of Meat*, eds. A.J. St. Angelo and M.E. Bailey. Academic Press, Orlando, pp. 237–266.

Baines, D.A. and Mlotkiewicz, J.A. (1984). In *Recent Advances in the Chemistry of Meat*, ed. A.J. Bailey, Roy. Soc. Chem., Special Publ. No. 47.

Barnicoat, C.R. and Shorland, F.B. (1952). New Zealand lamb and mutton. Part II. Chemical composition of edible tissues. *N. Z. J. Sci. Technol.*, **5**, 16–23.

Batcher, O.M., Brant, A.W. and Kunze, M.S. (1969). Sensory evaluation of lamb and yearling mutton flavors. *J. Food Sci.*, **34**, 272–274.

Bensadoun, A. and Reid, J.T. (1965). Effect of physical form, composition and level of intake of diet on the fatty acid composition of the sheep carcass. *J. Nutrition*, **87**, 239–244.

Berry, B.W., Cross, H.R. and Smith, G.C. (1989). Processing, chemical, sensory and physical properties of bacon chopped and formed, made from pork, beef, mutton and chevon. *J. Musc. Foods*, **1**, 45–57.

Block, E. (1985). The chemistry of garlic and onions. *Sci. Am.*, **252**, 94–99.

Bramblett, V.D., Judge, M.D. and Vail, G.E. (1963). Stress during growth. II. Effects on palatability and cooking characteristics of lamb meat. *J. Anim. Sci.*, **22**, 1064–1067.

Brennand, C.P. and Lindsay, R.C. (1982). Sensory discrimination of species-related meat flavors. *Lebensm. Wiss. u. Technol.*, **15**, 249–252.

Broad, T.E. and Davies, A.S. (1980). Pre- and postnatal study of carcass growth of sheep. 1. Growth of dissectable fat and its chemical components. *Anim. Prod.*, **31**, 63–71.

Brown, G.I. (1989). New Zealand tallow markets: a survey of opportunities. Meat Ind. Res. Inst. N.Z. (Hamilton, New Zealand). Publ. No. RM 185.

Busboom, J.R., Miller, G.J., Field, R.A., Crouse, J.D., Riley, M.L., Nelms, G.E. and Ferrell, C.L. (1981). Characteristics of fat from heavy ram and wether lambs. *J. Anim. Sci.*, **52**, 83–92.

Butler-Hogg, B.W. (1984). The growth of Clun and Southdown sheep: body composition and the positioning of total body fat. *Anim. Prod.*, **39**, 405–411.

Butler-Hogg, B.W. and Brown, A.J. (1986). Muscle weight distribution in lambs: a comparison of entire male and female. *Anim. Prod.*, **42**, 343–348.

Buttery, R.G., Ling, L.C., Teranishi, R. and Mon, T.R. (1977). Roasted lamb fat: basic volatile components. *J. Agric. Food Chem.*, **25**, 1227–1229.

Caporaso, F., Sink, J.D., Dimick, P.S., Mussinan, C.J. and Sanderson, A. (1977). Volatile flavor constituents of ovine adipose tissue. *J. Agric. Food Chem.*, **25**, 1230–1233.

Cassens, R.G., Greaser, M.L., Ito, T. and Lee, M. (1979). Reactions of nitrite in meat. *Food Technol.*, **33**(7), 46–55.

Chrystall, B.B. and West, J. (1991). Composition of New Zealand Foods. 4. Beef and lamb. Meat Ind. Res. Inst. N.Z. (Inc.) and N.Z. Govt., Dept. Sci. Ind. Res. (in preparation).

Cramer, D.A. (1983). Chemical compounds implicated in lamb flavour. *Food Technol.*, **37**(5), 249–257.

Cramer, D.A., Barton, R.A., Shorland, F.B. and Czochanska, Z. (1967). A comparison of the effects of white clover (*Trifolium repens*) and of perennial ryegrass (*Lolium perenne*) on fat composition and flavour of lamb. *J. Agric. Sci.*, **69**, 367–373.

Cramer, D.A., Pruett, J.B., Kattnig, R.M. and Schwartz, W.C. (1970a). Comparing breeds of sheep. I. Flavor differences. *Proc. Western Sect. Am. Soc. Anim. Sci.*, **21**, 267a–271a.

Cramer, D.A., Pruett, J.B., Swanson, V.B., Schwartz, W.C., Kattnig, R.M., Phillips, B.R. and Wookey, R.E. (1970b). Comparing breeds of sheep. II. Carcass characteristics. *Proc. Western Sect. Am. Soc. Anim. Sci.*, **21**, 267b–272b.

Cross, C.K. and Ziegler, P. (1965). A comparison of the volatile fractions from cured and uncured meat. *J. Food Sci.*, **30**, 610–614.

Crouse, J.D. (1983). The effects of breed, sex, slaughter weight, and age on lamb flavor. *Food Technol.*, **37**(5), 264–268.

Crouse, J.D., Field, R.A., Chant, G., Jr., Ferrell, C.L., Smith, G.M. and Harrison, V.L. (1978). Effect of dietary energy intake on carcass composition and palatability of different weight carcasses from ewe and ram lambs. *J. Anim. Sci.*, **47**, 1207–1218.

Crouse, J.D., Busboom, J.R., Field, R.A. and Ferrell, C.L. (1981). The effects of breed, diet, sex, location and slaughter weight on lamb growth, carcass composition and meat flavor. *J. Anim. Sci.*, **53**, 376–386.

Crouse, J.D., Ferrell, C.L., Field, R.A., Busboom, J.R. and Miller, G.J. (1982). The relationship of fatty acid composition and carcass characteristics to meat flavor in lamb. *J. Food Qual.*, **5**, 203–214.

Crouse, J.D., Ferrell, C.L. and Cross, H.L. (1983). The effects of dietary ingredient, sex and slaughter weight on cooked meat flavor profile of market lamb. *J. Anim. Sci.*, **57**, 1146–1153.

Czochanska, Z., Shorland, F.B., Barton, R.A. and Rae, A.L. (1970). A note on the effect of the length of the resting period before slaughter on the intensity of flavour and odour of lamb. *N.Z. J. Agric. Res.*, **13**, 662–663.

Dimick, P.S., Patton, S., Kinsella, J.E. and Walker, N.J. (1966). The prevalence of aliphatic delta-lactones or their precursors in animal fats. *Lipids*, **1**, 387–390.

Dransfield, E., Nute, G.R., MacDougall, D.B. and Rhodes, D.N. (1979). Effect of sire breed on eating quality of cross-bred lambs. *J. Sci. Food Agric.*, **3**, 805–808.

Dransfield, E., Nute, G.R., Hogg, B.W. and Walters, B.R. (1990). Carcass and eating quality of ram, castrated ram and ewe lambs. *Anim. Prod.*, **50**, 291–299.

Dwivedi, B.K. (1975). Meat flavor. *CRC Crit. Rev. Food Technol.*, **5**, 487–535.

Field, R.A., Williams, J.C., Ferrell, C.L., Crouse, J.D. and Kunsman, J.E. (1978). Dietary alteration of palatability and fatty acids in meat from light and heavy weight ram lambs. *J. Anim. Sci.*, **47**, 858–864.

Field, R.A., Williams, J.C. and Miller, G.J. (1983). The effects of diet on lamb flavor. *Food Technol.*, **37**(5), 258–263.

Field, R.A., Williams, J.C., Brewer, M.S., Cross, H.R. and Secrist, J.L. (1984). Influence of sex, NaCl, MSL, nitrite and storage on sensory properties of restructured lamb roasts. *J. Food Qual.*, **7**, 121–129.

Ford, A.L., Park, A.J. and McBride, R.L. (1975). Effect of a protected lipid supplement on flavor properties of sheep meats. *J. Food Sci.*, **40**, 236–239.

Fox, C.W., Eller, R., Sather, L. and McArthur, J.A.B. (1964). Effect of sire and breed on eating qualities from weanling lambs. *J. Anim. Sci.*, **23**, 596.

Fox, C.W., McArthur, J.A.B. and Sather, L. (1962). Effect of sire and breed on flavor scores from weanling lambs. *J. Anim. Sci.*, **21**, 665.

Garton, G.A., Hovell, F.D.D. and Duncan, W.R.H. (1972). Influence of dietary volatile fatty acids on the fatty-acid composition of lamb triglycerides, with special reference to the effect of propionate on the presence of branched-chain components. *Br. J. Nutr.*, **28**, 409–416.

Gill, C.O. (1988). CO_2 packaging—the technical background. *Proc. 25th Meat Ind. Res. Conf. (Hamilton, New Zealand)*, Meat Ind. Res. Inst. N.Z. Publ. No. 855, pp. 181–185.

Gordon, M. (1987). Novel antioxidants. *Food Sci. Technol. Today*, **1**(3), 172–173.

Graafhuis, A.E., Muir, P.D., Devine, C.E. and Young, O.A. (1991). The effect of ultimate pH on the meat quality of seven month and fifteen month ram lambs with the same carcass weight. *Meat Sci.*, in preparation.

Hansen, R.P. and Czochanska, Z. (1978). Fatty acid composition of the subcutaneous fats of lambs fed 71% barley grain, and a comparison with those of lambs grazed on pasture, or fed higher levels of barley grain. *N.Z. J. Sci.*, **21**, 85–90.

Hoffmann, G. and Meijboom, P.W. (1968). Isolation of two isometric 2,6-nonadienals, and two isomeric 4-heptenals from beef and mutton tallow. *J. Am. Oil Chemists' Soc.*, **45**, 468–474.

Horning, M.G., Martin, D.B., Karmen, A. and Vagelos, P.R. (1961). Fatty acid synthesis in adipose tissue. II. Enzymatic synthesis of branched chain and odd-numbered fatty acids. *J. Biol. Chem.*, **236**, 669–672.

Hornstein, I. and Crowe, P.F. (1960). Flavor studies on beef and pork. *J. Agric. Food Chem.*, **8**, 494–498.

Hornstein, I. and Crowe, P.F. (1963). Meat flavor: Lamb. *J. Agric. Food Chem.*, **11**, 147–149.

Hudson, J.E. and Loxley, R.A. (1983). The effect of pentose sugars on the aroma and flavour of mutton. *Food Technol. Australia*, **35**(4), 174–175.

Jacobs, J.A., Field, R.A., Botkin, M.P. and Riley, M.L. (1973). Effect of dietary stress on lamb carcass composition and quality. *J. Anim. Sci.*, **36**, 507–510.

Jacobson, M. and Koehler, H.H. (1963). Components of the flavor of lamb. *Agric. Food Chem.*, **11**, 336–339.

Johns, A.M., Birkinshaw, L.H. and Ledward, D.A. (1989). Catalysts of lipid oxidation in meat products. *Meat Sci.*, **25**, 209–220.

Johnson, A.R. and Vickery, J.R. (1964). Factors influencing the production of hydrogen sulphide from meat during heating. *J. Sci. Food Agric.*, **15**, 695–701.

Kemp, J.D., Shelley, J.M., Ely, D.G. and Moody, W.G. (1972). Effects of castration and slaughter weight on fatness, cooking losses and palatability of lamb. *J. Food Sci.*, **34**, 560–562.

Kemp, J.D., Mahyuddin, M., Ely, D.G., Fox, J.D. and Moody, W.G. (1981). Effect of feeding systems, slaughter weight and sex on organoleptic properties and fatty acid composition of lamb. *J. Anim. Sci.*, **51**, 321–330.

Kirton, A.H., Clarke, J.N. and Hickey, S.M. (1982). A comparison of the composition and carcass quality of Kelly and Russian castrate, ram, wether and ewe lambs. *Proc. N.Z. Soc. Anim. Prod.*, **42**, 117–118.

Kirton, A.H., Sinclair, D.P., Chrystall, B.B., Devine, C.E. and Woods, E.G. (1981). Effect of plane of nutrition on carcass composition and the palatability of pasture-fed lamb. *J. Anim. Sci.*, **52**, 285–291.

Kirton, A.H., Winger, R.J., Dobbie, J.L. and Duganzich, D.M. (1983). Palatability of meat from electrically stimulated carcasses of yearling and older entire-male and female sheep. *J. Food Technol.*, **18**, 639–649.

Klettner, P., Pöllein, H. and Ott, G. (1989). Processing of old sheep in the meat industry. *Fleischwirtsch.*, **69**, 1810–1812.

Kunsman, J.E. and Riley, M.L. (1975). A comparison of hydrogen sulfide evolution from cooked lamb and other meats. *J. Food Sci.*, **40**, 506–508.

Lingnert, H. and Eriksson, C.E. (1981). Antioxidant effect of Maillard reaction products. *Prog. Food Nutr. Sci.*, **5**, 453–466.

Locker, R.H. (1980). In *Developments in Meat Science–1*, ed. R. Lawrie. Applied Sci., London, pp. 181–193.

Locker, R.H. and Moore, V.J. (1977). Lamb, ham and bacon. *Food Technol. N.Z.*, **12**(4), 27–31.

Mabrouk, A.F. (1976). In *Phenolic, Sulphur and Nitrogen Compounds in Food Flavors*, eds. G. Charalambous and I. Katz. ACS Symp. Series 26, American Chemical Society, Washington, DC, pp. 146–183.

MacLeod, G. (1986). In *Developments in Food Flavours*, eds. G.G. Birch and M.G. Lindley. Elsevier Appl. Sci., London, pp. 191–223.

MacLeod, G. and Ames, J.M. (1986). Capillary gas chromatography–mass spectrometric analysis of cooked ground beef aroma. *J. Food Sci.*, **51**, 1427–1433.

MacLeod, G. and Seyyedain-Ardebili, M. (1981). Natural and simulated meat flavors (with particular reference to beef). *CRC Crit. Rev. Food Sci. Nutr.*, **14**, 309–437.

Macy, R.L., Naumann, H.D. and Bailey, M.E. (1964a). Water-soluble flavor and odor precursors of meat. II. Effects of heating on amino nitrogen constituents and carbohydrates in lyophilized diffusates from aqueous extracts of beef, pork, and lamb. *J. Food Sci.*, **29**, 142–148.

Macy, R.L., Naumann, H.D. and Bailey, M.E. (1964b). Water-soluble flavor and odor precursors of meat. I. Qualitative study of certain amino acids, carbohydrates, non-amino acid nitrogen compounds, and phosphoric acid esters of beef, pork, and lamb. *J. Food Sci.*, **29**, 136–141.

Meilgaard, M., Civille, G.V. and Carr, B.T. (1987). *Sensory evaluation techniques*. CRC Press, Boca Raton.

Mendenhall, V.T. and Ercanbrack, S.K. (1979). Effect of carcass weight, sex and breed on consumer acceptance of lamb. *J. Food Sci.*, **44**, 1063–1066.

Miller, G.J., Field, R.A. and Agboola, H.A. (1986). Lipids in subcutaneous tissues and longissimus muscles of feedlot and grass-fed ewes. *J. Food Qual.*, **9**, 39–47.

Misock, J.P., Campion, D.R., Field, R.A. and Riley, M.L. (1976). Palatability of heavy ram lambs. *J. Anim. Sci.*, **42**, 1440–1444.

Moody, W.G. (1983). Beef flavor—a review. *Food Technol.*, **37**(5), 227–232.

Munday, R. and Winterbourn, C.C. (1989). Reduced glutathione in combination with superoxide dismutase as an important biological defence mechanism. *Biochem. Pharmacol.*, **38**, 4349–4352.

Mussinan, C.J., Wilson, R.A., Katz, I., Hruza, A. and Vock, M.H. (1976). in *Phenolic, Sulfur and Nitrogen Compounds in Food Flavors*, eds. G. Charalambous and I. Katz. ACS Symp. Series 26. Am. Chem. Soc., Washington, DC, pp. 133–145.

Nicol, A.M. and Jagusch, K.T. (1971). The effect of different types of pasture on the organoleptic qualities of lambs. *J. Sci. Food Agric.*, **22**, 464–466.

Nixon, L.N. (1981). A comparison of the effects of grass and legume pasture on sheep meat flavour. *N.Z. J. Agric. Res.*, **24**, 277–279.

Nixon, L.N., Wong, E., Johnson, C.B. and Birch, E.J. (1979). Nonacidic constituents of volatiles from cooked mutton. *J. Agric. Food Chem.*, **27**, 355–359.

Ockerman, H.W., Emsen, H., Parker, C.F. and Pierson, C.J. (1982). Influence of type (wooled or hair) and breed on growth and carcass characteristics and sensory properties of lamb. *J. Food Sci.*, **47**, 1365–1371.

Park, R.J. and Minson, D.J. (1972). Flavour differences in meat from lambs grazed on tropical legumes. *J. Agric. Sci.*, **79**, 473–478.

Park, R.J., Corbett, J.L. and Furnival, E.P. (1972). Flavour differences in meat from lambs grazed on lucerne (*Medicago sativa*) or phalaris (*Phalaris tuberosa*) pastures. *J. Agric. Sci.*, **78**, 47–52.

Park, R.J., Ford, A., Minson, D.J. and Bacter, R.I. (1975). Lucerne-derived flavour in sheep meat as affected by season and duration of grazing. *J. Agric. Sci.*, **84**, 209–213.

Park, R.J., Ford, A.L. and Ratcliff, D. (1978a). Use of protected lipid supplement to modify the flavor of mutton. *J. Food Sci.*, **43**, 874–881.

Park, R.J., Ford, A.L. and Ratcliff, D. (1978b). A study of "sweet" flavor in lamb produced by feeding protected sunflower seed. *J. Food Sci.*, **43**, 1363–1367.

Paul, A.A. and Southgate, D.A.T. (1978). McCance and Widdowson's *The Composition of Foods*. Her Majesty's Stationery Office, London.

Paul, P.C., Torten, J. and Spurlock, G.M. (1964). Eating quality of lamb. *Food Technol.*, **18**(11), 121–124.

Pearson, A.M., Wenham, L.M., Carse, W.A., McLeod, K., Davey, C.L. and Kirton, A.H. (1973). Observations on the contribution of fat and lean to the aroma of cooked beef and lamb. *J. Anim. Sci.*, **36**, 511–515.

Pepper, F.H. and Pearson, A.M. (1969). Changes in hydrogen sulfide and sulfhydryl content of heated beef adipose tissue. *J. Food Sci.*, **34**, 10–12.

Peterson, G.V. (1983). Preslaughter and slaughter factors affecting meat quality in lambs. Ph.D. Thesis, Massey University, Palmerston North, N.Z.

Purchas, R.W., O'Brien, L.E. and Pendleton, C.M. (1979). Some effects of nutrition and castration on meat production from male Suffolk cross (Border Leicester–Romney cross) lambs. *N.Z. J. Agric. Res.*, **22**, 375–383.

Purchas, R.W., Johnson, C.B., Birch, E.J., Winger, R.J., Hagyard, C.J. and Keogh, R.G. (1986). Flavour studies with beef and lamb. Massey University, Palmerston North, N.Z.

Raymer, J.H. and Pellizzari, E.D. (1987). Toxic organic compound recoveries from 2,6-diphenyl-p-phenylene oxide porous polymer using supercritical carbon dioxide and thermal desorption methods. *Anal. Chem.*, **59**, 1043–1048.

Rhee, K.S. (1988). Enzymic and nonenzymic catalysis of lipid oxidation in muscle foods. *Food Technol.*, **42**(6), 127–132.

Rubin, L.J. and Shahidi, F. (1988). Lipid oxidation and the flavour of meat products. In *Proc. 34th Int. Congr. Meat Sci. Technol.*, ICMST, Brisbane, pp. 295–301.

Sato, K., Hegarty, G.R. and Herring, H.K. (1973). The inhibition of warmed-over flavor in cooked meats. *J. Food Sci.*, **38**, 398–403.

Schutte, L. (1974). Precursors of sulfur-containing flavor compounds. *CRC Crit. Rev. Food Technol.*, **4**, 457–505.

Shorland, F.B., Czochanska, Z., Moy, M., Barton, R.A. and Rae, A.L. (1970). Influence of pasture species on the flavour, odour and keeping quality of lamb and mutton. *J. Sci. Food Agric.*, **21**, 1–4.

Shorthose, W.R. (1989). *Dark-cutting in Cattle and Sheep. Proceedings of an Australian Workshop.* Australian Meat & Live-stock Research and Development Corporation, Report No. 89/2.

Smith, M. and Young, O.A. (1991). Spicing up sheepmeat. *Food Technol. N.Z.*, **26**(3), 25–29.

Spillane, C. and L'Estrange, J.L. (1977). The performance and carcass fat characteristics of lambs fattened on concentrate diets. *Ir. J. Agric. Res.*, **16**, 205–219.

Spurway, R.A. (1972). Flavour differences in sheep meat. *Food Technol. Australia*, **24**(12), 645.

St. Angelo, A.J., Vercellotti, J.R., Legendre, M.G., Vinnett, C.H., Kuan, J.W., James, C. and

Dupy, H.P. (1987). Chemical and instrumental analyses of warmed-over flavor of beef. *J. Food Sci.*, **52**, 1163–1168.

Suzuki, J. and Bailey, M.E. (1985). Direct sampling capillary GLC analysis of flavor volatiles from ovine fat. *J. Agric. Food Chem.*, **33**, 343–347.

Vimini, R.J., Field, R.A., Crouse, J.D. and Miller, G.J. (1984). Factors affecting melting point of subcutaneous fat from heavy ram and wether lambs. *Int. Goat Sheep Res.*, **2**, 105–113.

Wasserman, A.E. and Talley, F. (1968). Organoleptic identification of roasted beef, veal, lamb and pork as affected by fat. *J. Food Sci.*, **33**, 219–223.

Watanabe, K. and Sato, Y. (1968). Aliphatic γ- and δ-lactones in meat fats. *Agric. Biol. Chem.*, **32**, 1318–1324.

Weller, M., Galgan, M.W. and Jacobson, M. (1962). Flavor and tenderness of lamb as influenced by age. *J. Anim. Sci.*, **21**, 927–929.

Wenham, L.M. (1974). Studies in ewe mutton quality–palatability of beef and mutton patties. *N.Z. J. Agric. Res.*, **17**, 203–205.

Werkhoff, P., Bruning, J., Emberger, R., Guntert, M., Kopsel, M., Kuhn, W. and Surburg, H. (1990). Isolation and characterization of volatile sulfur-containing meat flavor compounds in model systems. *J. Agric. Food Chem.*, **38**, 777–791.

Wong, E. (1975). Mutton flavour. *Food Technol. N.Z.*, **10**(1), 13–15.

Wong, E. and Mabrouk, A.F. (1979). Isolation of precursors of mutton odor. *J. Agric. Food Chem.*, **27**, 1415–1416.

Wong, E., Nixon, L.N. and Johnson, C.B. (1975a). Volatile medium chain fatty acids and mutton flavor. *J. Agric. Food Chem.*, **23**, 495–498.

Wong, E., Johnson, C.B. and Nixon, L.N. (1975b). The contribution of 4-methyloctanoic (hircinoic) acid to mutton and goat meat flavour. *N.Z. J. Agric. Res.*, **18**, 261–266.

6 Umami flavour of meat

J.A. MAGA

6.1 Introduction

Many compounds have been shown to be present in the flavour fraction of food. However, the flavour chemist has to decide which compound or series of compounds are the major contributors to specific food flavours. This has been a long and difficult task, and as a result, the flavours of foods such as meats are still not completely understood.

Even more complex is the situation where certain compounds have been shown to intensify, modify or mask the flavours of certain foods. The fact that a specific compound or combination of compounds, when intentionally added or formed in foods by biological or thermal pathways, has the ability to change the perceived flavour properties of certain foods is a research area that is fascinating.

Over the years, various nomenclatures have been proposed for compounds that have the ability to modify flavour perception. These include terms such as flavour potentiators, flavour enhancers and umami. Currently, the scientific community appears to be adopting the name umami, defined as the taste of monosodium glutamate (MSG) and 5′-nucleotides such as 5′-inosinate (IMP) and 5′-guanylate (GMP).

The major objectives of this chapter are to define and discuss the properties of umami in model system studies, and to review the formation, identification, quantitation and stability, together with the sensory significance in beef, pork, chicken, turkey and lamb. The influence of meat aging/processing and compound synergism is also discussed, in this attempt to update the role of umami in meat flavour chemistry.

6.2 Definitions

Umami can be defined as the taste properties resulting from the natural occurrence or intentional addition of compounds such as monosodium glutamate (MSG) and certain 5′-nucleotides such as 5′-inosine monophosphate (IMP) and 5′-guanosine monophosphate (GMP). Other researchers have used terms such as ‘savoury’, ‘beefy’ and ‘brothy’ to describe the same taste sensations. These nucleotides have also been referred to respectively as inosinic and guanylic acids, 5′-inosine and 5′-guanylic acids, inosine 5′- and guanosine 5′-phosphates and disodium 5′-inosinate or disodium 5′-guanylate. IMP and GMP blends have also been marketed under the trade name Ribotide®.

Compounds of these types are especially interesting in that they have the ability to modify taste, even though they do not possess characteristic flavours of their own, especially at the low concentrations at which they affect food flavour. In using the above definition of flavour alteration without taste contribution, one could also consider that sodium chloride, if used at subthreshold levels, also possesses umami properties. This is an area that deserves research attention.

6.3 Historical background

Many cultures throughout the world have long used ingredients or food preparation techniques that result in the presence of umami compounds that intensify certain food flavours. Experience has taught cooks what it took scientists many years to discover.

It was not until the early 1900s that a specific compound that was proven to be responsible for an umami sensation was isolated. Ikeda (1909) was able to identify the compound monosodium glutamate (MSG) in the naturally occurring form in an extract from dried kombu or sea tangle, a type of seaweed. The importance of his discovery was soon evident because the commercial production of MSG for the intentional addition to foods began shortly thereafter. Apparently, Ikeda was the first to propose the name 'umami', which means 'deliciousness' in Japanese, for the taste sensation associated with MSG. Today MSG is produced in many countries and is consumed internationally.

A few years later, another food common to Oriental cuisine, dried bonito tuna, was the source for the identification of another umami compound, namely inosine monophosphate (IMP). It was reported in the initial study (Kodama, 1913) that the compound in question was the histidine salt of 5′-inosinic acid. However, it was later concluded that histidine was not a significant contributor to umami. In contrast to MSG, the commercialization of IMP was not begun until the early 1960s.

In the early 1960s another compound, guanosine monophosphate (GMP) was identified from another natural source, the Shiitake black mushroom (Nakajima *et al.*, 1961; Shimazono, 1964). It is quite interesting to note that the three commercially available umami compounds were first identified from natural sources. In fact, it has been postulated (Hashimoto, 1965) that all types of marine products possess umami compounds due to the high amounts of glutamic acid and nucleotides that they contain.

More recently, other umami compounds, including ibotenic and tricholomic acids, have been identified as naturally occurring compounds in other types of Japanese mushrooms (Takemoto and Nakajima, 1964; Takemoto *et al.*, 1964). Currently these umami compounds are not commercially available due to the effectiveness and ready availability of MSG, IMP and GMP. In the meantime, over 80 years of research on umami compounds has accumulated and their commercialization now represents an international industry worth several hundred million dollars.

6.4 Structural considerations

To date, umami compounds have been found to structurally contain either L-amino acids containing five carbon atoms or a purine ribonucleotide 5′-monophosphate having an oxy group in the 6-position. These structures are representative of MSG, IMP and GMP as shown in Figure 6.1.

In commercial practice the amino acid based compound is in the monosodium salt form whereas the nucleotides are in the disodium form. In addition, MSG can also be obtained in the potassium, ammonium or calcium forms for utilization in products where low sodium levels are desired.

The isomeric structure of umami compounds can also dramatically influence their taste properties. In the case of MSG, the D-form, which is not naturally occurring, has no umami properties, whereas the L-form, which is naturally occurring, does. The nucleotides IMP and GMP can occur as the 2′-, 3′- or 5′- forms but only the 5′-form is taste active.

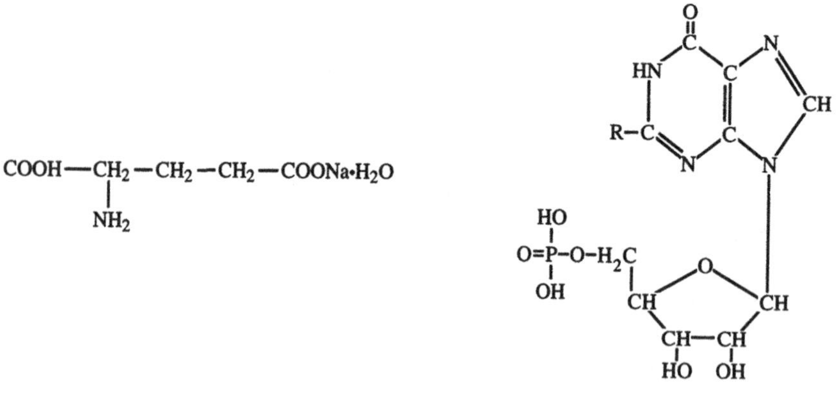

$COOH—CH_2(NH_2)—CH_2—CH_2—COONa{\cdot}H_2O$

IMP: R=H
GMP: $R=NH_2$

Figure 6.1 Umami structures.

In the case of MSG, it is interesting to note that L-glutamine, which has the same basic structure as MSG, has no umami taste. The umami properties of other structurally related amino acids have been investigated (Akabori, 1939) and it was concluded that the L-forms of α-amino dicarboxylic acids with four to seven carbons also have umami properties similar to those of L-glutamic acid. The flavour properties of various substituted nucleotides have also been reported (Yamazaki *et al.*, 1968a,b) and as expected, some structures do result in compounds possessing umami properties.

6.5 Stability

When in solution, MSG acts as an ampholyte and as such can exist in a number of ionic forms, in equilibrium, which are pH dependent. At various pHs forms such as glutamic acid hydrochloride, free glutamic acid, neutral MSG and basic disodium glutamate can exist. The influence of pH on these forms is summarized in Table 6.1, and as can be seen, the neutral MSG form is predominant over most of the pH scale, except in very acid conditions. The neutral MSG form in turn possesses the most potent umami sensation.

The thermal stability of IMP and GMP has been shown to be also pH dependent. As seen in Table 6.2, acidic conditions decrease compound stability and IMP appears to be slightly more stable than GMP. Their thermal degradation has been shown to follow first order kinetics with the major degradation products being nucleosides and phosphoric acid, which would suggest degradation via hydrolysis of the phosphoric ester bond (Matoba *et al.*, 1988).

Nucleotide stability at temperatures normally used in retorting also

Table 6.1 Ionic form distribution of MSG as influenced by pH

pH	Distribution (%)			
	R_1	R_2	R_3	R_4
8.0	0	0	96.9	3.1
7.0	0	0	99.8	0
6.0	0	1.8	98.2	0
5.5	0	5.3	94.7	0
5.0	0	15.1	84.9	0
4.5	0	36.0	64.0	0
4.0	0.9	63.1	36.0	0
3.5	4.0	80.9	15.1	0
3.0	13.4	81.3	5.3	0

Table 6.2 Stability of GMP and IMP to temperature and pH

Compound	pH	Temperature (°C)	Half-life (h)
GMP	4.0	100	6.4
	7.0	100	8.2
	9.0	100	38.5
IMP	4.0	100	8.7
	7.0	100	13.1
	9.0	100	46.2

A 10°C increase in temperature lowered all half-lives by one-third.
From Matoba *et al.* (1988).

Table 6.3 Aqueous hydrolysis rates for nucleotides at 121°C and various pH values

pH	Compound	Half-life (min)
3.0	IMP	32.4
	GMP	31.5
4.0	IMP	45.3
	GMP	62.4
5.0	IMP	45.9
	GMP	57.8
6.0	IMP	63.0
	GMP	41.0

From Shaoul and Sporns (1987).

presents problems, especially for larger-size cans where extended retorting times are required. The data summarized in Table 6.3 show that under highly acidic conditions half of either IMP or GMP can be lost in approximately 30 min. In contrast, it has been estimated that at pH 5 and at 23°C the half-life for IMP and GMP is 36 and 19 years, respectively (Shaoul and Sporns, 1987).

6.6 Synergism

One of the most fascinating aspects of umami compounds that is far from understood is their ability to act synergistically when used in combination with each other. This concept has been extensively reviewed (Maga, 1983) and thus will only be briefly touched upon. As shown in Table 6.4, if MSG is assigned a umami intensity of 1.0, the addition of an

Table 6.4 Synergistic flavour intensity of MSG–GMP

Ratio of MSG:GMP	Relative flavour intensity
1:0	1.0
1:1	30.0
10:1	18.8
20:1	12.5
50:1	6.4
100:1	5.5

From Ribotide® Product Data Sheet No. 3.

equal amount of GMP increases relative flavour intensity 30 times. Even addition of as little as 1% GMP to MSG significantly increases flavour intensity. Similar effects are noted for combinations of IMP and MSG.

6.7 Taste properties

The individual taste thresholds for various umami compounds, alone and in combination with each other, are summarized in Table 6.5. It can be seen that when used alone, GMP has a threshold an order of magnitude lower than that for MSG and IMP. The role of synergism as described above is also evident in Table 6.5 when all three compounds are utilized in equal proportions. Therefore, if naturally present in combination or intentionally added in combination, a relatively small total amount is required to elicit a umami response.

Relative to the basic tastes of sweet, salt, sour and bitter, the taste threshold of MSG is well within their range (Table 6.6). This is quite interesting in light of the fact that there are many studies attempting to relate the taste of MSG to the other basic tastes. As seen in Figure 6.2, various portions of the MSG molecule have been proposed to possess the

Table 6.5 Umami compound taste thresholds

Compound	Taste threshold (%)
MSG	0.012
IMP	0.014
GMP	0.0035
IMP + GMP	0.0063
IMP + GMP + MSG	0.000031

From Maga (1983).

Table 6.6 Detection thresholds in water of various taste substances

Compound	Threshold (%)
Sucrose	0.086
Sodium chloride	0.0037
Tartaric acid	0.00094
Quinine sulphate	0.000049
MSG	0.012

From Yamaguchi (1987).

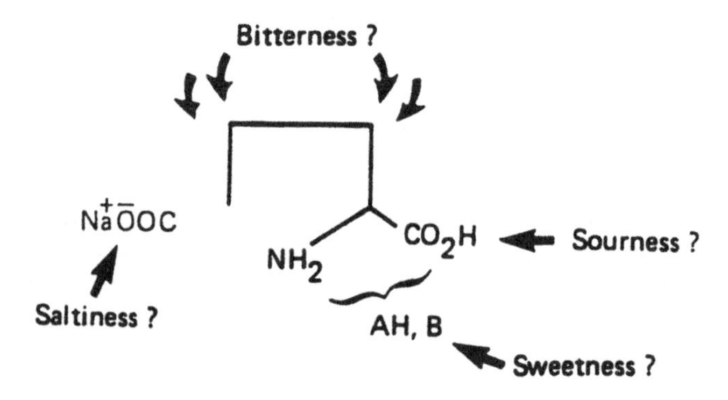

Figure 6.2 Proposed MSG taste properties. From Birch (1987).

potential to elicit other tastes. This approach has led Birch (1987) to propose that MSG not only possesses a umami taste but also a salty one (Table 6.7). In addition, he proposes that aspartic and glutamic acids also have a umami taste in combination with an acidic sensation.

Yamaguchi (1987) has attempted to locate the relative position of umami taste to the other basic tastes as well as to that of various foods, utilizing multi-dimensional plotting. As seen in Figure 6.3, umami falls outside the regions occupied by sweet, salt, sour and bitter tastes but is closely associated with tastes derived from various marine and meat products. These data can be interpreted in at least two different ways. First, the data clearly demonstrate that umami is neither associated with, nor is the result of, the four acknowledged basic tastes, from which one could conclude that it indeed is a separate and distinctive taste. On the other hand, one could also conclude that umami is a specialized form of taste associated with meat or marine-based foods.

Table 6.7 Compounds related to MSG and having umami taste

Compound	Tastes
MSG	Umami and salty
Aspartic acid	Acidic and umami
Glutamic acid	Acidic and umami

From Birch (1987).

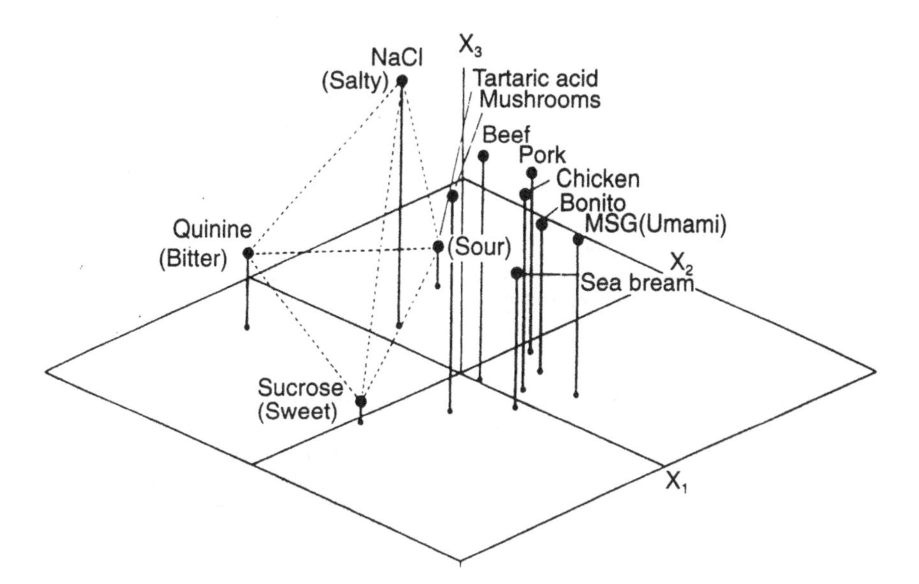

Figure 6.3 Relationship of umami taste to other tastes and foods. From Yamaguchi (1987).

6.8 Food occurrence

Umami compounds and their precursors are present in a wide variety of foods. This can perhaps be best appreciated by viewing the information summarized in Table 6.8. These foods include raw and processed, fermented and unfermented, as well as plant and animal sources. Therefore, those individuals who may be concerned with the intentional addition of umami compounds to various foods would be hard pressed to identify a varied diet that did not contain naturally occurring sources of the compounds in question.

If one looks at the free glutamate levels that are present in various

Table 6.8 Foods containing umami compounds

Beverages (beer, tea, wine)
Fruits (various)
Marine products (seaweed, fish, clams, crab, oysters, shrimp)
Meats (beef, chicken, lamb, pork)
Milk products (milk, cheese)
Plant proteins (barley, coconut, corn, cottonseed, peanut, soybean, wheat)
Vegetables (various)

From Maga (1983).

Table 6.9 Free glutamate levels in various foods

Food	Glutamate level (mg/100 g)
Tomato	140
Tomato juice	260
Grape juice	258
Broccoli	176
Parmesan cheese	1,200
Gruyère cheese	1,050

From Giacometti (1979).

Table 6.10 Glutamate levels in various animal proteins

Protein	Glutamate level (g/100 g)
α-Casein (milk)	22.5
β-Lactoglobulin (milk)	20.0
Actin (muscle)	14.8
Myosin (muscle)	21.0
Albumin (egg white)	16.5

From Giacometti (1979).

foods (Table 6.9), it can quickly be appreciated that significant amounts of glutamate may be present, especially in cheeses that are aged for long periods.

Glutamic acid is usually the major amino acid in protein, and as seen in Table 6.10, it represents approximately 20% of all amino acids in various animal protein sources. When these proteins are processed or intentionally hydrolysed in the manufacture of hydrolysed protein, glutamic acid is freed, which in turn can result in the formation of MSG.

If one was to rank the importance of umami compounds to the flavour properties of various foods, as shown in Table 6.11, it becomes quite

Table 6.11 Major taste-active compounds in foods

Compound	Meat	Vegetables	Fruit	Roots	Seeds
Inorganic ions	× ×	×	×	×	×
Amino acids	× × ×	× ×	×	× ×	× ×
Peptides and proteins	× × ×	× ×		×	× ×
Histidine dipeptides	× × ×				
Nucleotides	× × ×	×		× ×	× ×
Amines	× ×	×			
Sugars			× × ×	× ×	× ×
Phenols (simple)		× ×	× ×		×
Hydroxy compounds		×	× ×		
Polyphenolic compounds		× ×	× ×		×
Carbonyl compounds		× ×	× × ×		×
Esters			× ×		
Sulphur compounds	× ×	× ×			×
Acids		× ×	× × ×		
Furans		× ×	× ×	×	×
Lactones		× ×	× × ×		
N,S-heterocycles	×	× × ×	×		×

From Grill and Flynn (1987).
× of minor importance; × × somewhat important;
× × × very important.

apparent that they make major contributions to meat flavours. Therefore, the remainder of this review will be devoted to the role of umami compounds in meat flavour.

6.9 Umami compounds and meat flavour

Numerous early studies have demonstrated that the intentional addition of MSG and nucleotides significantly modifies meat flavour perception. For example, Girardot and Peryam (1954) showed that the addition of MSG to a variety of processed meats resulted in products that were preferred over the same products to which MSG was not added (Table 6.12). This was especially true in the case of chicken. Of the products they evaluated, meat loaf had the least improvement in flavour perception.

When using added IMP, other researchers (Kurtzman and Sjostrom, 1964) concluded that canned chicken-containing noodle soup was not flavour-enhanced (Table 6.13). Based on IMP thermal stability data presented earlier, perhaps most or all of the added IMP was degraded. However, other products evaluated, including canned beef noodle soup, did show improvement with IMP addition. For some unexplained reason, no improvement was found with canned beef hash.

Bauer (1983) clearly demonstrated that the aging of meat clearly influences resulting glutamic acid content. He aged beef and pork for 4 and 7

Table 6.12 Preference for processed meats containing MSG

Product	% Preferring MSG sample
Beef stew	61
Beef and gravy	61
Boned chicken	75
Hamburger	65
Meat loaf	53
Pork and gravy	66

From Giradot and Peryam (1954).

Table 6.13 Influence of IMP addition to the flavour of various meat products

Product	Effect
Beef bouillon	Enhanced
Beef noddle soup, canned	Enhanced
Chicken noodle soup, canned	Undecided
Canned luncheon meats	Enhanced
Corned beef hash	Not enhanced
Ham	Enhanced

From Kurtzman and Sjostrom (1964).

Table 6.14 Meat age versus glutamic acid content

Product	Age (days)	Glutamic acid content (mg/100 g)
Beef	4	9.3
	7	21.1
Pork	4	11.4
	7	16.0

From Bauer (1983).

days, and as seen in Table 6.14, in the case of both products, the longer-aged meat had significantly higher levels of glutamic acid. Apparently, microbial activity during aging resulted in partial protein hydrolysis thereby liberating free glutamic acid. However, as seen in Table 6.15, the levels of glutamate present in a range of processed meats are quite variable. The relatively low levels found in some of these products could perhaps be attributed to the degree of heat processing or low pH.

The free glutamate content in a variety of cooked meats has been summarized in Table 6.16, and as can be seen, most of the values being rela-

Table 6.15 Glutamate content of processed meats

Product	Glutamate content (%)
Dry sausages	0.03–0.25
Frankfurters	0.02–0.15
Liver sausages	0.01–0.29
Hams	0.01–0.11
Pickled tongue	0.07–0.17

From Gerhardt and Schulz (1985).

Table 6.16 Free glutamate content of various cooked meats

Product	Glutamate content (%)
Beef, tenderloin	0.042
Beef, shank	0.014
Beef, standing rib, medium rare	0.057
Juice from above	0.088
Beef, standing rib, well done	0.013
Bologna	0.004
Chicken	0.055
Duck	0.064
Frankfurters, boiled	0.001
Lamb	0.003
Mutton	0.008
Pork, loin	0.029

From Maga (1983).

Table 6.17 Nucleotide composition (%) of various meats

Nucleotide	Beef	Chicken
IMP	0.106–0.443	0.075–0.122
GMP	0.002	0.002
AMP	0.007	0.007–0.013
CMP	0.001	0.002
UMP	0.001	0.003

From Maga (1983).

tively low except for the two poultry products and the free run juice from beef rib roast. It is also interesting to note that cooking beef rib roast to well done compared to medium rare resulted in approximately an 80% reduction in the amount of free glutamate present. Other products that had low values included lamb and mutton as well as boiled frankfurters. In the case of the frankfurters, the low value could be attributed to one of two reasons; either their high fat content had a dilution effect on the

Table 6.18 Effect of meat storage on IMP levels

Meat and storage days	IMP level (μ mol/g meat)
Beef (12)	3.2
Pork (6)	6.7
Chicken (2)	7.2

From Kato and Nishimura (1987).

Table 6.19 Effect of meat storage on umami taste intensity

Meat and storage days	Number of samples judged to have a more intense umami taste		Significance
	Before storage	After storage	
Beef (12)	12	4	NS
Pork (6)	2	14	S
Chicken (2)	8	23	S

NS, not significant; S, significant.
From Kato and Nishimura (1987).

Table 6.20 Effect of storage on IMP levels in cooked meats

Meat and storage days	IMP level (μ mol/g meat)
Beef (12)	4.1
Pork (6)	10.5
Chicken (2)	10.9

From Kato and Nishimura (1987).

actual amount of meat present or the free glutamate was lost to the cooking water.

The nucleotide levels naturally found in beef and chicken are summarized in Table 6.17. These data indicate that beef actually has higher levels of IMP than does chicken whereas the levels of GMP are the same. Interestingly, chicken has somewhat higher levels of other nucleotides (AMP, CMP, UMP), which have limited umami properties, than beef.

Kato and Nishimura (1987) measured the IMP levels in beef, pork and chicken that had been aged for varying lengths of time, and as seen in Table 6.18, chicken had the highest IMP level even though it had only been aged two days. Pork had the next highest level while beef, although aged the longest, had the lowest amount of IMP. When a panel was asked to judge whether samples had a more intense umami taste before or after

Table 6.21 Beef broth GLC headspace peak area ratios versus umami additions

Peak area ratio*	Additive
1.66	0.05% MSG
2.30	IMP + GMP (0.05% total)
2.35	IMP + GMP + MSG (0.05% total)

*Control versus compound addition.
From Maga and Lorenz (1972).

Table 6.22 Flavour profile of cooked hamburger with 1% MSG

Descriptor	Score
Aroma	
Whole aroma	0
Meaty	0.2
Beefy	0
Acceptability	0.4
Basic taste	
Whole taste	0.6
Salty	0.3
Sweet	0.1
Sour	0.2
Bitter	0
Flavour characteristic	
Continuity	0.5
Mouthfulness	0.6
Impact	0.5
Mildness	0.5
Thickness	0.4
Other flavours	
Spicy	0.2
Oily	0
Meaty	0.2
Beefy	0.3
Overall preference	
Palatability	0.6

0, Same as control (No MSG); 1, Slight increase; 2, Marked increase.
From Yamaguchi and Kimizuka (1979).

aging, no difference was found for beef, which had the lowest IMP level, while storage significantly influenced umami taste intensity for pork and chicken (Table 6.19). Interestingly, cooking stored beef, pork and chicken (Table 6.20) resulted in levels of IMP that were 30% higher in all three products, as compared to their noncooked but aged counterparts.

Table 6.23 Umami content (mg/100 ml) in chicken stock

Compound	Amount
AMP	2.26
IMP	5.84
GMP	0
ATP	ND
Glutamic acid	15.00

From Yamaguchi (1987).
ND = not detected.

Table 6.24 Detection levels of umami compounds added to stocks

Stock	Detection level (%)	
	MSG	IMP
Beef	0.00625	0.025
Chicken	0.0125	0.00625

From Yamaguchi (1987).

Most researchers have exclusively investigated the influence of umami compounds on taste perception but one should also question if these compounds can influence odour perception. A limited number of articles has appeared addressing this issue. Maga and Lorenz (1972) prepared beef stock samples to which they added either no umami compounds, or a total of 0.05% of only MSG, IMP or GMP, or a combination of IMP, GMP and MSG. They sampled the headspace above the various stocks and measured total peak areas using gas chromatography. The stock sample which had no added compounds served as the control. As seen in Table 6.21, the addition of MSG alone as well as combinations of umami compounds significantly increased peak area ratios thereby indicating that these compounds can also influence aroma properties.

Yamaguchi and Kimizuka (1979) performed a flavour profile analysis on cooked hamburgers to which 1% MSG had been added as compared to a no additive control. As seen in Table 6.22, they observed a slight increase in the 'meaty' and 'acceptability' descriptors associated with the aroma portion of the profile of hamburgers containing added MSG. It is also interesting to note that added MSG also intensified many of the other sensory properties.

Yamaguchi (1987) has conducted extensive research on umami compounds in meat stocks. For example, she found that the major nucleotide in chicken stock was IMP (Table 6.23). Interestingly, no GMP was

Table 6.25 Taste intensities (0–100) of various 1% meat proteins with added umami compounds

Meat	No added umami	Added umami compound		
		MSG (0.015%)	IMP (0.010%)	GMP (0.004%)
Beef	64	83	80	89
Pork	60	68	70	76
Lamb	41	46	45	48
Chicken	53	84	86	90
Turkey	49	61	60	65

From Maga (1987).

Table 6.26 Percent increase in taste intensity caused by umami addition to meats

Meat	Added umami compound		
	MSG	IMP	GMP
Beef	30	25	39
Pork	13	17	27
Lamb	12	10	17
Chicken	58	62	70
Turkey	24	22	33

From Maga (1987).

detected in this study. In addition to IMP, relatively high levels of glutamic acid were found. In comparing beef and chicken stocks, she reported (Table 6.24) significant differences between minimum detectable levels for MSG and IMP. Twice the amount of MSG was required before it was detected in chicken stock as compared to beef stock, whereas only one-third the amount of IMP required for beef stock, needed to be added to chicken stock before it was detected.

Maga (1987) attempted to evaluate the role of MSG, IMP and GMP on the taste intensities of various purified meat proteins including beef, pork, lamb, chicken and turkey as compared to the same meat proteins containing no added compounds. From Table 6.25 it can be seen that with no additions, beef and pork had the highest intensity ratings while lamb had the lowest. In all cases, the addition of any of the three compounds increased taste intensity although the increase for lamb was minimal. In looking at percent increases over the control (Table 6.26) it is apparent that the most affected meat protein was chicken. Also, not all additives were equally effective. Therefore, these data would indicate that the functionality and perhaps mechanism of interaction between umami compounds and meat proteins vary with protein source.

6.10 Conclusions

The literature clearly demonstrates that umami compounds have the ability to alter the taste and possibly aroma properties of a wide range of foods independent of the four basic tastes. Umami compounds react synergistically therefore reducing the total amount required to elicit a response.

They and their precursors occur naturally in a wide range of foods and their effectiveness can be improved by their intentional addition to most foods. Acidic conditions and high processing temperatures minimize the effectiveness of 5′-nucleotide-based umami compounds. Umami compounds are very effective in contributing to meat flavour, especially chicken and exhibit minimum effectiveness with lamb.

References

Akabori, S. (1939). Relationships between the taste of amino acids and their structures. *J. Japan Biochem. Soc.*, **14**, 185–197.

Bauer, F. (1983). Free glutamic acid in meats products. *Ernahrung*, **7**, 688.

Birch, G.G. (1987). Structure, chirality, and solution properties of glutamates in relation to taste. In *Umami: A Basic Taste*, ed. Y. Kawamura and M.R. Kare. Marcel Dekker, New York, 173–184.

Gerhardt, U. and Schulz, W. (1985). The presence of free glutamic acid in foods, with special reference to meat products. *Fleischwirtschaft*, **65**, 1483–1486.

Giacometti, T. (1979). Free and bound glutamate in natural products. In *Glutamic Acid: Advances in Biochemistry and Physiology*, ed. L.J. Filer, S. Garattini, M.R. Kare, W.A. Reynolds and R.J. Wurtman. Raven Press, New York, 25–34.

Girardot, N.F. and Peryam, D.R. (1954). MSG's power to perk up foods. *Food Eng.*, **26**(12), 71–74.

Grill, H.J. and Flynn, F.W. (1987). Behavioral analysis of oral stimulating effects of amino acid and glutamate compounds on the rat. In *Umami: A Basic Taste*, ed. Y. Kawamura and M.R. Kare. Marcel Dekker, New York, 461–480.

Hashimoto, Y. (1965). Taste-producing substances in marine products. In *The Technology of Fish Utilization*, ed. R. Kreuzery. Elsevier, London, 57–65.

Ikeda, K. (1909). On a new seasoning. *J. Tokyo Chem.*, **30**, 820–826.

Kato, H. and Nishimura, T. (1987). Taste components and conditioning of beef, pork, and chicken. In *Umami: A Basic Taste*, ed. Y. Kawamura and M.R. Kare. Marcel Dekker, New York, 289–306.

Kodama, S. (1913). On a procedure for separating inosinic acid. *J. Tokyo Chem.*, **34**, 751–755.

Kurtzman, C.H. and Sjostrom, L.B. (1964). The flavor-modifying properties of disodium inosinate. *Food Technol.*, **18**, 221–225.

Maga, J.A. (1983). Flavor potentiators. *CRC Crit. Rev. Food Sci. Nutr.*, **18**, 231–312.

Maga, J.A. (1987). Organoleptic properties of umami substances. In *Umami: A Basic Taste*, ed. Y. Kawamura and M.R. Kare. Marcel Dekker, New York, 255–269.

Maga, J.A. and Lorenz, K. (1972). The effect of flavor enhancers on direct headspace gas–liquid chromatography profiles of beef broth. *J. Food Sci.*, **37**, 963–964.

Matoba, T., Kuchiba, M., Kimura, M. and Hasegawa, K. (1988). Thermal degradation of flavor enhancers, inosine 5′-monophosphate, and guanosine 5′-monophosphate in aqueous solution. *J. Food Sci.*, **53**, 1156–1159, 1170.

Nakajima, N., Ishikawa, K., Kamada, M. and Fujita, E. (1961). Food chemical studies on 5′-ribonucleotides. I. On the 5′-ribonucleotides in foods. Determination of the 5′-ribonucleotides in various stocks by ion exchange chromatography. *J. Agric. Chem. Soc. Japan*, **35**, 797–804.

Ribotide® Product Data Sheet No. 3. The synergistic effect between disodium 5′-inosinate, disodium 5′-guanylate and monosodium glutamate. Takeda Chemical Industries, Ltd., Tokyo, Japan.

Shaoul, O. and Sporns, P. (1987). Hydrolytic stability at intermediate pHs of the common purine nucleotides in food, inosine-5′-monophosphate, guanosine-5′-monophosphate and adenosine-5-monophosphate. *J. Food Sci.*, **52**, 810–812.

Shimazono, H. (1964). Distribution of 5′-ribonucleotides in foods and their application to foods. *Food Technol.*, **18**, 294–298.

Takemoto, T. and Nakajima, T. (1964). Studies on the constituents of indigenous fungi. I. Isolation of the flycidal constituent from *Tricholoma muscarium*. *J. Pharm. Soc. Japan*, **84**, 1183–1185.

Takemoto, T., Yokobe, T. and Nakajima, T. (1964). Studies on the constituents of indigenous fungi. II. Isolation of the flycidal constituent from *Amanita strobiliformis*. *J. Pharm. Soc. Japan*, **84**, 1186–1189.

Yamaguchi, S. (1987). Fundamental properties of umami in human taste sensation. In *Umami: A Basic Taste*, ed. Y. Kawamura and M.R. Kare. Marcel Dekker, New York, 41–73.

Yamaguchi, S. and Kimizuka, A. (1979). Psychometric studies on the taste of monosodium glutamate. In *Glutamic Acid: Advances in Biochemistry and Physiology*, ed. L.J. Filer, S. Garattini, M.R. Kare, W.A. Reynolds and R.J. Wurtman. Raven Press, New York, 35–54.

Yamazaki, A., Kumashiro, I. and Takenishi, T. (1968a). Synthesis of 2-alkyl-thioinosine 5′-phosphate and N′-methylated guanosine 5′-phosphates. *Chem. Pharm. Bull.*, **16**, 338–343.

Yamazaki, A., Kumashiro, I. and Takenishi, T. (1968b). Synthesis of some N′-methyl-2-substituted inosines and their 5′-phosphates. *Chem. Pharm. Bull.*, **16**, 1561–1565.

7 Lipid-derived off-flavours in meat—formation and inhibition

J.I. GRAY and A.M. PEARSON

7.1 Introduction

Lipid oxidation is a major deterioration reaction in foods which often results in a significant loss of quality. Oxidative deterioration of fats, oils and lipid-containing foods can result in the development of rancid off-flavours as well as the loss of the desirable characteristic flavour notes. It is also well known that oxidative reactions can cause discoloration of pigments, especially haem and carotenoids, destruction of certain nutrients and the possible formation of toxic by-products (Pearson *et al.*, 1983; Addis, 1986). Generally, these changes in quality prevent consumer acceptance of oxidized food products.

The oxidation of fats and oils occurs at sites of unsaturation and is generally accepted to proceed according to a free radical chain mechanism (Labuza, 1971). The overall mechanism of lipid oxidation consists of three phases: initiation, propagation and termination. Hydroperoxides, the primary oxidation products, function to make the propagation reactions autocatalytic, while secondary oxidation products such as aldehydes, ketones and acids are responsible for off-flavours and odours.

The scientific literature pertaining to lipid oxidation in meat and other biological systems is often confusing and in addition there remain many unanswered questions. There is considerable debate regarding the catalytic role of haem and nonhaem iron in lipid oxidation, the nature of the initiating species involved and the relative importance of nonenzymatic and enzymatic catalysts. This chapter will address these issues as well as the inhibition of lipid oxidation in meats. The term lipid oxidation is used consistently throughout the text, although it is recognized that the term lipid peroxidation is more generally favoured, particularly in the biochemical literature. Special attention will also be directed toward the mechanisms by which nitrite functions as an antioxidant in cured meats, and the relationship between dietary vitamin E and meat quality.

7.2 Role of lipids in meat flavour

Lipids can contribute both desirable and undesirable flavours to meats. Among the desirable flavours are the characteristic species-associated flavours and aromas that occur in beef and pork (Hornstein and Crowe, 1960), in beef and whale meat (Hornstein *et al.*, 1963) and in lamb (Hofstrand and Jacobson, 1960; Hornstein and Crowe, 1963). These flavours and aromas are generally accepted as desirable by at least some segment of the consuming public, although others may object to certain flavours, as is the case with lamb (Hofstrand and Jacobson, 1960).

Lipids may contribute to the desirable flavour of cooked meat in other ways. During cooking, they undergo thermally induced oxidation to give a range of volatile products which can contribute to meat aroma, but which may also react with components from the lean tissue to give other flavour compounds. In addition, they may also act as a solvent for aroma compounds accumulated during production, processing and cooking of meat (Mottram and Edwards, 1983). The importance of the interaction of lipids in the Maillard reaction to the formation of meat flavour compounds was confirmed recently in a series of studies by Mottram and co-workers. Farmer *et al.* (1989) demonstrated that phospholipid had a marked effect on the overall odour of reaction mixtures containing ribose and cysteine and on the nature of the individual aromas detected. Without phospholipid, sulphur-containing heterocyclic Maillard products predominated, while the inclusion of phospholipid added lipid degradation products as well as certain compounds specific to the interaction of lipid in the Maillard reaction. The most likely routes by which lipid could interact in the Maillard reaction were summarized by Farmer and Mottram (1990) as follows: (1) the reaction of carbonyl compounds from lipids with the amino groups of cysteine and ammonia produced by its Strecker degradation; (2) the reaction of the amino group in phosphatidylethanolamine with sugar-derived carbonyl compounds; (3) the interaction of free radicals from oxidized lipids in the Maillard reaction; and (4) the reaction of hydroxy and carbonyl lipid oxidation products with free hydrogen sulphide.

Development of oxidative off-flavours (rancidity) has long been recognized as a serious problem during the holding or storage of meat products for subsequent consumption. Rancidity in meat begins to develop soon after death and continues to increase in intensity until the meat product becomes unacceptable to consumers. The rate at which these off-flavours develop varies greatly and depends on a number of factors which are reviewed in detail elsewhere (Pearson *et al.*, 1977; Gray and Pearson, 1987). When uncured meat is stored after cooking, flavour changes occur rapidly. The term 'warmed-over flavour' was used by Tims and Watts (1958) to describe the rapid development of an oxidized flavour in refrigerated cooked meats which can develop within 48 h at 4°C. This is in marked contrast to rancidity in uncooked meat that is not normally apparent until it has been stored for several months. Warmed-over flavour has also been used to describe the flavour problem in unheated products such as mechanically separated and restructured meats, in which the muscle membranes are broken down by mechanical means (Pearson *et al.*, 1977; Gray and Pearson, 1987).

The terms rancidity and warmed-over flavour are often used interchangeably as lipid oxidation has been generally considered the major contributor to warmed-over flavour (Pearson *et al.*, 1977). However, the term 'meat flavour deterioration' has been suggested to better describe the complex series of chemical reactions that contribute to an overall increase in off-flavour notes and a loss in desirable meat flavour quality (Spanier *et al.*, 1988). Recently, it has been demonstrated that warmed-over flavour is not due solely to lipid oxidation (St. Angelo *et al.*, 1988), but may also be associated with the deterioration of the desirable beef flavour itself. Many of these reactions involve free radicals (St. Angelo *et al.*, 1990). From a sensory point of view, these chemical reactions result in the formation of undesirable cardboardy and painty flavours and decreases in the desirable cooked beef brothy flavour. Love (1988) indicated that the use of these more descriptive terms in place of ones such as 'warmed-over flavour' or 'stale' should enable researchers in the warmed-over flavour area to obtain more useful information from sensory studies.

7.3 Lipid oxidation in meats

7.3.1 Role of haem and nonhaem iron as catalysts

The rate and extent of lipid oxidation in meat are dependent on a number of factors, the most important being the level of polyunsaturated fatty acids present in a particular muscle system (Allen and Foegeding, 1981). Younathan and Watts (1960) demonstrated that the lipids involved in flavour deterioration in cooked meat were the unsaturated fatty acids of the lean tissue or cellular lipids which exist primarily as phospholipids. The development of rancid off-flavours varies with species, most likely because of differences in phospholipid content and fatty acid composition. Susceptibility of phospholipids to oxidation is due in part to their high content of polyunsaturated fatty acids, particularly linoleic and arachidonic acids (Igene and Pearson, 1979). Phospholipids from beef are approximately 15% more unsaturated than the triacylglycerols. Their pronounced susceptibility to oxidation may also be due to their close association in membranes with tissue catalysts of oxidation. Grinding and chopping disrupt the membranes and expose the phospholipids to oxygen, enzymes, haem pigments and metal ions which can cause the rapid development of rancidity even in fresh raw meat (Pearson *et al.*, 1977; Asghar *et al.*, 1988).

Because lipid oxidation is an important problem in red meat, there has been a great deal of interest in identifying the catalysts that promote the oxidative process (Love, 1983). Traditionally, lipid oxidation has been attributed to haem catalysts such as haemoglobin, myoglobin and cytochromes. Fox and Benedict (1987) summarized the role of haem pigment catalysis of oxidative reactions beginning with the initial work of Robinson (1924). Numerous studies have confirmed that haem compounds function as pro-oxidants when in contact with purified lipids (Brown *et al.*, 1963; Hirano and Olcott, 1971). Tappel (1962) proposed that the most probable mechanism of catalysis involved the formation of a coordinate complex between the haem compound and lipid hydroperoxide, followed by homolytic scission of the hydroperoxide to form free radicals. In this mechanism, there would be no change in the valence of haem iron.

Meat contains nonhaem iron as well as haem iron and there has been considerable interest in the possibility that the former plays an important role in accelerating lipid oxidation in meat (Love, 1983). Sato and Hegarty (1971) first questioned the catalytic role of haem by demonstrating that the addition of myoglobin or haemoglobin to meat which had been exhaustively washed to remove the haem pigments had little effect on lipid oxidation. However, ferrous chloride and ferric chloride both promoted lipid oxidation, the ferrous ions being more active. Love and Pearson (1974) and Igene *et al.* (1979) working with similar systems, also concluded that myoglobin was not the principal pro-oxidant in cooked meat. Igene *et al.* (1979) reported that the addition of a pigment extract to cooked beef muscle residue enhanced lipid oxidation. However, treatment of the extract with ethylenediaminetetraacetic acid reduced its pro-oxidant activity. When hydrogen peroxide was added to the extract to destroy the haem pigments, increased catalytic activity was observed. These investigators further reported that over 90% of the iron in the fresh meat pigment was present as bound haem iron. Cooking destroyed the haem molecule and increased the concentration of nonhaem iron in the extract from 8.7% of the total iron in the unheated extract to 27.0% in extracts heated to 70°C. It was concluded that the increased rate of lipid oxidation in cooked meat is due, in part, to the release of nonhaem iron during cooking which catalyses oxidation rather than the meat pigments *per se*. The nonhaem iron increase in cooked meat has also been reported by Schricker and Miller (1983) and Rhee and Ziprin (1987). However, as correctly pointed out by Rhee (1988), the greater susceptibility of cooked meat toward lipid oxidation may also be due to the disruption of meat tissues by cooking, thus bringing the lipid substrates and catalysts into closer contact.

Table 7.1 Rates of lipid oxidation catalysed by metmyoglobin and iron in raw and heated water-extracted muscle systems from several species at 4°C for 5 days[a]

Muscle	Increase in TBARS/day[b]					
	Control		MetMb[c]		Fe^{2+}	
	Raw	Heated	Raw	Heated	Raw	Heated
Fish	0.14	0.16	0.28	2.70	2.71	4.37
Turkey	0.13	0.15	0.27	2.18	1.89	3.84
Chicken	0.13	0.15	0.26	1.98	1.77	3.73
Pork	0.12	0.13	0.23	1.82	1.42	3.27
Beef	0.09	0.10	0.16	1.19	0.62	2.13
Lamb	0.07	0.09	0.12	0.86	0.44	1.09

[a]Adapted from Tichivangana and Morrissey (1985), with permission.
[b]Mean TBARS/day of four replicates carried out in duplicate.
[c]Metmyoglobin (5 mg/g), Fe^{2+} (5 mg/kg).

Somewhat similar studies by Tichivangana and Morrissey (1985) revealed that only in raw fish was metmyoglobin a significant pro-oxidant, whereas significant increases in thiobarbituric acid (TBA)-reactive substances (TBARS, also referred to as TBA numbers or values and usually expressed as mg malonaldehyde per kg muscle tissue) were seen with pork, chicken, turkey and fish in the presence of ferrous ions (Table 7.1). In all cooked systems, oxidation was much faster than in the raw state. Ferrous ions at the 1 mg/kg level were found to be highly catalytic in cooked muscle, while copper(II) and cobalt(II) at a similar concentration were more effective than metmyoglobin (5 mg/g).

Recently, Kanner and his associates have demonstrated a direct involvement of ferric haem pigments in lipid oxidation in muscle tissues (Kanner and Harel, 1985; Harel and Kanner, 1985a,b). Their results suggest that the haem pigments may only be effective catalysts in the presence of hydrogen peroxide. The addition of metmyoglobin or hydrogen peroxide to the sarcosomal fraction isolated from turkey dark meat had little effect on the rate of oxidation. When the two reactants were added together, the rate of oxidation was significantly enhanced (Harel and Kanner, 1985a). These investigators proposed that the hydrogen peroxide-activated metmyoglobin is essentially a ferryl species called the porphyrin cation radical in which iron has an oxidation number of four. How lipid oxidation is initiated by this radical species will be discussed later.

There is, however, some evidence that the catalytic effect of the metmyoglobin–hydrogen peroxide system may be due, at least in part, to release of iron from the haem pigment by hydrogen peroxide. Rhee *et al.* (1987) reported that raw water-extracted beef muscle treated with met-

Table 7.2 TBARS and nonhaem iron content of raw and cooked water-extracted beef muscle residue samples stored at 4°C as affected by the addition of metmyoglobin or metmyoglobin–hydrogen peroxide[a]

Treatment[b]	Raw				Cooked			
	TBARS[d]			Nonhaem iron content[d]	TBARS			Nonhaem iron content[d]
Days	0	3	6	0	0	3	6	0
None (control)	0.27	0.28	0.29	1.03	0.64	1.24	1.37	1.12
MetMb	0.49	0.60	0.52	2.22	0.89	1.50	1.82	2.25
MetMb + H_2O_2								
1:0.1[c]	1.79	1.16	1.04	2.60	1.23	1.77	2.55	2.79
1:0.25	2.63	3.35	4.13	–	1.36	2.26	3.06	–
1:0.5	0.64	2.08	2.80	9.43	1.13	2.64	4.00	8.95
1:1	0.87	2.10	2.69	10.70	2.34	3.97	4.85	11.54
1:2	–	–	–	11.84	–	–	–	12.19

[a]Adapted from Rhee *et al.* (1987), with permission.
[b]The level of MetMb added was 4 mg/g residue or 0.26 M based on the moisture content (83%) of the meat residue mixtures.
[c]Molar ratio of MetMb to H_2O_2.
[d]Iron concentration expressed as mg/kg sample; TBARs, mg malonaldehyde/kg sample.

myoglobin–hydrogen peroxide in molar ratios of 1:0.5, 1:1 and 1:2 had significantly higher nonhaem iron contents than samples treated with metmyoglobin alone (Table 7.2). They concluded that activated metmyoglobin was the primary catalyst of oxidation in the raw system, although the nonhaem iron released from the haem pigment was also contributory. When the beef muscle systems were heated, the rate of lipid oxidation was slow in the system initially treated with metmyoglobin alone, but was rapid in those systems containing metmyoglobin and hydrogen peroxide. The rate of lipid oxidation increased steadily and progressively as the relative amount of hydrogen peroxide increased (Table 7.2). These investigators surmised that haem iron (activated metmyoglobin) may initiate lipid oxidation in cooked meat, as it may do so in uncooked meat, but nonhaem iron plays a greater role in accelerating lipid oxidation in cooked meat than in uncooked meat. Recently, Kanner *et al.* (1988a,b), in studies with membrane model systems, also concluded that lipid oxidation in ground turkey muscle is effected by free iron and other transition metal ions.

The exhaustively washed muscle fibre systems were developed to give a clearer understanding of the respective roles of haem and nonhaem iron as catalysts of lipid oxidation (Sato and Hegarty, 1971). However, several recent studies with such systems have provided some interesting and often contradictory results. Johns *et al.* (1989) demonstrated that haemoglobin, when added to emulsions containing refined lard, egg white and corn starch at levels similar to the haemoprotein contents in fresh meat, was a more effective catalyst of lipid oxidation than inorganic iron compounds at levels approximating those present in meat. Similarly, when added to and evenly distributed in washed muscle fibres at levels comparable to those found in meat, haemoglobin was again a strong catalyst, whereas all forms of inorganic iron appeared to have little pro-oxidant activity (Table 7.3). They concluded that the conflicting results relative to the roles of haem pigments and inorganic iron in lipid oxidation found in the literature (Sato and Hegarty, 1971; Love and Pearson, 1974; Tichivangana and Morrissey, 1985) may be partly due to the difficulty of evenly dispersing the catalysts in the washed fibres and that hydrogen peroxide, formed by autoxidation of the oxyhaem pigment, may be necessary for the methaem pigments to be effective catalysts. Monahan *et al.* (1993) investigated some of the subtle differences between the muscle fibre systems of Love and Pearson (1974), Tichivangana and Morrissey (1985) and Johns *et al.* (1989) and determined that the mode of addition of the reactants to the washed muscle fibres (in dry form or in solution) had no effect on their catalytic activity. Furthermore, similar results were obtained for haemoglobin (mixture of met and oxy pigments) and myoglobin (met form), thus casting doubt on the suggestion that hydrogen peroxide, produced via the oxidation of haemoglobin, activated the methaemoglobin into an effective catalyst (Johns *et al.*, 1989). It was further demonstrated that the pro-oxidant effects of haem proteins and inorganic iron increased with increasing catalyst concentration, and were similar in heated model systems when present at equimolar iron concentrations.

Table 7.3 TBARS of washed, cooked muscle fibres containing various potential catalysts during aerobic storage at 2°C

Catalyst	Concentration (mg Fe/kg fibre)	Time (days)		
		0	4	7
None	–	1.17 ± 0.02	1.37 ± 0.05	–
$FeSO_4$	1	0.86 ± 0.02	1.91 ± 0.05	2.85 ± 0.06
Hb	16.5	1.01 ± 0.04	4.01 ± 0.08	6.20 ± 0.06
None	–	0.80 ± 0.01	1.48 ± 0.01	2.11 ± 0.17
$FeSO_4$	1	0.92 ± 0.02	1.79 ± 0.06	2.18 ± 0.03
Hb	16.5	1.04 ± 0.02	4.67 ± 0.10	5.66 ± 0.18
None	–	0.79 ± 0.02	1.10 ± 0.02	0.81 ± 0.03
$FeSO_4$	1	1.57 ± 0.18	1.70 ± 0.02	1.25 ± 0.08
Hb	16.5	1.58 ± 0.16	3.07 ± 0.05	3.12 ± 0.02

From Johns *et al.* (1989), with permission.

The inadequacy of the muscle fibre model system to fully represent the intact muscle was further demonstrated by Kanner *et al.* (1991) who contended that the aqueous extraction of muscle fibres might also have extracted several essential compounds such as enzymes, hydrogen peroxide, and reducing compounds or chelating agents which may affect the overall catalysis of muscle lipid peroxidation *in situ.* Asghar *et al.* (1988) had also proposed that the removal of endogenous hydrogen peroxide from the muscle tissue through the washing process might explain why earlier studies had revealed metmyoglobin to be an inferior catalyst to ferrous iron in such systems. Kanner *et al.* (1991) evaluated the effect of the cytosolic extract (soluble fraction) on lipid oxidation of the water-extracted muscle residue (insoluble fraction) stimulated by hydrogen peroxide-activated myoglobin and iron ion-dependent enzymatic and nonenzymatic reactions. They demonstrated that the cytosolic fraction contains reducing compounds that stimulate iron redox cycle-dependent lipid oxidation, but totally inhibit the catalysis by activated haem proteins. Clearly these results and the apparent contradictory reports on the relative roles of haem and nonhaem catalysts in model systems necessitate further studies in this area.

7.3.2 Ferritin as a catalyst of lipid oxidation in meat

Ferritin is a soluble iron storage protein found in the liver, spleen and skeletal muscle, has a molecular mass of 450,000 Da and contains 4,500 ions of iron per molecule of protein. Ferritin releases iron as Fe^{2+} in the presence of reducing agents such as ascorbate, superoxide anion, and thiols (Boyer and McCleary, 1987). It is also well established that iron released from ferritin by reducing agents catalyses lipid oxidation *in vitro* (Gutteridge *et al.*, 1983; Thomas *et al.*, 1985).

Several studies have focused on ferritin as a source of catalytic iron ions for lipid oxidation in muscle tissue. Apte and Morrissey (1987a) evaluated the effects of haemoglobin and ferritin on lipid oxidation in raw and cooked muscle systems and demonstrated that intact ferritin is not a catalyst, i.e. ferritin does not contribute to lipid oxidation in raw muscle. However, these investigators demonstrated in this and a subsequent study (Apte and Morrissey, 1987b) that ferritin is highly catalytic in cooked systems. They concluded that heating denatured the ferritin molecule, probably releasing free iron which then catalysed lipid oxidation and the development of off-flavours. On the other hand, Decker and Welch (1990) reported that physiological concentrations of ferritin-bound iron found in beef muscle catalysed lipid oxidation *in vitro* and heating (100°C for 15 min) increased the ability of ferritin to promote oxidation. These data suggest that ferritin could be involved in the development of off-flavours in both cooked and uncooked muscle foods. Model system studies by Seman *et al.* (1991) also provided data which suggested that ferritin may be at least partially responsible for catalysing lipid oxidation in muscle foods. Kanner and Doll (1991) reached a similar conclusion from their study of ferritin-catalysed oxidation of turkey muscle membrane lipids.

7.4 Initiation of lipid oxidation in muscle tissue

7.4.1 *Initiators of the oxidation reaction*

Although the studies described in the previous section have provided some understanding of the mechanism of lipid oxidation in meats, they have also added somewhat to the uncertainty surrounding the respective roles of haem and nonhaem iron in the oxidation process. This uncertainty is further accentuated by the interchangeable use of the terms *catalysts*, *initiators* or *promoters* in the meat science literature, particularly as they refer to iron salts and other iron complexes as initiating lipid oxidation. Although the whole subject of lipid oxidation and the factors influencing the oxidation reaction in meats has been the focus of much research for several decades, there still remain several unanswered questions. For example, what are the primary catalysts or species that initiate lipid oxidation in raw muscle tissue? Is lipid oxidation in meat systems initiated by enzymatic or nonenzymatic oxidation systems? These and other issues have been addressed recently in two excellent reviews by Kanner *et al.* (1987) and Halliwell and Gutteridge (1990) and these should be consulted for further information.

Much of the information pertaining to lipid oxidation in muscle tissues deals with hydroperoxide-dependent lipid oxidation. Pure lipid hydroperoxides are fairly stable at physiological temperatures, but in the presence of transition metal complexes, especially iron salts, their decomposition is greatly accelerated. Ferrous ion will cause fission of O—O bonds to form very reactive alkoxy radicals for the propagation reaction, whereas the iron(III) complex can form both peroxy and alkoxy radicals (Ingold, 1962).

$$\mathrm{LOOH} + \mathrm{Fe^{2+}} \rightarrow \mathrm{LO^{\bullet}} + \mathrm{OH^{-}} + \mathrm{Fe^{3+}}$$
$$\mathrm{LOOH} + \mathrm{Fe^{3+}} \rightarrow \mathrm{LOO^{\bullet}} + \mathrm{H^{+}} + \mathrm{Fe^{2+}}$$

(further reaction to give alkoxy radical)

The washed muscle fibre systems described in the previous section were prepared by several extractions of ground muscle tissue and thus would be expected to contain lipid hydroperoxides before the addition of haem and nonhaem iron. It is also likely that the microsomal preparations used by several investigators (Harel and Kanner, 1985b; Buckley *et al.*, 1989;

Asghar *et al.*, 1991) also contained preformed lipid hydroperoxides because of the grinding/homogenization of the muscle tissues used in these studies. Halliwell and Gutteridge (1990) pointed out that the microsomal fraction prepared from many animal tissues is a heterogeneous collection of vesicles from the plasma membrane and endoplasmic reticulum obtained by high speed centrifugation of cell homogenates. Membrane fractions isolated from disrupted cells should contain some lipid hydroperoxide. Thus, the addition of iron salts and haem compounds to the washed fibre and microsome systems is doing no more than stimulating further lipid oxidation by breaking down lipid hydroperoxides.

A primary question remains then as to what initiates lipid oxidation in uncooked muscle tissue and other biological systems when no preformed lipid hydroperoxide is present, i.e. what species have sufficient reactivity to abstract a hydrogen atom from a methylene ($-CH_2-$) group in a polyunsaturated fatty molecule? The lipid oxidation process consists of three generally recognized phases:

Initiation $$LH \xrightarrow{\text{Initiator}} L^{\bullet}$$

Propagation $$L^{\bullet} + O_2 \xrightarrow{\text{Fast}} LOO^{\bullet}$$

$$LH + LOO^{\bullet} \xrightarrow{\text{Slow}} LOOH + L^{\bullet}$$
$$LOOH \longrightarrow LO^{\bullet} + {}^{\bullet}OH$$

Termination $$L^{\bullet} + L^{\bullet} \longrightarrow L\text{—}L$$
$$L^{\bullet} + LOO^{\bullet} \longrightarrow LOOL$$
$$LOO^{\bullet} + LOO^{\bullet} \longrightarrow LOOL + O_2$$

where $L^{\bullet}$ is the alkyl radical, LOO is the peroxy radical and $LO^{\bullet}$ is the alkoxy radical. The direct interaction of oxygen with unsaturated fatty acids is extremely slow because the initiation of oxidation requires an active form of oxygen (Hsieh and Kinsella, 1989). Because molecular oxygen in the ground state is paramagnetic (i.e. contains two unpaired electrons) and unsaturated fatty acids in the ground state are diamagnetic (i.e. contain no unpaired electrons), there is a spin barrier which prevents the direct addition of triplet oxygen to singlet state unsaturated fatty acid molecules (Asghar *et al.*, 1988). This spin restriction of molecular oxygen can be overcome by any of a number of initiation mechanisms: (1) electronic excitation of the triplet state dioxygen to the singlet state, as happens in photooxidation (Foote, 1985); (2) partially reduced or activated oxygen species such as hydrogen peroxide, superoxide anion or hydroxyl radicals; (3) active oxygen–iron complexes (ferryl radical); and (4) iron-mediated homolytic cleavage of the hydroperoxides which generate organic free radicals (Kanner *et al.*, 1987; Hsieh and Kinsella, 1989).

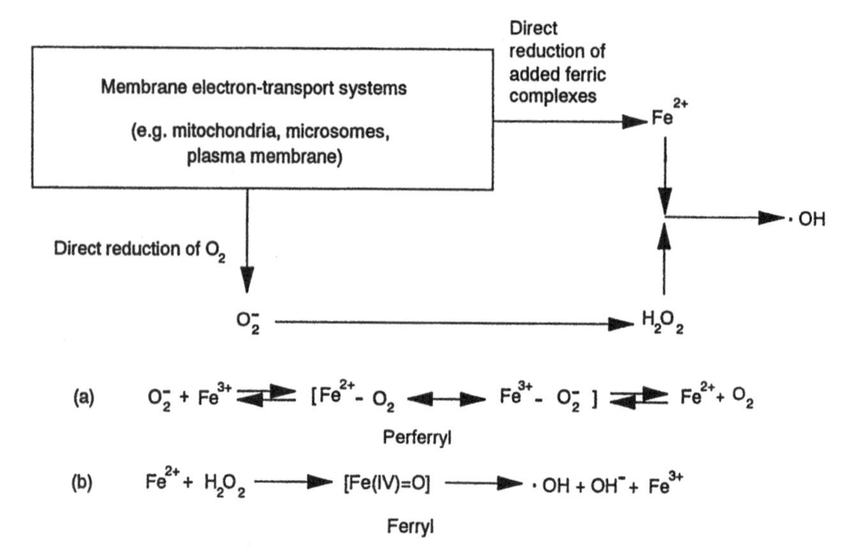

Figure 7.1 Independent generation of •OH and other reactive species in biological membrane systems. Reproduced from Gutteridge and Halliwell (1990), with permission.

Active oxygen species in biological systems can be generated via non-enzymatic and enzymatic mechanisms (Fridovich, 1976). Figure 7.1 summarizes the iron-dependent generation of the hydroxyl radical (Gutteridge and Halliwell, 1990). *In vivo* sources of the hydroxyl radical (•OH) include homolytic fission of the —O—O— bond in hydrogen peroxide, which is catalysed by heat, radiation and the Fenton–Haber–Weiss reaction (Kanner *et al.*, 1987; Halliwell and Gutteridge, 1990). The hydroxyl radical is extremely reactive and can readily initiate lipid oxidation:

$$LH + {}^{\bullet}OH \longrightarrow L^{\bullet} + H_2O$$

Figure 7.1 also indicates that reactive species other than •OH can also be formed in biological systems. Aust and his associates (Svingen *et al.*, 1979; Tien and Aust, 1982) extensively studied lipid oxidation in liver microsomes involving NADPH–cytochrome P-450 reductase and xanthine oxidase. They concluded that cytochrome P-450 reductase catalyses the transfer of an electron from NADPH to Fe^{3+} to generate Fe^{2+}, which can form a complex with oxygen to produce perferryl or ferryl radicals. They assumed that these radicals have the ability to abstract hydrogen from polyunsaturated fatty acids to initiate oxidation. In the case of xanthine oxidase, superoxide anion ($O_2^{\bullet -}$) produced by this enzyme reduces Fe^{3+} to Fe^{2+}, which then can produce the ferryl species via its interaction with hydrogen peroxide.

It has also been reproducibly observed in many membrane systems that the initial rate of oxidation measured when Fe^{2+} is added, is increased if Fe^{3+} is also present. Minotti and Aust (1987) proposed that a ferrous–dioxygen–ferric complex is the actual species responsible for the initiation reaction. This complex originates either from Fe^{3+} in the presence of reducing agents such as NADPH, superoxide anion, ascorbate or thiol compounds, or from Fe^{2+} in the presence of an oxidizing agent such as hydrogen peroxide. Complete reduction of the ferric ion, however, would inhibit lipid peroxidation, while complete oxidation of the ferrous ion would have the same effect. However, several recent studies have indicated that other metal ions such as Pb^{2+} and Al^{3+} can replace Fe^{3+} in stimulating Fe^{2+}-dependent peroxidation, thus suggesting that a specific Fe^{2+}–Fe^{3+}–O_2 cannot be required for the initiation of oxidation (Halliwell and Gutteridge, 1990).

As mentioned previously, it is generally accepted that haem proteins catalyse the propagation step and are not truly initiators of lipid oxidation. However, recent studies have indicated that the interaction of hydrogen peroxide with metmyoglobin will produce an activated haem protein which is capable of initiating membrane lipid oxidation (Harel and Kanner, 1985a,b). The activated species appears to be a porphyrin cation radical which is also a ferryl species and not a hydroxyl radical. The proposed sequence of events to explain the initiation and propagation of lipid oxidation in meat systems is as follows:

(1) Autoxidation and oxygen activation of oxymyoglobin to metmyoglobin and hydrogen peroxide.
(2) Activation of metmyoglobin by hydrogen peroxide to form the porphyrin cation radical, in which iron has an oxidation number of four ($P^{+} - Fe^{4+} = 0$).
(3) Initiation of lipid oxidation by the porphyrin cation radical via two-electron reduction of the catalyst.

Thus, hydrogen peroxide which is generated by muscle tissue (Harel and Kanner, 1985a) seems to play an important role in the formation of the primary pool of catalysts that may be involved in the initiation of lipid oxidation in muscle tissues (Kanner *et al.*, 1987).

7.4.2 *Enzymatic lipid oxidation*

While it is generally accepted that lipid oxidation in muscle foods is non-enzymatic in nature, there is evidence that there are enzymatic lipid oxidation systems associated with muscle microsomes (Love, 1983; Rhee, 1988). The lipid oxidation in subcellular fractions reported by various investigators is mediated by Fe^{2+} iron maintained enzymatically in the reduced form by NADPH. Hultin and his associates (Lin and Hultin, 1976; Player and Hultin, 1977) showed that chicken breast or leg muscle microsomes produced malonaldehyde when incubated with appropriate components (NADH or NADPH, ADP and iron) of an oxidation system. Similar lipid oxidation systems associated with beef and pork muscle microsomes were reported by Rhee *et al.* (1984) and Rhee and Ziprin (1987), respectively. An enzymatic lipid oxidation system in trout muscle mitochondria was identified by Luo and Hultin (1986) and was found to be similar to that in fish muscle microsomes in terms of cofactor requirements and optimal pH.

However, the use of the term 'enzymatic lipid oxidation' to describe the lipid oxidation systems associated with muscle microsomes and mitochondria has been questioned recently by Halliwell and Gutteridge (1990). It is their view that the term enzymatic lipid oxidation should be reserved for the enzymes lipoxygenase and cyclooxygenases, which catalyse the controlled oxidation of unsaturated fatty acids to give hydroperoxides and endoperoxides that are stereospecific. Although an enzyme is involved, microsomal lipid oxidation in the presence of NADPH and Fe^{3+} (often added as complexes of ADP) is essentially nonenzymatic in nature, the enzyme serving only to reduce the iron complexes to Fe^{2+}. Similarly, enzyme-generated superoxide anion (e.g. by xanthine plus xanthine oxidase) can increase the rate of membrane lipid oxidation in reaction mixtures containing Fe^{3+} by reducing Fe^{3+} to Fe^{2+}. These investigators conclude that the function of the enzyme is similar to that of ascorbic acid in nonenzymatic oxidation.

Recently, it has been demonstrated that lipoxygenases in gill and skin tissues of several species of fish, catalyse the oxidation of polyunsaturated fatty acids to produce unstable hydroperoxides. The hydroperoxides, following carbon–carbon cleavage of the hydroperoxide group, are potential precursors of many compounds, such as hexanal, 4-heptenal and 2,4-heptadienal (Hsieh *et al.*, 1988). These carbonyls are sources of oxidative off-flavours that can adversely affect consumer acceptability of fish (Josephson *et al.*, 1984). Two lipoxygenases, 12-lipoxygenase (German and Kinsella, 1985) and 15-lipoxygenase (German and Creveling, 1990), have been identified in fish gill tissue and provide the potential for development of many different volatile compounds from the same fatty acid precursor (German and Creveling, 1990). The presence of a 15-lipoxygenase in chicken muscle has also been reported by Grossman *et al.* (1988), who suggested that the enzyme may be responsible for some of the oxidative changes occurring in fatty acids of chicken meat during frozen storage. The latter supports the earlier findings of Decker and Schanus (1986a,b) which suggested that both haem and nonhaem catalysts and enzyme systems might be involved in catalysis of lipid oxidation in raw chicken dark meat.

7.5 Prevention of lipid oxidation in meats

Many studies have indicated that lipid oxidation in meat products can be effectively controlled or, at least, minimized by the use of antioxidants. These compounds can range from commercial phenolic antioxidants to more exotic compounds isolated from natural food products. Gray and Pearson (1987) and Bailey (1988) have reviewed the applications of antioxidants in meat systems and these will not be addressed in any detail in this chapter. The major focus will be on recent studies on the mechanism of nitrate as an antioxidant, stabilization of meat lipids through vitamin E-supplementation of animal diets, and the use of spice extracts in the processing of meat systems.

7.5.1 Antioxidant role of nitrite

In the processing of cured meats, sodium nitrite is used as a curing adjunct because of its ability to produce the characteristic cured meat flavour and colour and its antibotulinal activity (Gray and Pearson, 1984). In addition, nitrite is an effective antioxidant and has been found to retard oxidation in cooked ground beef when used at the 50 mg/kg level and to completely eliminate oxidation at the 2000 mg/kg level (Sato and Hegarty, 1971). Fooladi *et al.* (1977) confirmed these and other observations by demonstrating that nitrite (156 mg/kg) inhibited warmed-over flavour in cooked meat with a two-fold reduction in TBARS for beef and chicken and a five-fold reduction for pork. Addition of nitrite also inhibited the development of rancidity in raw beef and chicken during storage at 4°C for 48 h. Ramarathnam *et al.* (1991) recently reported that the concentration of hexanal, a major lipid oxidation product, was approximately 400 times greater in uncured pork than in cured pork. These results confirm many previous observations that hexanal formation in meats is inhibited by the presence of nitrite (Cross and Ziegler, 1965; Swain, 1972; MacDonald *et al.*, 1980).

The mechanism by which nitrite prevents or retards the oxidation of meat lipids is not fully understood. Literature data suggest that more than one mechanism is involved. Zipser *et al.* (1964) proposed that nitrite formed a complex with the iron porphyrins in heat-denatured meat, thus inhibiting haem-catalysed lipid oxidation. Pearson *et al.* (1977) suggested that it was more likely that nitrite stabilizes the lipid-containing membranes, thereby preventing oxidation. Igene *et al.* (1985) concluded that nitrite may inhibit oxidative reactions by three possible mechanisms: (1) by forming a strong complex with the haem pigments, thus preventing the release of ferrous iron with its attendant catalysis of the propagation stage of lipid oxidation; (2) by stabilizing the polyunsaturated lipids within the membranes; and (3) by serving to 'tie-up' or 'chelate' metal ions such as ferrous ions, thus rendering them unavailable for catalysis of lipid oxidation. Evidence was presented to indicate that all three mechanisms may be operative in cured meats, although the first mechanism appeared to be the most important. Morrissey and Tichivangana (1985) also postulated that nitrite probably formed inactive 'chelates' or complexes with nonhaem iron, copper and cobalt and in this manner inhibited catalytic activity and prevented oxidation. In a subsequent series of experiments, Apte and Morrissey (1987a,b) demonstrated that nitrite could inhibit the catalysis of lipid oxidation in heated water-washed muscle fibres by ferritin, myoglobin, haemoglobin and free iron. They concluded that the complexing of either free or low molecular weight iron fractions by nitric oxide may be the critical reaction controlling lipid oxidation in cured meats.

Recently, Freybler *et al.* (1989) provided further evidence that nitrite functions via several co-operative mechanisms. Mitochondrial and microsomal fractions isolated from cured pork were significantly more stable than their counterparts from uncured pork (nitrite-free) when exposed to the catalyst system of metmyoglobin and hydrogen peroxide (Figure 7.2). Phospholipids from the cured pork samples also oxidized less rapidly than those from the uncured samples when subjected to the same oxidation conditions. When the latter phospholipids were reacted with dinitrogen

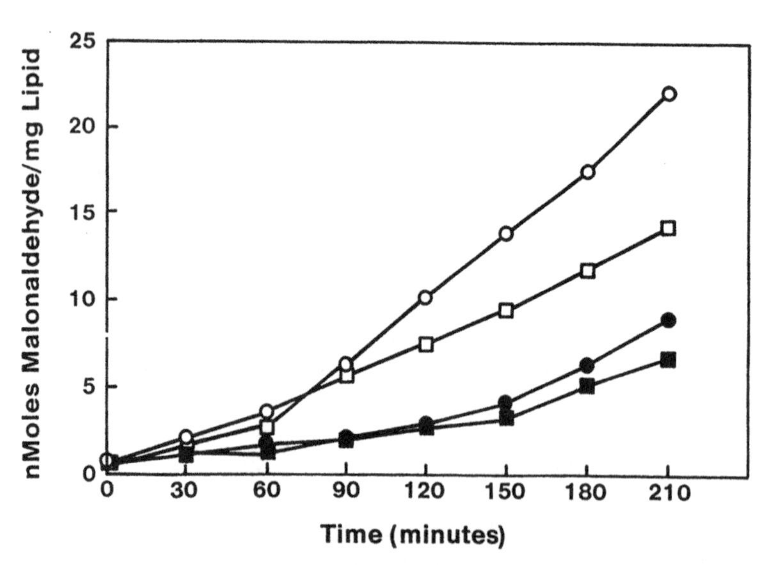

Figure 7.2 Metmyoglobin/hydrogen peroxide-catalysed lipid oxidation in microsomes and mitochondria from cured and uncured pork. □ = Microsomes from uncured pork; ■ = microsomes from cured pork; ○ = mitochondria from uncured pork; ● = mitochondria from cured pork. Reproduced from Freybler *et al.* (1989).

trioxide, their oxidative stability was significantly enhanced. To confirm that the stabilization of the lipids in cured meats was due to the interaction of nitrite or dinitrogen trioxide with the double bonds of unsaturated fatty acids, the lipid samples were heated with a secondary amine in order to form the corresponding *N*-nitrosamine. Results clearly indicated that phospholipids extracted from cured pork were capable of nitrosating morpholine, whereas those from the uncured samples were not. These data support the results of Liu *et al.* (1988) and imply that nitrite reacts with unsaturated lipids to form nitro-nitrosite derivatives, thus stabilizing the lipids against oxidative changes.

In a second series of experiments, Freybler *et al.* (1989) provided additional data in support of nitrite stabilization of haem pigments. Water-extracted pork samples were treated with metmyoglobin and nitric oxide myoglobin, with and without hydrogen peroxide, and stored at 4°C. Lipid oxidation was significantly ($p < 0.05$) higher in samples containing metmyoglobin/hydrogen peroxide compared to control samples and those containing nitric oxide myoglobin and metmyoglobin (Table 7.4). Raw samples treated with nitric oxide myoglobin, alone or in combination with hydrogen peroxide, showed no increase in lipid oxidation over the 72 h

Table 7.4 Effect of metmyoglobin and nitric oxide myoglobin on the oxidative stability of lipids in water-extracted muscle

Storage time (h)	Control	TBARS (mg malonaldehyde/kg muscle)				
		H_2O_2	NOMb	NOMb/ H_2O_2	MetMb	MetMb/ H_2O_2
Raw						
0	0.20	0.38	0.27	0.22	0.49	0.38
24	0.31	0.35	0.37	0.43	0.71	0.82
48	0.31	0.37	0.38	0.43	0.80	1.11
72	0.32[a]	0.40[a]	0.39[a]	0.45[a]	0.87[b]	1.18[c]
Short-term heating						
0	0.32	0.39	0.32	0.32	0.55	0.66
24	0.89	0.88	0.46	0.51	1.25	1.47
48	1.23	0.97	0.56	0.63	1.60	2.02
72	1.48[c]	1.11[b]	0.56[a]	0.74[a]	1.95[d]	2.49[e]
Prolonged heating						
0	0.62	0.81	0.37	0.34	0.63	0.68
24	1.30	1.23	0.59	0.49	1.57	1.66
48	1.41	1.39	0.75	0.79	1.96	2.39
72	1.48[b]	1.57[b]	0.90[a]	0.93[a]	2.35[c]	3.01[d]

Adapted from Freybler *et al.* (1989).
Short-term heating samples heated to 70°C and immediately cooled in an ice bath.
Long-term heating samples heated to 70°C and held for 30 min, then cooled in an ice bath.
Means followed by different superscripts within rows are significantly ($p < 0.05$) different.

storage period. TBARS for the uncooked samples containing nitric oxide myoglobin and nitric oxide myoglobin/hydrogen peroxide after 72 h were 0.39 and 0.45 respectively. Short- and long-term heating did not accelerate lipid oxidation in either sample, which implies that nitric oxide myoglobin acted as a specific antioxidant in these systems. Iron analyses also revealed that the nonhaem iron content did not increase when the model system containing nitric oxide myoglobin was heated (Table 7.5). In contrast, when metmyoglobin was added to the muscle fibres and subjected to the same heat treatments, there was a significant increase in the nonhaem iron content. These results confirm that nitrite stabilizes haem pigments in cured meats, thus preventing the release of free iron as a consequence of exposure to heat and hydrogen peroxide.

The observation that nitric oxide myoglobin possesses antioxidant properties supports the previous results of Kanner *et al.* (1980) and Morrissey and Tichivangana (1985). Kanner *et al.* (1984) proposed that the antioxidant effect of this compound and other nitroxides was due to quenching of substrate free radicals. Morrissey and Tichivangana (1985), however, suggested that the specific contribution of nitric oxide myoglobin as an antioxidant may be relatively insignificant in most cured meats.

7.5.2 *Stabilization of meat lipids with nitrite-free curing mixtures*

While the beneficial effects of nitrite in cured meats are many, it also has the disadvantage of contributing to the formation of *N*-nitrosamines, particularly in cooked bacon (Gray, 1976). Thus, there has been some interest in identifying a suitable substitute for nitrite in the preparation of cured meat products. One approach has been to develop a multicomponent system in which individual components are used to reproduce the colour and flavour imparted by nitrite, and to reproduce its antioxidant and antimicrobial effects. Shahidi *et al.* (1988) formulated several nitrite-free meat

Table 7.5 Effect of heating and addition of hydrogen peroxide on the release of iron from metmyoglobin and nitric oxide myoglobin added to water-extracted muscle

Heat	mg Iron/kg muscle					
	Control	H_2O_2	NOMb	NOMb/ H_2O_2	MetMb	MetMb/ H_2O_2
Raw	1.79[a]	1.45[a]	1.54[a]	1.64[a]	2.02[b]	3.87[b]
Short-term heating	1.33[a]	1.54[a]	1.57[a]	1.70[a]	3.12[b]	4.98[c]
Prolonged heating	1.36[a]	1.49[a]	1.63[a]	1.79±[a]	4.44[b]	5.14[c]

Adapted from Freybler *et al.* (1989).
Means followed by different superscripts within rows are significantly different at ($p < 0.05$).

curing mixtures which included salt, sugar, ascorbates, an antioxidant and/or a chelator, an antimicrobial agent and dinitrosyl ferrohaemochrome. They imparted to meat a similar oxidative stability to that of sodium nitrite. Butylated hydroxyanisole and *t*-butylhydroquinone were the best antioxidants and polyphosphates; ethylenediaminetetraacetic acid and diethyltriaminepentaacetic acid were the superior chelators. O'Boyle *et al.* (1990) applied such a nitrite-free curing system to wieners and produced a product which, in almost all aspects, could not be distinguished from its nitrite-cured counterparts.

7.5.3 Vitamin E and meat quality

The presence or absence of vitamin E in animal tissues can influence the stability of lipids in meats during storage (Pearson *et al.*, 1977). Consequently, many studies have focussed on the effects of dietary vitamin E on lipid stability in muscle foods. Summaries of these activities have been provided by Pearson *et al.* (1977) and Gray and Pearson (1987). Recently, several studies have been conducted on the basis that the incorporation of α-tocopherol into the cell membranes will stabilize the membrane lipids and consequently enhance the quality of meat during storage. This hypothesis was based on the premise that lipid oxidation is initiated at the membrane level (Asghar *et al.*, 1988).

Buckley *et al.* (1989) evaluated the effects of dietary vitamin E and oxidized oil on the oxidative stability of membrane lipids in pig muscles and on the oxidative stability of pork chops and ground pork during refrigerated and frozen storage. Their results clearly demonstrated that dietary treatments had a major impact on the stability of pork products, those from pigs fed the vitamin E supplemented diets (200 IU vitamin E/kg feed) being more stable than those from the control pigs (Table 7.6). The negative effect of oxidized dietary oil on product stability was very pronounced as demonstrated by the rapid increase in TBARS during storage.

To further study the relationship between membrane lipid stability and the oxidative stability of meat during storage, Asghar *et al.* (1991) evaluated the influence of three levels of vitamin E (10, 100 and 200 IU/kg feed) in the diets of pigs on the subcellular deposition of α-tocopherol in the muscle and on certain quality characteristics of pork (oxidative stability of lipids, colour and drip loss) during storage. Data in Table 7.7 show the beneficial effect of dietary vitamin E on the oxidative stability of pork chops when exposed to fluorescent light for 10 days. Even though lipid oxidation increased in all cases during storage, pork chops from the pigs receiving the greatest level of vitamin E (200 IU/kg feed) exhibited the smallest increase in TBARS. In addition, increased colour stability and decreased drip loss were observed during the same storage period.

Table 7.6 Effect of vitamin E supplementation (200 IU/kg feed) and oxidized oil in swine diets on the stability of pork chops during refrigerated storage, as measured by the TBA procedure

Treatment	Days of storage at 4°C							
	Fluorescent Light				Dark			
	0	3	6	9	0	3	6	9
Control	0.08[a]	0.44[b]	0.84[b]	2.58[b]	0.08[a]	0.51[b]	0.72[b]	0.42[b]
Short-term vitamin E (4 week feeding)	0.07[a]	0.16[c]	0.80[c]	0.07[a]	0.07[a]	0.32[c]	0.26[c]	0.29[c,d]
Long-term vitamin E (10 week feeding)	0.08[a]	0.13[c]	0.46[d]	0.76[c]	0.08[a]	0.32[c]	0.24[c]	0.21[d]
Oxidized corn oil[y]	0.08[a]	1.28[a]	3.98[a]	4.98[a]	0.08[a]	1.12[a]	3.35[a]	3.78[a]

Adapted from Buckley *et al.* (1989) with permission.
Corn oil was oxidized to PV = 300, and represented 3% of feed for duration of feeding trial.
TBARS in columns having different superscripts are significantly different ($p < 0.05$).

Table 7.7 Effect of vitamin E supplementation on the quality of pork chops stored at 4°C under fluorescent light

Storage time (days)	Vitamin E supplementation (IU/kg feed)		
	10	100	200
	TBARS (mg malonaldehyde/kg meat)		
0	0.28	0.27	0.27
3	1.54	0.56	0.35
6	2.96	0.94	0.58
10	5.17	2.96	1.93
	Hunter *a* value		
0	10.7	11.6	11.5
3	10.3	11.0	11.7
6	7.3	9.2	10.2
10	7.2	7.9	8.5
	Percent drip loss		
3	19.0	16.2	10.2
6	20.1	19.5	12.2
10	21.3	21.2	14.1

Adapted from Asghar *et al.* (1991) with permission.
Pork loins were initially frozen at −20°C for 3 months before cutting into pork chops for the refrigerated storage study.

The concentrations of α-tocopherol in the mitochondrial and microsomal fractions of the *Longissimus dorsi* muscle increased significantly with increasing levels of dietary vitamin E. The differences in the concentrations of α-tocopherol in the subcellular membranes were clearly reflected in the enhanced stability of membrane-bound lipids when micro-

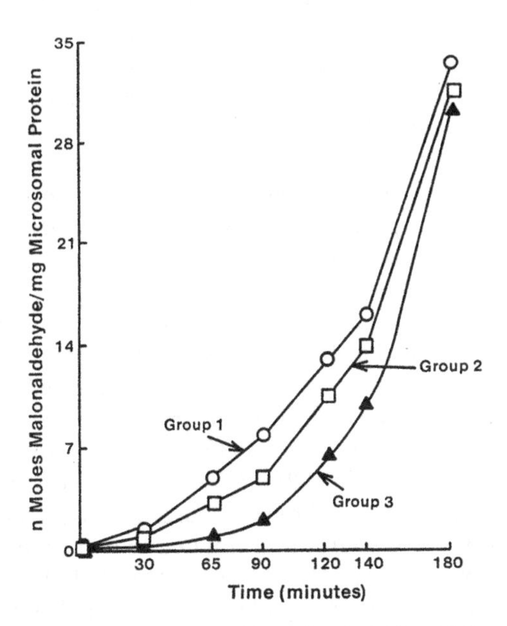

Figure 7.3 Metmyoglobin/hydrogen peroxide-catalysed lipid oxidation in microsomal fractions isolated from the *Longissimus dorsi* muscles of pigs fed vitamin E-supplemented diets. Group 1, 10 IU vitamin E/kg feed; Group 2, 100 IU/kg feed; Group 3, 200 IU/kg feed. Reproduced from Asghar *et al.* (1991), with permission.

somal fractions were subjected to a metmyoglobin/hydrogen peroxide stress (Figure 7.3). These results confirmed the previous findings of Buckley *et al.* (1989) that the stabilization of the membrane-bound lipids through the incorporation of α-tocopherol into the membranes has a positive effect on the oxidative stability of meat during storage.

Results of these studies were confirmed by Monahan *et al.* (1990a) who demonstrated that feeding a supplement of vitamin E (200 IU/kg feed) to pigs from the time of weaning to slaughter significantly increased the α-tocopherol content of a number of tissues and reduced the susceptibility of these tissues to oxidative deterioration. Monahan *et al.* (1990b) further concluded that the incorporation of vitamin E into the membrane lipids via the diet may be one of the most effective means of extending the shelf-life of restructured and precooked meat products.

The influence of dietary vitamin E on beef and veal quality has also been investigated recently. Faustman *et al.* (1989a,b) reported improved lipid and pigment stability in beef from Holstein steers whose diets were supplemented with 370 IU vitamin E/animal/day for approximately 43 weeks. They concluded that the additional vitamin E might be absorbed by the animals and incorporated into the cellular membranes to exert its antioxidant effects. Engeseth (1990) supplemented the whole milk diets of male Holstein calves with vitamin E (500 IU/day) from birth to slaughter (12 weeks) and reported increased oxidative stability of raw and cooked muscle during refrigerated storage. Dietary supplementation resulted in higher concentrations of α-tocopherol in the muscles (6.1 mg/kg) relative to the control animals (1.1 mg/kg). Membrane concentrations followed similar trends and this was reflected in the enhanced stability of the membrane lipids when exposed to metmyoglobin/hydrogen peroxide.

This later study also addressed the oxidative stability of cholesterol in cooked meat during storage. Interest in cholesterol oxidation has increased as a result of the many possible associations of cholesterol oxide consumption with various health disorders (Smith, 1981; Addis and Park, 1989). Cholesterol oxidation products have been reported in a variety of foods including dairy products, French fried potatoes, egg and egg products, and meats (Missler *et al.*, 1985; Park and Addis, 1986, 1987; Sander *et al.*, 1989; Pie *et al.*, 1991). Cholesterol in lean meat tissues is principally located in the membranes (Dugan, 1987; Hoelscher *et al.*, 1988), thereby placing cholesterol in a lipid environment that is very susceptible to oxidation. Engeseth (1990) evaluated cholesterol oxide development in veal muscle (raw and cooked) and concluded that vitamin E supplementation effectively controlled the rate of cholesterol oxidation. As with other indices of oxidation (e.g. TBARS), oxide development was greater in the cooked samples.

Park and Addis (1987) also noted greater cholesterol oxide development with increasing degrees of rancidity (as determined by TBARS) in meats. A comparison of their data with that of Engeseth (1990) is presented in Table 7.8, where the total cholesterol oxidation products are expressed as a percentage of the cholesterol concentration in a variety of cooked meats. The greater percentage for turkey relative to the other meats may be due to the higher degree of unsaturation of the fatty acids in the phospholipid

Table 7.8 Total cholesterol oxidation products, expressed as a percentage of total cholesterol in various cooked meat samples stored at 4°C

Time (days)	Turkey[a]	Beef[b]	Beef[c]	Veal[d]	Veal–Vit E[d]
0	ND	ND	0.30	0.85	0.78
3	0.20	0.45	–	–	–
4	–	–	1.69	1.53	0.53
8	2.90	1.74	–	–	–

[a,b]Park and Addis (1987).
[c,d]Engeseth (1990).
ND, Not detected; detection limit, 1 ng.

fraction of turkey (Jantawat and Dawson, 1980). Thus, it appears that membrane lipid composition affects the stability of cholesterol in muscle tissue and that dietary vitamin E supplementation assists in stabilizing cholesterol in the membranes.

7.5.4 Spice extracts as antioxidants

Many spices and herbs possess antioxidant activity (Gerhardt and Blat, 1984). Rosemary is particularly effective and its activity is due to the presence of a number of phenolic compounds such as carnosol, rosmanol, rosmariquinone and rosmaridiphenol (Wu *et al.*, 1982; Inatani *et al.*, 1982, 1983; Houlihan *et al.*, 1984, 1985). The antioxidant activity of rosemary extracts has been evaluated in several food products as well as in accelerated oxidation tests using chicken fat and lard as substrates. Barbut *et al.* (1985) added 200 mg/kg rosemary oleoresin to turkey breakfast sausage prepared with 75% hand deboned turkey and 25% mechanically deboned turkey meat. Sensory aroma and taste scores and TBARS for samples stored at 4°C for up to 16 days demonstrated that rosemary oleoresin was comparable to a commercial blend of butylated hydroxyanisole, butylated hydroxytoluene and citric acid in suppressing lipid antioxidation.

The antioxidant efficacy of rosemary oleoresin in ground beef patties has been recently evaluated by St. Angelo *et al.* (1990). Ground beef patties treated with metal chelators, free radical scavengers, rosemary oleoresin and sodium alginate were examined by chemical and sensory means with regard to warmed-over flavour. Results indicated that the first two groups of compounds could inhibit warmed-over flavour development. It was concluded that the free radical scavengers were perhaps serving two functions, i.e. they retarded/inhibited lipid oxidation (preventing painty and cardboardy flavour notes), as well as protein degradation (maintaining desirable cooked beef brothy flavour notes). At low concentrations, oil-soluble rosemary oleoresin appeared to be an effective inhibitor of warmed-over flavour. Lai *et al.* (1991) and Stoick *et al.* (1991) also demonstrated that rosemary oleoresin, when used in combination with sodium tripolyphosphate, effectively controlled lipid oxidation in restructured chicken nuggets and beef steaks, respectively, during frozen storage.

7.6 Future research needs

In spite of the numerous publications on the subject, many questions relating to the mechanism of lipid oxidation and its prevention need further investigation. The recent reviews of Kanner *et al.* (1987), Hsieh and Kinsella (1989) and Halliwell and Gutteridge (1990) clearly emphasize the ubiquitous nature and pervasiveness of lipid oxidation in foods and *in vivo*, and the challenges that face researchers to provide some resolution to these questions. Meat technologists/chemists are now addressing many of the same issues which have confused and challenged biochemists for decades. However, it is critical that researchers involved in studying lipid oxidation in meat and meat products use terminology which is consistent and adequately describes the problem being addressed. For example, the term initiation can only be applied to the removal or abstraction of the initial proton from the unsaturated fatty acid and should not be used interchangeably with other terms that really apply to the decomposition of preformed hydroperoxides.

Some of the research needs pertaining to the oxidation of lipids in biological systems have been described by Halliwell and Gutteridge (1990). In addition, a review of the meat science literature indicates that there is a very real need for further research into the mechanism of warmed-over flavour development in muscle foods. More conclusive evidence showing the relative contributions of haem and nonhaem iron to rancidity development in precooked and restructured meats is needed in order to design systems that will minimize or prevent oxidative deterioration of the meat products. Similarly, more detailed studies are necessary to evaluate the importance of lipoxygenase enzymes in initiating oxidative reactions in meat systems. Evidence to date suggests that the role of lipoxygenases may be greater than was previously believed.

Further research is needed to establish the mechanism of cholesterol oxidation in meat products, even though Park and Addis (1987) have indicated that cooked meat contributes only a small portion of the cholesterol oxidation products in the human diet. Furthermore, the mechanism of salt-catalysed oxidation of meat lipids is not well understood. Knowledge of the independent pro-oxidative influences of salt, trace metals, ions and haem pigments is necessary to clarify the mechanism(s) involved in rancidity development in meat systems.

Lipid oxidation in meat systems can be minimized by the use of synthetic antioxidants (Gray and Pearson, 1987). However, more studies are needed to assess the application of naturally occurring antioxidants in controlling warmed-over flavour in precooked meats and restructured meat products during storage. This is especially important in view of the negative public reaction to synthetic chemicals in foods.

Acknowledgments

The authors wish to express their appreciation to Ms Marlene Green and Barbara Bommarito for editorial assistance.

References

Addis, P.B. (1986). Occurrence of lipid oxidation products in food. *Food Chem. Toxicol.*, **24**, 1021–1030.

Addis, P.B. and Park, S.W. (1989). Role of lipid oxidation products in atherosclerosis. In *Food Toxicology: A Perspective on the Relative Risks*, ed. S.L. Taylor and R.A. Scanlan. Marcel Dekker, Inc., New York, pp. 297–330.

Allen, C.E. and Foegeding, E.A. (1981). Some muscle characteristics and interactions in muscle foods—a review. *Food Technol.*, **35**, 253–257.

Apte, S. and Morrissey, P.A. (1987a). Effect of haemoglobin and ferritin on lipid oxidation in raw and cooked muscle systems. *Food Chem.*, **25**, 127–134.

Apte, S. and Morrissey, P.A. (1987b). Effect of water-soluble haem and non-haem iron complexes on lipid oxidation of heated muscle systems. *Food Chem.*, **26**, 213–222.

Asghar, A., Gray, J.I., Buckley, D.J., Pearson, A.M. and Booren, A.M. (1988). Perspectives on warmed-over flavor. *Food Technol.*, **42**(6), 102–108.

Asghar, A., Gray, J.I., Booren, A.M., Gomaa, E., Abouzied, M.N., Miller, E.R. and Buckley, D.J. (1991). Influence of supranutritional dietary vitamin E levels on subcellular deposition of alpha-tocopherol in the muscle and on pork quality. *J. Sci. Food Agric.*, **57**, 31–41.

Bailey, M.E. (1988). Inhibition of warmed-over flavor with emphasis on Maillard reaction products. *Food Technol.*, **42**, 123–126.

Barbut, S., Josephson, D.B. and Maurer, A.J. (1985). Antioxidant properties of rosemary oleoresin in turkey sausage. *J. Food Sci.*, **50**, 1356–1359.

Boyer, R.F. and McCleary, C.J. (1987). Superoxide ion as a primary reductant in ascorbate-mediated ferritin iron release. *Free Radical Biol. Med.*, **3**, 389–395.

Brown, W.D., Harris, L.S. and Olcott, H.S. (1963). Catalysis of unsaturated lipid oxidation by iron photoporphyrin derivatives. *Arch. Biochem. Biophys.*, **101**, 14–20.

Buckley, D.J., Gray, J.I., Asghar, A., Price, J.F., Crackel, R.L., Booren, A.M., Pearson, A.M. and Miller, E.R. (1989). Effects of dietary antioxidants and oxidized oil on membranal lipid stability and pork product quality. *J. Food Sci.*, **54**, 1193–1197.

Cross, C.K. and Ziegler, P. (1965). A comparison of the volatile fractions from cured and uncured meat. *J. Food Sci.*, **30**, 610–614.

Decker, E.A. and Schanus, E.G. (1986a). Catalysis of linoleate oxidation by soluble chicken muscle proteins. *J. Am. Oil Chem. Soc.*, **63**, 101–104.

Decker, E.A. and Schanus, E.G. (1986b). Catalysis of linoleate oxidation by nonheme and heme-soluble chicken muscle proteins. *J. Agric. Food Chem.*, **34**, 991–994.

Decker, E.A. and Welch, B. (1990). Role of ferritin as a lipid oxidation catalyst in muscle food. *J. Agric. Food Chem.*, **38**, 674–677.

Dugan, L.R., Jr. (1987). Fats. In *The Science of Meat and Meat Products*, Part 2, 3rd ed., ed. J.F. Price and B.S. Schweigert, Food and Nutrition Press, Inc., Westport, CT, pp. 507–530.

Engeseth, N.J. (1990). Membranal lipid oxidation in muscle tissue—mechanism and prevention. Ph.D. dissertation, Michigan State University, East Lansing, USA.

Farmer, L.J. and Mottram, D.S. (1990). Interaction of lipid in the Maillard reaction between cysteine and ribose: the effect of a triglyceride and three phospholipids in the volatile products. *J. Sci. Food Agric.*, **53**, 505–525.

Farmer, L.J., Mottram, D.S. and Whitfield, F.B. (1989). Volatile compounds produced in Maillard reactions involving cysteine, ribose and phospholipid. *J. Sci. Food Agric.*, **48**, 347–368.

Faustman, C., Cassens, R.G., Schaefer, D.M., Buege, D.R. and Scheller, K.K. (1989a). Vitamin E supplementation of Holstein steer diets improves sirloin steak color. *J. Food Sci.*, **54**, 485–486.

Faustman, C., Cassens, R.G., Schaefer, D.M., Buege, D.R., Williams, S.N. and Scheller, K.K. (1989b). Improvement of pigment and lipid stability in Holstein steer beef by dietary supplementation with vitamin E. *J. Food Sci.*, **54**, 858–862.

Fooladi, M.H., Pearson, A.M., Coleman, T.H. and Merkel, R.A. (1977). The role of nitrite in preventing development of warmed-over flavour. *Food Chem.*, **4**, 283–292.

Foote, C.S. (1985). Chemistry of reactive oxygen species. In *Chemical Changes in Food During Processing*, ed. T. Richardson and J.W. Finley. AVI Publishing Co., Westport, CT, pp. 17–32.

Fox, J.B., Jr. and Benedict, R.C. (1987). The role of heme pigments and nitrite in oxidative processes in meat. In *Warmed-Over Flavor of Meat*, ed. A.J. St. Angelo and M.E. Bailey. Academic Press, Inc., New York, pp. 119–139.

Freybler, L.A., Gray, J.I., Asghar, A., Booren, A.M., Pearson, A.M. and Buckley, D.J. (1989). Mechanism of nitrite stabilization of meat lipids and heme pigments. In *Proceedings of the 35th International Congress of Meat Science and Technology*, Vol. 3, pp. 903–908.

Fridovich, I. (1976). Oxygen radicals, hydrogen peroxide and oxygen toxicity. In *Free Radicals in Biology*, ed. W.A. Pryor. Academic Press, New York, Vol. 1, pp. 239–277.

Gerhardt, U. and Blat, P. (1984). Dynamische messmethode zur ermittlung der fettstabilitat. Einfluss von gewurzen und zusatzstoffen. *Fleischwirtschaft*, **64**, 484–486.

German, J.B. and Creveling, R.K. (1990). Identification and characterization of a 15-lipoxygenase from fish gills. *J. Agric. Food Chem.*, **38**, 2144–2147.

German, J.B. and Kinsella, J.E. (1985). Lipid oxidation in fish tissue. Enzymatic initiation via lipoxygenase. *J. Agric. Food Chem.*, **36**, 680–683.

Gray, J.I. (1976). N-Nitrosamines and their precursors in bacon—a review. *J. Milk Food Technol.*, **39**, 686–692.

Gray, J.I. and Pearson, A.M. (1984). Cured meat flavor. *Adv. Food Res.*, **29**, 1–86.

Gray, J.I. and Pearson, A.M. (1987). Rancidity and warmed-over flavour. In *Advances in Meat Research, Vol. 3—Restructured Meat and Poultry Products*, ed. A.M. Pearson and T.R. Dutson. Van Nostrand Reinhold Company, New York, pp. 221–269.

Grossman, S., Bergman, M. and Sklan, D. (1988). Lipoxygenase in chicken muscle. *J. Agric. Food Chem.*, **36**, 1268–1270.

Gutteridge, J.M.C. and Halliwell, B. (1990). The measurement and mechanism of lipid peroxidation in biological systems. *TIBS*, **15**, 129–135.

Gutteridge, J.M.C., Halliwell, B., Treffry, A., Harrison, P.M. and Blake, D. (1983). Effect of ferritin-containing fractions with different iron loading on lipid peroxidation. *Biochem. J.*, **209**, 557–560.

Halliwell, B. and Gutteridge, J.M.C. (1990). Role of free radicals and catalytic metal ions in human disease: an overview. *Methods in Enzymol.*, **186**, 1–84.

Harel, S. and Kanner, J. (1985a). Hydrogen peroxide generation in ground muscle tissue. *J. Agric. Food Chem.*, **33**, 1186–1188.

Harel, S. and Kanner, J. (1985b). Muscle membranal lipid peroxidation initiated by H_2O_2-activated metmyoglobin. *J. Agric. Food Chem.*, **33**, 1188–1192.

Hirano, Y. and Olcott, H.S. (1971). Effect of heme compounds on lipid oxidation. *J. Am. Oil Chem. Soc.*, **48**, 523–524.

Hoelscher, L.M., Savell, J.W., Smith, S.B. and Cross, H.R. (1988). Subcellular distribution of cholesterol within muscle and adipose tissues of beef loin steaks. *J. Food Sci.*, **53**, 718–722.

Hofstrand, J. and Jacobson, M. (1960). The role of fat in the flavor of lamb and mutton as tested with broths and with depot fats. *Food Res.*, **25**, 706–711.

Hornstein, I. and Crowe, P.F. (1960). Meat flavor chemistry: studies on beef and pork. *J. Agric. Food Chem.*, **8**, 494–498.

Hornstein, I. and Crowe, P.F. (1963). Meat flavor: lamb. *J. Agric. Food Chem.*, **11**, 147–149.

Hornstein, I., Crowe, P.F. and Sulzbacher, W.L. (1963). Flavor of beef and whale meat. *Nature (London)*, **199**, 1252–1254.

Houlihan, C.M., Ho, C.T. and Chang, S.S. (1984). Elucidation of the chemical structure of a novel antioxidant rosmaridiphenol, isolated from rosemary. *J. Am. Oil Chem. Soc.*, **61**, 1036–1039.

Houlihan, C.M., Ho, C-T. and Chang, S.S. (1985). The structure of rosmariquinone—a new antioxidant isolated from *Rosmarinus officinalis* L. *J. Am. Oil Chem. Soc.*, **62**, 96–98.

Hsieh, R.J. and Kinsella, J.E. (1989). Oxidation of polyunsaturated fatty acids: mechanisms, products, and inhibition with emphasis on fish. *Adv. Food Nutr. Res.*, **33**, 233–341.

Hsieh, R.J., German, J.B. and Kinsella, J.E. (1988). Lipoxygenase in fish tissue. Some properties of the 12-lipoxygenase from trout gill. *J. Agric. Food Chem.*, **36**, 680–685.

Igene, J.O. and Pearson, A.M. (1979). Role of phospholipids and triglycerides in warmed-over flavor development. *J. Food Sci.*, **44**, 1285–1290.

Igene, J.O., King, J.A., Pearson, A.M. and Gray, J.I. (1979). Influence of heme pigments, nitrite and non-heme iron on development of warmed-over flavor (WOF) in cooked meat. *J. Agric. Food Chem.*, **27**, 838–842.

Igene, J.O., Yamauchi, K., Pearson, A.M. and Gray, J.I. (1985). Mechanisms by which nitrite inhibits the development of warmed-over flavour (WOF) in cured meat. *Food Chem.*, **18**, 1–18.

Inatani, R., Nakatani, N., Fuwa, H. and Seto, H. (1982). Structure of a new antioxidative phenolic diterpene isolated from rosemary (*Rosmarinus officinalis* L.). *Agric. Biol. Chem.*, **46**, 1661–1666.

Inatani, R., Nakatini, N. and Fuwa, H. (1983). Antioxidative effect of the constituents of rosemary (*Rosmarinus officinalis* L.) and their derivatives. *Agric. Biol. Chem.*, **47**, 521–528.

Ingold, K.U. (1962). Metal catalysis. In *Symposium on Foods: Lipids and Their Oxidation*, ed. H.W. Schultz, E.A. Day and R.O. Sinnhuber. AVI Publishing Co., Westport, CT, pp. 93–121.

Jantawat, P. and Dawson, L.E. (1980). Composition of lipids from mechanically deboned poultry meats and their composite tissues. *Poultry Sci.*, **59**, 1043–1052.

Johns, A.M., Birkinshaw, L.H. and Ledward, D.A. (1989). Catalysts of lipid oxidation in meat products. *Meat Sci.*, **25**, 209–220.

Josephson, D.B., Lindsay, R.C. and Stuiber, D.A. (1984). Variations in the occurrences of enzymatically derived volatile aroma compounds in salt- and freshwater fish. *J. Agric. Food Chem.*, **32**, 1344–1347.

Kanner, J. and Doll, L. (1991). Ferritin in turkey muscle tissue. A source of catalytic iron ions for lipid peroxidation. *J. Agric. Food Chem.*, **39**, 247–249.

Kanner, J. and Harel, S. (1985). Initiation of membranal lipid peroxidation by activated met-myoglobin and methemoglobin. *Arch. Biochem. Biophys.*, **237**, 314–321.

Kanner, J., Ben-Gera, I. and Berman, S. (1980). Nitric oxide–myoglobin as an inhibitor of lipid oxidation. *Lipids*, **15**, 944–948.

Kanner, J., Harel, S., Shegalovich, J. and Berman, S. (1984). Anti-oxidative effect of nitrite in cured meat products. Nitrite oxide–iron complexes of low molecular weight. *J. Agric. Food Chem.*, **32**, 512–515.

Kanner, J., German, J.B. and Kinsella, J.E. (1987). Initiation of lipid peroxidation in biological systems. *CRC Crit. Rev. Food Sci. Nutr.*, **25**(4), 317–364.

Kanner, J., Shegalovich, I., Harel, S. and Hazan, B. (1988a). Muscle lipid peroxidation dependent on oxygen and free metal ions. *J. Agric. Food Chem.*, **36**, 409–412.

Kanner, J., Hazan, B. and Doll, L. (1988b). Catalytic 'free' iron ions in muscle foods. *J. Agric. Food Chem.*, **36**, 412–415.

Kanner, J., Salan, M.A., Harel, S. and Shegalovich, I. (1991). Lipid peroxidation of muscle food. The role of the cytosolic fraction. *J. Agric. Food Chem.*, **39**, 242–246.

Labuza, T.P. (1971). Kinetics of lipid oxidation in foods. *CRC Crit. Rev. Food Technol.*, **2**, 355–404.

Lai, S.M., Gray, J.I., Smith, D.M., Booren, A.M., Crackel, R.L. and Buckley, D.J. (1991). Effects of oleoresin rosemary, tertiary butylhydroquinone and sodium tripolyphosphate on the development of oxidative rancidity in restructured chicken nuggets. *J. Food Sci.*, **56**, 616–620.

Lin, T.S. and Hultin, H.O. (1976). Enzymic lipid peroxidation in microsomes of chicken skeletal muscle. *J. Food Sci.*, **41**, 1488–1489.

Liu, R.H., Conboy, J.J. and Hotchkiss, J.H. (1988). Nitrosation by nitro-nitrosite derivatives of olefins: a potential mechanism for N-nitrosamine formation in fried bacon. *J. Agric. Food Chem.*, **36**, 984–987.

Love, J.D. (1983). The role of heme iron in the oxidation of lipids in red meats. *Food Technol.*, **37**(7), 116–120, 122.

Love, J.D. (1988). Sensory analysis of warmed-over flavor in meat. *Food Technol.*, **42**(6), 140–143.

Love, J.D. and Pearson, A.M. (1974). Metmyoglobin and non-heme iron as prooxidants in cooked meats. *J. Agric. Food Chem.*, **22**, 1032–1034.

Luo, S.W. and Hultin, H.O. (1986). An enzymic-catalyzed lipid peroxidation system in trout muscle mitochondria. Paper presented at 46th Annual Meeting of Institute of Food Technologists, June 15–18, Dallas, TX.

MacDonald, B., Gray, J.I. and Kakuda, Y. (1980). The role of nitrite in cured meat flavor. II. Chemical analyses. *J. Food Sci.*, **45**, 889–892.

Minotti, G. and Aust, S.D. (1987). The requirement for iron(III) in the initiation of lipid peroxidation by iron(II) and hydrogen peroxide. *J. Biol. Chem.*, **262**, 1098–1104.

Missler, S.R., Wasilchuk, B.A. and Merritt, C. Jr. (1985). Separation and identification of cholesterol oxidation products in dried egg preparations. *J. Food Sci.*, **50**, 595–598.

Monahan, F.J., Buckley, D.J., Morrissey, P.A., Lynch, P.B. and Gray, J.I. (1990a). Effect of dietary alpha-tocopherol supplementation on alpha-tocopherol levels in porcine tissues and on susceptibility to lipid peroxidation. *Food Sci. Nutr.*, **42F**, 203–212.

Monahan, F., Buckley, D.J., Gray, J.I., Morrissey, P.A., Asghar, A., Hanrahan, T.J. and Lynch, P.B. (1990b). Effect of dietary vitamin E on the stability of raw and cooked pork. *Meat Sci.*, **27**, 99–108.

Monahan, F.J., Crackel, R.L., Gray, J.I., Buckley, D.J. and Morrissey, P.A. (1993). Catalysis of lipid oxidation in muscle model systems by haem and inorganic iron. *Meat Sci.*, **34**, 95–106.

Morrissey, P.A. and Tichivangana, J.Z. (1985). The antioxidant activities of nitrite and nitrosylmyoglobin in cooked meats. *Meat Sci.*, **14**, 175–190.

Mottram, D.S. and Edwards, R.A. (1983). The role of triglycerides and phospholipids in the aroma of cooked beef. *J. Sci. Food Agric.*, **34**, 517–522.

O'Boyle, A.R., Rubin, L.J., Diosady, L.L., Aladin-Kassam, N., Comer, F. and Brightwell, W. (1990). A nitrite-free curing system and its application to the production of wieners. *Food Technol.*, **44**(5), 88–104.

Park, S.W. and Addis, P.B. (1986). Identification and quantitative estimation of oxidized cholesterol derivatives in heated tallow. *J. Agric. Food Chem.*, **34**, 653–659.

Park, S.W. and Addis, P.B. (1987). Cholesterol oxidation products in some muscle foods. *J. Food Sci.*, **52**, 1500–1503.

Pearson, A.M., Love, J.D. and Shorland, F.B. (1977). 'Warmed-over' flavor in meat, poultry and fish. *Adv. Food Res.*, **23**, 1–74.

Pearson, A.M., Gray, J.I., Wolzak, A.M. and Horenstein, N.A. (1983). Safety implications of oxidized lipids in muscle foods. *Food Technol.*, **37**(7), 121–129.

Pie, J.E., Spahis, K. and Seillan, C. (1991). Cholesterol oxidation in meat products during cooking and frozen storage. *J. Agric. Food Chem.*, **39**, 250–254.

Player, T.J. and Hultin, H.O. (1977). Some characteristics of the NAD(P)H-dependent lipid peroxidation system in the microsomal fraction of chicken breast muscle. *J. Food Biochem.*, **1**, 153–171.

Ramarathnam, N., Rubin, L.J. and Diosady, L.L. (1991). Studies on meat flavor. 1. Qualitative and quantitative differences in uncured and cured pork. *J. Agric. Food Chem.*, **39**, 344–350.

Rhee, K.S. (1988). Enzymic and nonenzymic catalysis of lipid oxidation in muscle foods. *Food Technol.*, **42**(6), 127–132.

Rhee, K.S. and Ziprin, Y.A. (1987). Lipid oxidation in retail beef, pork and chicken muscles as affected by concentrations of heme pigments and nonheme iron and microsomal enzymic lipid peroxidation activity. *J. Food Biochem.*, **11**, 1–15.

Rhee, K.S., Dutson, T.R. and Smith, G.C. (1984). Enzymic lipid peroxidation in microsomal fractions from beef skeletal muscle. *J. Food Sci.*, **49**, 675–679.

Rhee, K.S., Ziprin, Y.A. and Ordonez, G. (1987). Catalysis of lipid oxidation in raw and cooked beef by metmyoglobin–H_2O_2, nonheme iron, and enzyme systems. *J. Agric. Food Chem.*, **35**, 1013–1017.

Robinson, M.E. (1924). Hemoglobin and methemoglobin as oxidative catalysts. *Biochem. J.*, **18**, 255–264.

Sander, B.D., Smith, D.E., Addis, P.B. and Park, S.W. (1989). Effects of prolonged and adverse storage conditions on levels of cholesterol oxidation products in dairy products. *J. Food Sci.*, **54**, 874–879.

Sato, K. and Hegarty, G.R. (1971). Warmed-over flavor in cooked meats. *J. Food Sci.*, **36**, 1098–1102.

Schricker, B.R. and Miller, D.D. (1983). Effects of cooking and chemical treatment on heme and nonheme iron in meat. *J. Food Sci.*, **48**, 1340–1343, 1349.

Seman, D.L., Decker, E.A. and Crum, A.D. (1991). Factors affecting catalysis of lipid oxidation by a ferritin-containing extract of beef muscle. *J. Food Sci.*, **56**, 356–358.
Shahidi, F., Rubin, L.J. and Wood, D.F. (1988). Stabilization of meat lipids with nitrite-free curing mixtures. *Meat Sci.*, **22**, 73–80.
Smith, L.L. (1981). *Cholesterol Autoxidation*, Plenum Press, New York.
Spanier, A.M., Edwards, J.V. and Dupuy, H.P. (1988). The warmed-over flavor process in beef. A study of meat proteins and peptides. *Food Technol.*, **42**(6), 110–118.
St. Angelo, A.J., Vercellotti, J.R., Dupuy, H.P. and Spanier, A.M. (1988). Assessment of beef flavor quality: a multidisciplinary approach. *Food Technol.*, **42**(6), 133–138.
St. Angelo, A.J., Crippen, K.L., Dupuy, H.P. and James, C., Jr. (1990). Chemical and sensory studies of antioxidant-treated beef. *J. Food Sci.*, **55**, 1501–1505, 1539.
Stoick, S., Gray, J.I., Booren, A.M. and Buckley, D.J. (1991). The oxidative stability of restructured beef steaks processed with an oleoresin rosemary, tertiary butylhydroquinone, and sodium tripolyphosphate. *J. Food Sci.*, **56**, 597–600.
Svingen, B.A., Buege, J.A., O'Neal, F.O. and Aust, S.D. (1979). The mechanism of NADPH-dependent lipid peroxidation. *J. Biol. Chem.*, **254**, 5892–5899.
Swain, J.W. (1972). Flavor constituents of pork cured with and without nitrite. Ph.D. Dissertation, University of Missouri, Columbia, USA.
Tappel, A.L. (1962). Heme compounds and lipoxidase as biocatalysts. In *Symposium in Foods: Lipids and Their Oxidation*, ed. H.W. Schultz, E.A. Day and R.O. Sinnhuber. AVI Publishing Co., Westport, CT, pp. 122–138.
Thomas, C.E., Moorhouse, L.A. and Aust, S.D. (1985). Ferritin and superoxide-dependent lipid peroxidation. *J. Biol. Chem.*, **260**(6), 3275–3280.
Tichivangana, J.Z. and Morrissey, P.A. (1985). Metmyoglobin and inorganic metals as pro-oxidants in raw and cooked muscle systems. *Meat Sci.*, **15**, 107–116.
Tien, M. and Aust, S.D. (1982). Comparative aspects of several model lipid peroxidation systems. In *Lipid Peroxides in Biology and Medicine*, ed. K. Yagi. Academic Press, New York, pp. 23–39.
Tims, M.J. and Watts, B.M. (1958). Protection of cooked meats with phosphates. *Food Technol.*, **12**, 240–243.
Wu, T.W., Lee, M.H., Ho, C.T. and Chang, S.S. (1982). Elucidation of the chemical structures of natural antioxidants isolated from rosemary. *J. Am. Oil Chem. Soc.*, **59**, 339–345.
Younathan, M.T. and Watts, B.M. (1960). Oxidation of tissue lipids in cooked pork. *Food Res.*, **25**, 538–543.
Zipser, M.W., Kwon, T.W. and Watts, B.M. (1964). Oxidative changes in cured and uncured frozen cooked pork. *J. Agric. Food Chem.*, **12**, 105–109.

8 Lipid oxidation in meat by-products: effect of antioxidants and Maillard reactants on volatiles

T. CHERAGHI and J.P. ROOZEN

8.1 Introduction

In the food industry the dry rendering operation is the most economical and sanitary way of disposing of meat trimmings and offal. After drying, the defatted adipose tissue obtained (meat-meal) is a relatively inexpensive protein source, which contains about 70% protein and 15% crude fat (Greenberg, 1981). Utilization of this meat by-product is limited because of its high concentration of unsaturated fatty acids (50%), which makes it very susceptible to rancidity and off-flavour development (Hsieh and Kinsella, 1989). Many natural and processed foods undergo flavour changes during storage, resulting in the formation of characteristic off-flavours associated with lipid oxidation. There are many ways of preventing or delaying oxidation in food, for example, exclusion of oxygen, addition of antioxidants and/or reducing agents. The volatile components of cooked meats have been investigated by many researchers, and several reviews on meat flavours have appeared in the last decade (MacLeod and Seyyedain-Ardebili, 1981; Shahidi *et al.*, 1985). However, volatiles from meat by-products have not been thoroughly investigated.

Reactions between sugars and amino compounds were shown to yield products with different antioxidant effects, for example a rather small effect for the often-used combination of cysteine and xylose (Lingnert and Erikson, 1980). The presence of Maillard reaction intermediates resulted in a marked decrease in the formation of aldehydes associated with lipid oxidation.

The goal of this project was to study the volatile compounds of meat-meal, to evaluate the effect of antioxidants on the formation of these volatiles and finally, to investigate the effect of added Maillard reactants on the volatile composition of processed meat-meal water mixtures.

8.2 Materials and methods

8.2.1 Materials

The meat-meal used is commercially available defatted adipose tissue. For investigation of the effect of antioxidants on the oxidation rate of meat-meal, citric or ascorbic acid was added to meat-meal as follows:

150 g meat-meal + 450 ml water (control),
additional 15 mg citric acid (0.01% CA),
or additional 75 mg ascorbic acid (0.05% AA).

After mixing and freeze-drying for 24 h the dry matter was pulverized and stored in paper bags at an ambient temperature of about 23°C and humidity approximately 60%.

Sample preparation for processed Maillard meat-meal water mixtures (PMMWM) was as follows: 80 mg meat-meal was added to 165 ml boiling water and then cooked at 90°C for 2.5 min. Afterwards 150 ml water was added to each sample, of which: (1) was control (no Maillard reactants added); (2) was CYS (2.92 g cysteine added); (3) was XYL (3.62 g xylose added); and (4) was C/X (2.92 g cysteine + 3.62 g xylose added). After mixing, each sample was poured into 450 ml tin cans and then sealed, processed at 127°C for 35 min, cooled and stored at ambient conditions.

All reagents used were of the purest grade available. Glassware was soaked for 24 h in a concentrated chromic acid solution before use.

8.2.2 *Chemical analysis of meat-meal*

The analysis of fatty acids was carried out according to the method outlined by IUPAC (1979). A gas chromatograph (Carlo Erba Strumentazione, model Fractovap 4200) equipped with a flame ionization detector (FID) and a microbore capillary column (15 m × 0.53 mm i.d.; DB-225, J & W) was used for the analysis. The amino acid composition of meat-meal was determined by gas chromatography according to the method described by Zumwalt *et al.* (1987). An OV-1 column (25 m × 0.25 mm i.d.; J. & W) and an FID detector were used for this analysis. Lipid oxidation was determined by the *p*-anisidine value (PAV) and peroxide value (PV) methods described by IUPAC (1979). Additionally, direct and distillation thiobarbituric acid methods (TBARS) were used (Roozen, 1987). Moisture content was determined using the Karl Fischer titration method (Roozen and Pilnik, 1970).

8.2.3 *Analysis of volatiles in dry meat-meal*

Volatile compounds from meat-meal were collected using a dynamic headspace technique described by MacLeod and Ames (1987). Tubes were packed with exactly 0.1 g Tenax-TA 60/80 mesh (Chrompack, Netherlands) and conditioned for 4 h at 250°C before use. Volatiles were collected for 4 h using nitrogen gas at a flow rate of 15 ml/min. Tubes were backflushed for 15 min to remove any residual water from the Tenax-TA tubes. A Carlo Erba gas chromatograph model 6000 equipped with a purge and trap injector unit (Chrompack, Netherlands) and FID detector was used. The Carbowax 20 M bonded phase column (60 m × 0.25 mm i.d.; Supelco, USA) had a column flow of 0.75 ml/min (He). Volatiles were desorbed from the Tenax-TA tubes for 10 min at 180°C and collected in a capillary cold trap, which was then heated for 4 min at 250°C. The splitless mode was used. Volatiles were identified using combined capillary gas chromatography and mass spectrometry. A Pye 204 splitless capillary gas chromatograph was equipped with a DB 1701 (30 m × 0.32 mm i.d. × 0.25 μm film; J&W Betron), column flow, 1 ml/min of He. The temperature program was 30°C for 5 min, then 4°C/min to 220°C and held there for 16 min. A Micromass MM 7070F (VG, Winsford, UK) mass spectrometer was used in the EI mode, at 70 eV.

8.2.4 Analysis of volatiles in processed Maillard meat-meal water mixtures

The effect of lipid oxidation on Maillard reactions was investigated with meat-meal at two levels of oxidation. Vacuum-packed meat-meal was considered fresh and meat-meal exposed to air for 110 days was considered oxidized. Volatiles were extracted with a continuous stream distillation extraction technique. Conditions of the modified Likens–Nickerson apparatus were: 200 g sample was mixed with 100 ml distilled water and steam distilled for 3 h. The solvent was 25 ml 2-methyl butane, which was kept at 40°C. The extract was then concentrated to 1 ml using Kuderna–Danisch apparatus and then further concentrated to 0.6 ml using a gentle flow of nitrogen. Gas chromatography and mass spectrometry were carried out as mentioned before.

8.3 Results and discussion

Fatty acid composition of the meat-meal is presented in Table 8.1, which resembles the composition of pork fat. The high content of unsaturated fatty acids (58%) makes this meat by-product very susceptible to lipid oxidation. Table 8.1 also shows the major amino acids, of which alanine (9.8%), glycine (23.1%), proline (13.7%), hydroxyproline (9.7%), lysine (6.1%) and glutamic acid (10.3%) clearly indicate that the origin of this meat-meal is from connective and adipose tissues (Ward *et al.*, 1977).

The shelf-life of meat-meal is rather short when the product is stored in bulk containers or paper bags. Rancid odours develop within a few weeks, which indicates deterioration by lipid oxidation (PV ≈ 20 meq/kg of extracted fat). Exclusion of oxygen by vacuum packaging in plastic–aluminium laminates prevents lipid oxidation to a large extent (Table 8.2). PV and PAV give more distinct differences than TBARS, which might be caused by the monounsaturated nature of the majority of the fatty acids (Table 8.1) or by a delay in the secondary lipid oxidation reaction of the

Table 8.1 Fatty acid and amino acid composition of meat-meal[a]

Fatty acid		Amino acid	
Compound	Weight %	Compound	Weight %
C12:0	0.2	Alanine	9.8
C14:0	1.6	Glycine	23.1
C16:0	23	Valine	3.9
C16:1	2.3	Threonine	2.7
C17:0	0.7	Serine	3.4
C18:0	15	Leucine	5.3
C18:1	40	Proline	13.7
C18:2	10	Hydroxyproline	9.7
C18:3	1.5	Aspartic acid	3.6
C20:0	1.2	Phenylalanine	4.0
C20:1	2.0	Glutamic acid	10.3
C20:2	2.1	Lysine	6.1
		Tyrosine	1.8
		Arginine	2.7

[a]Fat content 11.5%, protein content 75%, moisture content 2.7%.

Table 8.2 Lipid oxidation in meat-meal stored in vacuum-pack or paper bags for 110 days at room temperature

Methods	Vacuum pack	Paper bag
p-Anisidine value	10.5	117
Peroxide value	3	240
TBARS (direct)	7	7
TBARS (distillation)	0	1

TBARS = thiobarbituric acid reactive substances.

polyunsaturated fatty acids to form malondialdehyde (Kim and LaBella, 1987).

The effect of antioxidants on lipid oxidation in meat-meal was tested by the analysis of major aldehydes using dynamic headspace collection of volatiles and gas chromatography (Figure 8.1). Both the heavy metal ion chelating agent citric acid and the oxygen scavenger ascorbic acid diminish the total amount of aldehydes by 70%. However, both antioxidants are less effective than oxygen exclusion by vacuum-packaging immediately after production of the meat-meal ('fresh' in Figure 8.1). In that case the total amount of aldehydes decreased by 95%.

The processed Maillard meat-meal water mixtures (PMMWM) were evaluated for their concentration of volatiles, in particular the four major aldehydes (Figure 8.2). Their concentrations in the 'fresh' and control sample were not changed much by the preparation procedure mentioned earlier (1 part meat-meal in 3 parts PMMWM). When the Maillard reac-

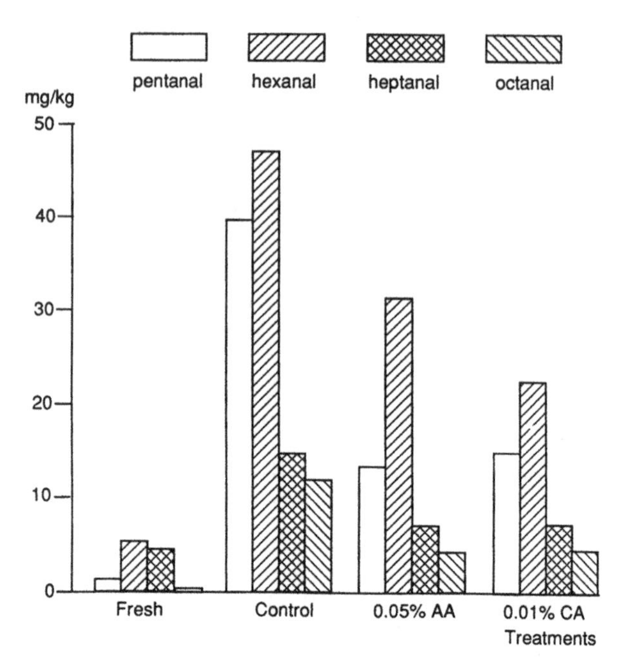

Figure 8.1 Effect of packaging and antioxidants on the content of four aldehydes in meat-meal stored for 110 days at room temperature. Fresh = stored vacuum-packed (P.V. = 3 meq/kg of extracted fat); Control = stored in paper bag (P.V. = 240 meq/kg of extracted fat); AA = ascorbic acid; CA = citric acid.

tants, cysteine and xylose, were present in combination in PMMWM, the concentration of aldehydes associated with oxidation decreased by 31%. Xylose was as effective as cysteine/xylose, but cysteine alone had no significant effect on the reduction in total aldehydes. Meat-meal seems to contain sufficient amino groups for starting up the Maillard reaction with xylose. The intermediates formed can then react with the aldehydes from the lipid oxidation reaction. Addition of cysteine does not change much in this respect, especially when cysteine is oxidized into cystine immediately after mixing.

Dynamic headspace volatile components from the meat-meal and Likens–Nickerson extracts of PMMWM were analysed by combined gas chromatography and mass spectrometry. More than 60 different compounds were identified in these samples, of which the aldehydes, alcohols, ketones, pyrazines, sulphur compounds and furans are presented in Table 8.3. 2- and 3-Methyl butanal, pentanal, hexanal, heptanal, octanal, nonanal and decanal were detected at relatively high concentrations and they are known to be the secondary products of lipid oxidation at moderate temperature, i.e. the scission products of monohydroperoxides of unsatu-

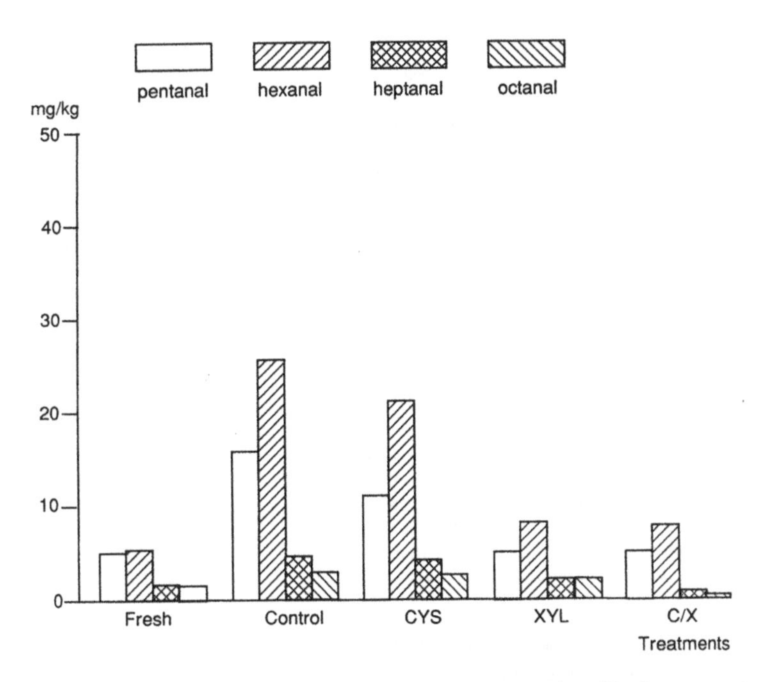

Figure 8.2 The content of four aldehydes determined in processed Maillard meat-meal water mixtures. Fresh = stored vacuum-packed (P.V. = 3 meq/kg of extracted fat); Control = stored in paper bag (P.V. = 240 meq/kg of extracted fat); CYS =cysteine (2.92 g) added; XYL = xylose (3.63 g) added; C/X = cysteine (2.92 g) + xylose (3.62 g) added.

rated fatty acids (Chan, 1987). In Figure 8.3 the two reconstructed ion chromatograms differ mainly at the higher retention times. This difference can only be caused by the use of 'fresh', versus 'old' meat-meal for PMMWM. Volatile heterocyclic sulphur compounds (probably 3-thiazolines, Table 8.4) were 10 times higher in the 'fresh' PMMWM than in the 'old' PMMWM. A rather simple explanation is the oxidation of cysteine into cystine in the old meat-meal PMMWM as discussed before. In that case cysteine is not available in PMMWM for producing fragments like hydrogen sulphide, acetaldehyde and ammonia, which react with acetoin from sugar–amine browning reactions to give 3-thiazolines (Lindsay, 1985).

Meat-meal as indicated in this paper, is a by-product of the meat industry. The product has a good balance of protein and fat content for nutritional purposes, and together with its water holding capacity, it could be a potential ingredient in pet food. Antioxidants (both reducing and chelating agents) are often added to products with a high content of unsaturated fatty acids to prevent the formation of undesirable volatiles. Our results show that the level of aldehydes associated with oxidation can be

Table 8.3 Volatile constituents of stored meat-meal and processed Maillard meat-meal water mixtures identified by gas chromatography–mass spectrometry

Aldehydes	Ketones	Alcohols
2-Methylpropanal	2-Butanone	2-Methylpropanol
Butanal	2-Pentanone	1-Butanol
2-Methylbutanal	2-Heptanone	1-Penten-1-ol
3-Methylbutanal	2-Octanone	1-Pentanol
Pentanal	2-Nonanone	1-Hexanol
Hexanal	3-Octene-2-one	1-Octanol
Heptanal	2-Decanone	1-Nonanol
Octanal	3,5-Octadien-2-one	1-Penten-3-ol
2-Octenal	2,3-Pentanedione	1-Octen-3-ol
Nonanal		1-Heptanol
2-Nonenal		2-Ethyl-1-hexanol
Decanal		2-Octen-1-ol
2-Decenal		3-Methyl-1-butanol
Benzaldehyde		1-Nonen-3-ol
Pyrazines	**Sulphur compounds**	**Furans**
2,5-Dimethylpyrazine	Dimethyl trisulphide	2-Ethylfuran
2-Methyl-5-ethylpyrazine	Methyl thiophene	2-Butylfuran
Trimethylpyrazine	2-Pentyl thiophene	2-Pentylfuran
Dimethyl-ethylpyrazine	Dimethyl disulphide	2-Hexylfuran

reduced by addition of ascorbic or citric acid, but that vacuum-packaging is much more effective. The stage of oxidation in meat-meal is extremely important for its application: PMMWM with 'fresh' meat-meal has a stronger meaty aroma than with 'old' meat-meal, probably associated with the presence of heterocyclic sulphur compounds (Cramer, 1983; Gasser and Grosch, 1988). In Figure 8.3 the 3-thiazoline peaks in the 'fresh' chromatogram compared with their negligible presence in the 'old' one are of practical interest.

The aliphatic aldehydes, alcohols and ketones are considered to contribute primarily to the fatty odour and flavour, whereas the pyrazines and some of the sulphur-containing compounds are more closely associated with the basic meaty flavour (Table 8.3). Many carbonyl compounds formed by lipid oxidation reactions are not important contributors, whereas certain aldehydes and ketones of Maillard reactions contribute to the desirable roasted flavour quality of a product, e.g. 3-methylbutanal, which is the Strecker degradation product of leucine formed during roasting (MacLeod and Coppock, 1977).

It is clear that lipids act as off-flavour precursors in meat-meal, but their exact role and the extent to which they affect flavour needs further study. It is generally agreed that fats influence the flavour of a product in two ways: (1) oxidation, principally of the unsaturated fatty acids, which results in the formation of carbonyl compounds present in sensory significant amounts; and (2) deposition of fat-soluble compounds, which volatilize upon heating and strongly affect flavour.

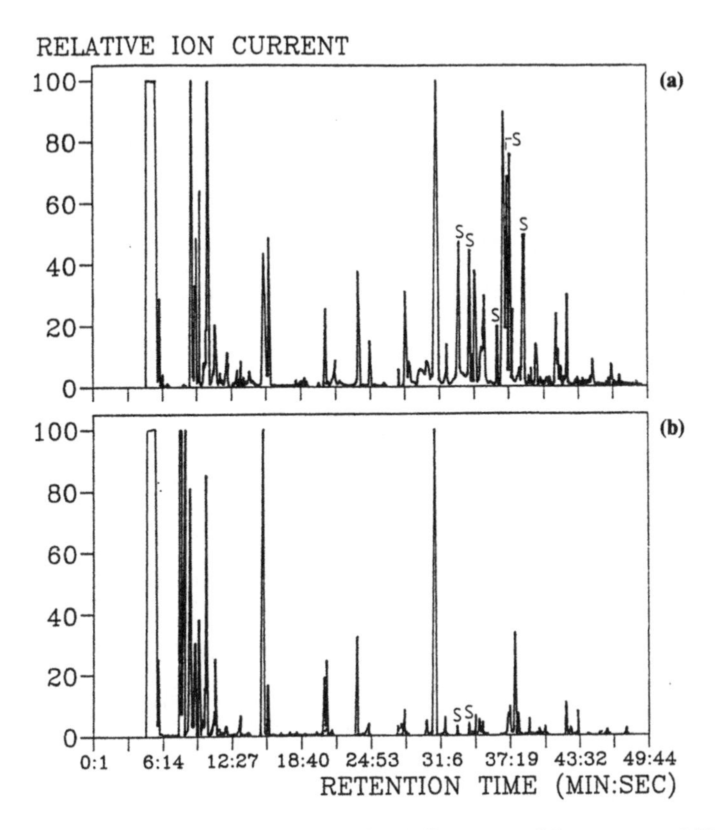

Figure 8.3 Reconstructed ion chromatograms of volatiles extracted from processed Maillard water mixtures with meat-meal stored in (a) vacuum-packs (fresh) or (b) paper bags (old) for 110 days. Conditions of preparation, extraction and analysis are described in Materials and methods. *S* = heterocyclic sulphur–nitrogen compound (3-thiazoline?).

Table 8.4 Listing of mass spectral data of two 3-thiazolines found and 2,4,5-trimethylthiazoline[a]

M/Z (relative intensity) of main fragments of mass spectrum
41 (35), 42 (55), 60 (30), 68 (25), 69 (100), 102 (83), 114 (27), 128 (23), 143 (57)
42 (59), 55 (38), 68 (30), 69 (56), 88 (100), 114 (26), 129 (52)
[a]42 (82), 55 (46), 68 (33), 69 (58), 88 (100), 114 (28), 129 (45)

[a]Mussinan *et al.* (1976).

References

Chan, H.W.-S. (1987). *Autoxidation of unsaturated lipids*. Academic Press Inc., London.

Cramer, D.A. (1983). Chemical compounds implicated in lamb flavor. *Food Technol.*, **37**, 249–257.

Gasser, U. and Grosch, W. (1988). Identification of volatile flavour compounds with high aroma values from cooked beef. *Zeitschrift fur Lebensmittel-Untersuchung und -Forschung*, **186**, 489–494.

Greenberg, M.J. (1981). Characterisation and comparison of flavor volatiles in meat by-products. In *Flavour '81*, ed. P. Schreier. Walter de Gruyter & Co., Berlin, pp. 599–608.

Hsieh, R.J. and Kinsella, J.E. (1989). Oxidation of polyunsaturated fatty acids: mechanisms, products, and inhibition with emphasis on fish. In *Advances in Food and Nutrition Research 33*, ed. J.E. Kinsella. Academic Press Inc., San Diego, pp. 233–341.

IUPAC (1979). *Standard Methods for the Analysis of Oils, Fats and Derivatives*. Sixth edition, part 1. International Union of Pure and Applied Chemistry, Pergamon Press, Oxford.

Kim, R.S. and LaBella, F.S. (1987). Comparison of analytical methods for monitoring autoxidation profiles of authentic lipids. *J. Lipid Res.*, **28**, 1110–1117.

Lindsay, R.C. (1985). Flavors. In *Food Chemistry*, ed. O.R. Fennema. Marcel Dekker, Inc., New York, p. 618.

Lingnert, H. and Eriksson, C.E. (1980). Antioxidative Maillard reaction products. 1. Products from sugars and amino acids. *J. Food Process. Preserv.*, **4**, 161–172.

MacLeod, G. and Coppock, B.M. (1977). A comparison of the chemical composition of boiled and roasted aromas of heated beef. *J. Agric. Food Chem.*, **25**(1), 113–117.

MacLeod, G. and Seyyedain-Ardebili, M. (1981). Natural and simulated meat flavors (with particular reference to beef). *Crit. Rev. Food Sci. Nutr.*, **14**, 309–437.

MacLeod, G. and Ames, J. (1986). 2-Methyl-3-(methylthio)furan: a meaty character impact aroma compound identified from cooked beef. *Chem. Ind.*, **3**, 175–177.

Mussinan, C.J., Wilson, R.A., Katz, I., Hruza, A. and Vock, M.H. (1976). Identification and flavor properties of some 3-oxazolines and 3-thiazolines isolated from cooked beef. In *Phenolic, Sulfur, and Nitrogen Compounds in Food Flavours*, ed. G. Charalambous and I. Katz. ACS Symposium series 26, American Chemical Society, Washington, DC, pp. 133–145.

Roozen, J.P. (1987). Effects of types I, II and III antioxidants on phospholipid oxidation in a meat model for warmed-over flavour. *Food Chem.*, **24**, 167–185.

Roozen, J.P. and Pilnik, W. (1970). Über die Stabilität adsorbierter Enzyme in wasserarmen Systemen. I. Die Stabilität von Peroxydase bei 25°C. *Lebensmittel-Wissenschaft und Technologie*, **3**, 37–40.

Shahidi, F., Rubin, L.J. and D'Souza, L.A. (1985). Meat flavor volatiles: a review of the composition, techniques of analysis, and sensory evaluation. *Crit. Rev. Food Sci. Nutr.*, **24**(2), 141–243.

Ward, A.G. and Courts, A. (1977). *The Science and Technology of Gelatin*. Academic Press Inc., London.

Zumwalt, R.W., Desgres, J., Kuo, K.C., Pautz, J.E. and Gehrke, C.W. (1987). Amino acid analysis by capillary gas chromatography. *J.A.O.A.C.*, **70**(2), 253–262.

9 Maillard reactions and meat flavour development

M.E. BAILEY

9.1 Introduction

9.1.1 Meat flavour

The flavour of raw fresh meat is bland, metallic and slightly salty, whereas desirable meat flavour is apparent only after heating. As with the flavour of most foods, both non-volatile and volatile components are essential. The non-volatile ingredients consist of taste compounds, flavour enhancers, or precursors for reaction products which may be responsible for desirable volatile compounds.

Obviously, thermal degradation is responsible for the formation of volatile components which are formed from water-soluble precursors such as thiamine, glycogen, glycoproteins, nucleotides, nucleosides, free sugars, amino acids, peptides, sugar phosphates, amines and organic acids. These natural precursors react in meat products during heating in primary reactions to form intermediate products which can further react with other degradation products to form a complex mixture of volatiles responsible for meat flavour.

Almost 1000 volatile compounds have been identified from meat or from model systems consisting of meat ingredients, and one might predict that many more thousands of volatile compounds will be identified from these systems in the future. Many reaction mechanisms have been proposed for the formation of these numerous compounds but the primary ones involve: (1) the degradation of vitamins, particularly thiamine; (2) the thermal degradation of carbohydrates and amines; and (3) the Maillard reaction, including Strecker degradation.

The importance of the chemistry of meat flavour to academia and the food industry is somewhat reflected in the number of reviews that have appeared in the literature since 1980 (Ho, 1980; Shibamoto, 1980; Katz, 1981; MacLeod and Seyyedain-Ardebili, 1981; Lawrie, 1982; Bailey, 1983; Cramer, 1983; IFT Symposium, 1983; Moody, 1983; Baines and Mlotkiewicz, 1984; MacLeod, 1984; MacLeod, 1986; Shahidi *et al.*, 1986; Hornstein and Wasserman, 1987; Parliment *et al.*, 1989; Rhee, 1989; Shahidi, 1989; Mottram, 1991), along with many other associated articles, particularly those concerned with the Maillard reaction and thermal reaction flavours.

Although many meat flavour compounds can be made by heating thiamine or a mixture of carbohydrates, ammonia and hydrogen sulphide, the Maillard reaction remains a focal point for the formation of most recognized meat flavour compounds. The water-soluble precursors of desirable meat flavour are well suited for the Maillard reaction. Wood and Bender (1957) and Bender *et al.* (1958) were the first to thoroughly examine the water-soluble extracts of boiled beef, which they believed contained precursors of meat flavour. Bender *et al.* (1958) concluded, 'it is beyond reasonable doubt that development of the brown colour and meaty flavour characteristics of these extracts is a result of the Maillard reaction. Hornstein and Crowe (1960) first demonstrated that lyophilized diffusate from cold water extracts of raw beef and pork produced meaty odours upon heating.

Macy *et al.* (1964a,b) extended these studies of the diffusates of beef, pork and lamb and demonstrated that amino acids, sugars, sugar phosphates, nucleotides and nucleosides all decreased in concentration during heating at 100°C for 1 h. Wasserman and Spinelli (1970) confirmed these results after heating beef diffusates in water at 125°C for 1 h. Wasserman (1979) also concluded that Maillard browning was important for the formation of desirable meat-flavoured compounds.

9.2 The Maillard reaction

The Maillard reaction (nonenzymatic browning) involves the reaction of aldehydes with amines and through numerous reactions, food flavour compounds and dark pigments (melanoidins) are formed. Louis-Camille Maillard, at the University of Nancy, published a series of papers on the reaction (Maillard, 1916), and scientists continue to study the many ramifications of the reaction.

The importance and complexity of the Maillard reaction is revealed by the large number of different review articles published. The most important early reviews related to food were those of Hodge (1953; 1967), Anet and Reynolds (1957) and Reynolds (1963; 1965). Many other reviews have been written on the chemistry of the Maillard reaction since 1970 (Hodge *et al.*, 1972; Williams, 1976; Hurrell and Carpenter, 1977; Mabrouk, 1979; Paulsen and Pflughaupt, 1980; Feather, 1981; Mauron, 1981; Nursten, 1981; Feeney and Whitaker, 1982; Hurrell, 1982; Vernin and Parkanyi, 1982; Namiki and Hayashi, 1983; Feather, 1985; Danehy, 1986; Heath and Reineccius, 1986; Nursten, 1986; Yaylayan and Sporns, 1987; Namiki, 1988; Baltes *et al.*, 1989; Ledl *et al.*, 1989; O'Brien and Morrissey, 1989; Eskin, 1990; Ledl, 1990), and there have been four symposia (Eriksson, 1981; Waller and Feather, 1983; Fujimaki *et al.*, 1986; Finot *et al.*, 1990).

Factors affecting the progress of the Maillard reaction include temperature, time, moisture content, pH, concentration and nature of the reactants. The initial reactions between the amines and aldehydes have been studied by numerous investigators and are reasonably well understood. The sugar reacts reversibly with an amine group to form a glycosylamine. The glucosylamines from aldose sugar undergo Amadori rearrangement to yield Amadori compounds such as 1-amino-1-deoxy-2-ketoses; the reaction between ketoses (fructose) and amines usually involves the formation of ketosylamines, followed by the Heynes rearrangement to form 2-amino-2-deoxy-aldolases (Reynolds, 1965).

The most widely accepted mechanisms for degradation pathways of Amadori compounds by dehydration and fission were suggested by the brilliant work of Hodge (1953) and the actual products formed depends on the basicity of the amine, the pH of the reaction mixture and the temperature.

Hodge (1967) emphasized that, although browning flavours in most foods are most likely to arise from the Maillard reaction, many other food components, particularly sugars, could interact to form carboxylic and other flavour constituents. Much recent evidence supports his ideas that oxidized components of lipids such as aldehydes, α,β-unsaturated ketones and acrolein participate in browning reactions to produce meat flavour compounds. Most importantly, he discussed mechanisms whereby sugar carmelization at high temperature produced many of the important intermediates in meat flavour that can likewise form by Maillard reaction at lower temperatures.

In an excellent paper, Vernin and Parkanyi (1982) summarized the Maillard reaction involving sugars and α-amino acids and outlined how reductones and dehydroreductones could be degraded into a number of flavour compounds by dehydration, retro-aldolization and Strecker degradation. The end products were *N,S,O*-heterocyclics. They also outlined how Amadori and Heynes intermediates were rearranged to form reductones. The primary reductones were 3-deoxyosones formed by 1,2-enolization of the carbonyl-amine compounds, 1-deoxysones and their equilibrium 1-deoxyreductones formed by 2,3-enolization of the carbonyl-amine compounds (Figure 9.1). Apparently the 3-deoxyosones are sufficiently stable to be separated by HPLC and characterized (Weenan, 1991). These compounds can be isolated and reacted to form many flavour intermediates, including 2-furfural derivatives, which can react with ammonia and hydrogen sulphide to form *N,S,O*-heterocyclics (Barone and Chanon, 1982).

A simplified version of the formation of rearrangement products from Amadori compounds and their possible degradation by dehydration and retro-aldolization (fission) is shown in Figure 9.2. The structures labelled flavour compounds are important intermediates that can react with other Maillard reaction degradation products to form meat flavour compounds. A modified version of the suggested reaction routes (Baltes *et al.*, 1989)

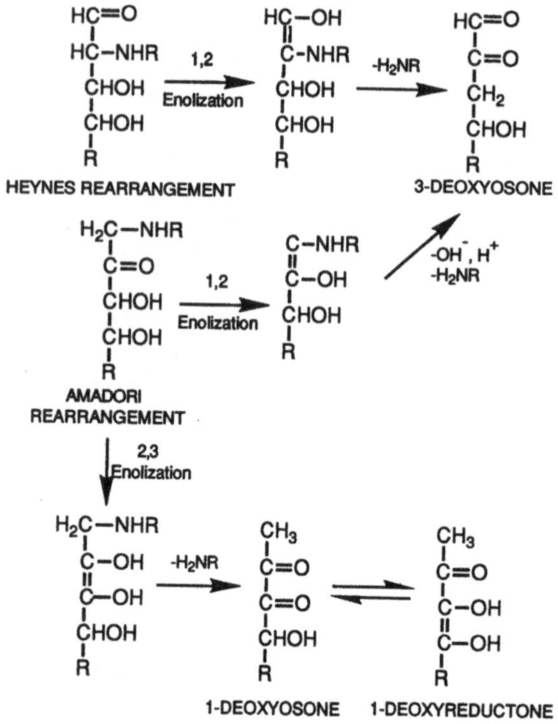

HEYNES REARRANGEMENT

3-DEOXYOSONE

AMADORI REARRANGEMENT

1-DEOXYOSONE 1-DEOXYREDUCTONE

Figure 9.1 Rearrangement of Amadori and Heynes intermediates from Maillard reaction of carbonyl–amine compounds to form reductones. Adapted from Vernin and Parkanyi (1982).

for the degradation via 1-deoxyosones to important meat flavour intermediates is presented in Figure 9.3. The compounds formed are key precursors involved in reactions responsible for meat flavour.

This pathway shows the formation of a number of cyclic oxygen intermediates important to meat flavour production. These compounds are 4-hydroxy-5-methyl-3(2*H*)-furanone from pentoses, 4-hydroxy-2,4-dimethyl-3(2*H*)-furanone (furaneol) from hexoses, along with isomaltol, maltol, 4-hydroxymaltol and 2-hydroxy-3-methyl cyclo-pentene-2-one (cyclotene). The most important reaction product of the 1-deoxyosone pathway is 5-hydroxy-5,6-dihydromaltol. This compound represents about 30% of volatiles formed by heating glucose at 120°C and it disappears at higher temperatures with the appearance of cyclotene and furaneol (Baltes *et al.*, 1989). Cyclotene is also formed by condensation of hydroxyacetone (Vernin and Parkanyi, 1982).

Feather (1981) evaluated the evidence for the existence of the 1-deoxyosone and its enolic form and concluded that it might be involved in the formation of isomaltol and maltol by cyclization–dehydration under basic conditions as hypothesized by Hodge (1953, 1967). A similar pathway for

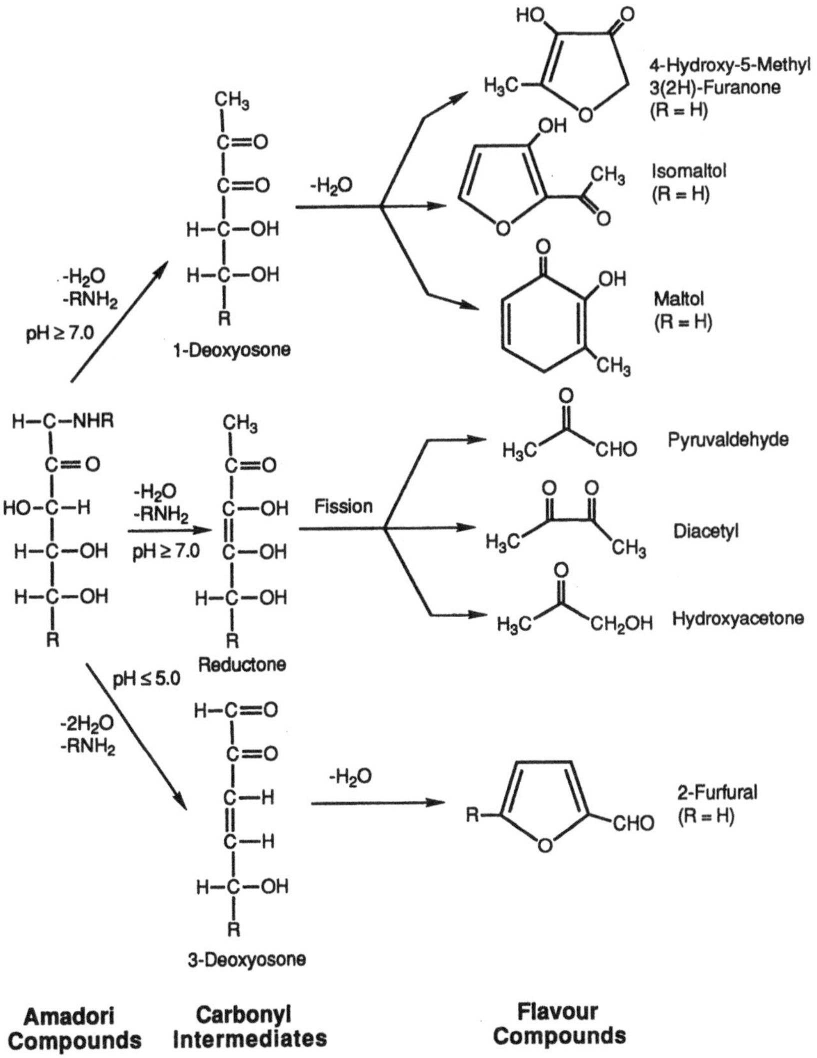

Figure 9.2 Deamination and dehydration of Amadori compounds to form important meat flavour intermediates.

the formation of maltol and isomaltol was recently confirmed by Ledl (1990).

Under the appropriate conditions for formation, it seems likely that 4-hydroxy-5-methyl-3(2*H*)-furanone could be formed by condensation of pentose with amine, formation of Amadori compound and dehydration via 2,3-enolization to form the 1-deoxyosone. Hicks and Feather (1975)

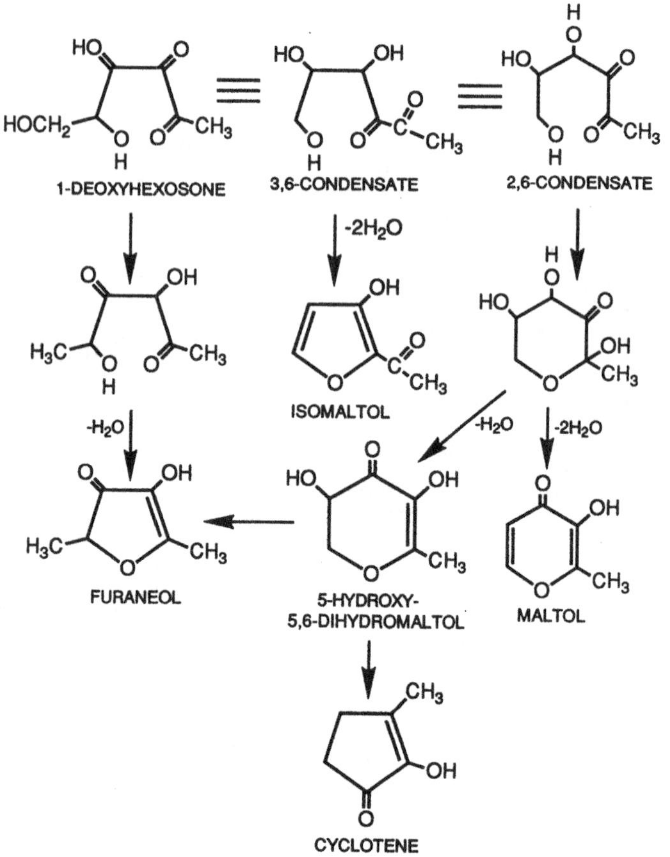

Figure 9.3 Formation of specific meat flavour intermediates by cyclization and dehydration of 1-deoxyhexosones. Modified from Baltes (1989).

conclusively showed that the pentose-derived Amadori compound 1-benzylamino-1-deoxy-D-*thrio*-pentalose dehydrates to 4-hydroxy-5-methyl-3(2*H*)-furanone. Hicks *et al.* (1974) also found that the furanone was formed from a 1-amino-1-deoxy-D-fructuronic acid by decarboxylation. It has also been formed by heating D-xylose (Severin and Seilmeier, 1968), D-ribose and D-ribose phosphate (Peer and van den Ouweland, 1968). Furaneol (4-hydroxy-2,5-dimethyl-3(2*H*)-furanone) was concluded to be involved in meat flavour by Tonsbeek *et al.* (1968), who isolated it along with 4-hydroxy-5-methyl-3(2*H*)-furanone from beef. Hexoses form 5-methyl-furfurals and furaneol, while pentoses form furfural and 4-hydroxy-5-methyl-3(2*H*) furanone, although there is some evidence that the furaneol can lose formaldehyde to yield the furanone during the

Maillard reaction (Ledl, 1990). Furaneol is also formed by heating precursors in yeast and cereal extracts (Schieberle, 1991). Ching (1979) identified three 3(2*H*) furanones by heating low molecular weight dialysable diffusate at 155°C in a closed system for 30 min. Thus, the compounds can be formed with low levels of sugars from extracts of meat.

1,2-Enolization of the Amadori compound is favoured by acid conditions, and following deamination and rearrangement, yields 3-deoxyosones as intermediates. These degrade to yield 5-hydroxymethyl-2-furfural for hexoses and 2-furfural from pentoses. These compounds are also produced by acid decomposition of sugars or caramelization (Feather and Harris, 1973). The furfural derivatives are very reactive with ammonia and hydrogen sulphide to form heterocyclic compounds. Barone and Chanon (1982), using a computer program simulating organic synthesis in the Maillard reaction, proposed more than 1000 structures from this reaction. Heterocyclics predicted in the reaction include furans, pyrroles, thiophenes, oxazoles, thiazoles, imidazoles, pyrans, pyridines and pyrazines, many of which have been identified by Shibamoto (1977).

The 1-deoxyreductone (equilibrium product of 1-deoxyosone) in basic medium ($pH > 5.0$) can degrade by retroaldo reactions to yield a number of very reactive carbonyl compounds such as pyruvic aldehyde, diacetyl, dihydroxyacetone, glyoxal and hydroxyacetol and acetic acid (Hodge, 1967; Feather, 1985). These compounds are particularly reactive with amino acids in Strecker degradation.

A fourth arm of the Maillard reaction is Strecker degradation by oxidation of the amino acids. This is undoubtedly one of the most important steps in the formation of meat flavour. This reaction involves the interaction of dicarbonyls such as diacetyl, pyruvaldehyde, hydroxyacetone and hydroxydiacetyl to degrade amino acids to aldehydes with one less carbon atom than the original amino acid, carbon dioxide, α-amino ketones, ammonia and hydrogen sulphide. Some degradation products include acetaldehyde from α-alanine, isobutyraldehyde from valine, isovaleraldehyde from leucine, 2-methylbutanal from isoleucine, benzaldehyde from phenylglycine and acetaldehyde from cysteine.

9.3 The Maillard reaction and meat flavour compounds

The above evidence reveals that the Maillard reaction is responsible for the formation of many meat flavour compounds, both in the cooking of meat and in model systems during formation of synthetic meat flavours. These compounds consist predominantly of *N,S,O*-heterocyclic compounds and other sulphur-containing constituents. The list of compounds includes furans, pyrazines, pyrroles, thiophenes, thiazoles (thiazolines), imidazoles, pyridines, oxazoles and cyclic ethylene sulphides.

Some of the many reviews on meat flavour chemistry were enumerated above and several of the contributors at this symposium have continued their discussions of meat flavour formation by reporting mechanisms involving the Maillard reaction. Only a few aspects of this broad topic can be considered here.

9.3.1 Low molecular weight precursors of meat flavour

The concentrated diffusate from water extracted meat (Wood, 1961; Hornstein and Crowe, 1963; Macy *et al.*, 1964a,b; Wasserman and Gray, 1965; Wasserman and Spinelli, 1972; Ching, 1979; Einig, 1983; Shin-Lee, 1988) is a natural starting place for the study of meat flavour chemistry. The heating of this non-proteinaceous meat flavour concentrate in water gives an aroma of boiled beef upon boiling and a stronger odour of broiled steak when heated to higher temperatures (150°C). This concentrate contains many precursors essential for meat flavour production as sugars, amino acids, sugar phosphates and nucleic acid components.

Macy *et al.* (1964a,b) demonstrated that many of these precursors from meat diminish in concentration during heating. Ching (1979), Einig (1983) and Shin-Lee (1988) separated and identified many volatile compounds formed by heating diffusate under various conditions. These compounds have been enumerated previously (Bailey and Einig, 1989), and the important compounds are summarized in Table 9.1.

Although 167 compounds were identified, the most important volatiles related to meat flavour were furanones, ketones, sulphur compounds (sulphides, thiophenes and thiazoles), pyrones, pyrroles and pyrazines.

Table 9.1 Important meat flavour compounds identified from heated beef diffusate[a]

Class of compound	Examples of compounds within classes
Furanones	2-Methyl-4,5-dihydro-3(2*H*)-furanone 2,5-Dimethyl-4-hydroxy-3(2*H*)-furanone 4-Hydroxy-5-methyl-3(2*H*)-furanone
Ketones	2-Hydroxy-3-methyl-2-cyclopentenone (cyclotene) Other cyclic ketones Acyclic ketones
Sulphur compounds	Sulphides, disulphides, trisulphides Thiophenes Thiazoles
Pyrones	3,5-Dihydroxy-2-methyl-4*H*-pyran-4-one
Pyrazines	37 Pyrazines, including five cyclopenta-pyrazines
Pyrroles	Eight pyrroles, including two acetyl pyrroles

[a]Compounds identified by Bailey and Einig (1989).

The latter compounds were a prominent group of volatiles found by heating beef diffusates.

9.3.2 *Pyrazines*

Pyrazines constitute a major class of volatiles formed from the Maillard reaction, particularly if Strecker degradation is considered as part of these reactions. Reactant conditions such as moisture content, temperature, pH and time are important in pyrazine formation (Vernin and Parkanyi, 1982). Van den Ouweland *et al.* (1989) surmised that a low percentage of volatiles identified from natural beef aroma have their origin from the Maillard reaction and that 50% of these are pyrazine derivatives.

A number of mechanisms have been published for the formation of pyrazines and an important pathway would be the condensation of dicarbonyl compounds formed by Strecker degradation to form alkyl pyrazines. Cyclopentapyrazines can be formed by the condensation of cyclic ketones such as cyclotene (2-hydroxy-3-methyl-cyclopentanone). Pathways proposed by Vernin and Parkanyi (1982) and Flament *et al.* (1976) are shown in Figure 9.4.

There is recent evidence that 3-deoxyglucosone can be a primary precursor for pyrazines. Data have been presented that it is degraded by retro-aldolization and a 2,4-scission to form pyruvaldehyde, which can be involved in the formation of dimethyl-pyrazine by Strecker degradation (Weenan, 1991).

(a) $R^1-C=O$, $R^2-C=O$ + Amino Acids $\xrightarrow[{[O]}]{-2H_2O}$ ALKYL PYRAZINES

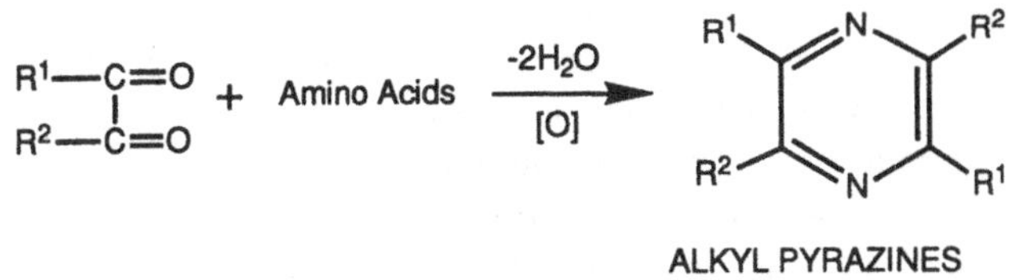

(b) $R^1-C=O$, $R^2-C=O$ + CYCLOTENE $\xrightarrow[{-2H_2O,\ [O]}]{\text{Amino acids}}$ CYCLOPENTAPYRAZINES

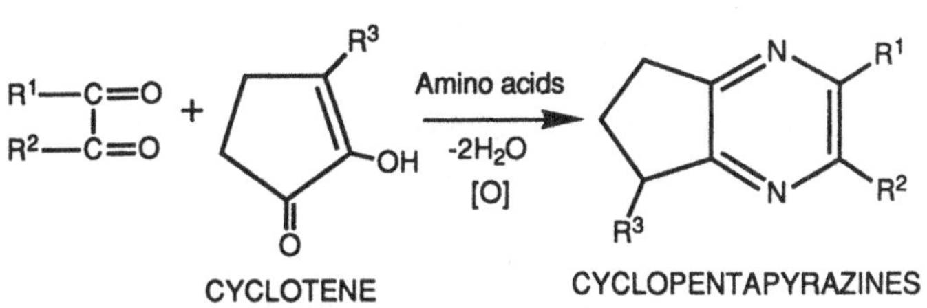

Figure 9.4 Reaction pathways proposed for the formation of (a) alkyl and (b) cyclic pyrazines. Modified from Vernin and Parkanyi (1982).

Pyrazines have been found in all meat species following cooking. Shahidi *et al.* (1986), who have done an excellent job in cataloguing volatile constituents from cooked meat, listed 48 pyrazines from beef, 36 from pork and 16 from lamb. Bailey and Einig (1989) listed 37 pyrazines identified in heated systems of low molecular weight diffusates from beef. Twenty-five were alkyl pyrazines and five were cyclopentapyrazines. The latter group of compounds are interesting since they have been reported to have roasted, grilled and other animal notes from roast meat (Ohloff and Flament, 1978). Bodrero *et al.* (1981) obtained a strong relationship between sensory scores for meaty odour and a reaction mixture of 2-acetyl-3-methylpyrazine and H_2S. Mottram *et al.* (1984) found 27 pyrazines in well-done grilled pork. Pyrazines accounted for 77% of the total volatiles found in pork and these compounds appear to be extremely important constituents of meat cooked at high temperature.

9.3.3 Sulphur compounds

Meat would indeed have an entirely different flavour in the absence of sulphur compounds. Large quantities of H_2S are produced during the heating of meat, which become more evident when the volatiles from meat cookery are filtered through acid to remove the ammonia. Under these conditions, the H_2S concentration is very prominent. Most researchers agree that sulphur compounds are the most important volatiles formed during meat cookery and sulphur precursors are used in essentially all synthetic meat flavour mixtures.

MacLeod (1986) has done a superlative job of identifying and describing the many volatile compounds in the literature described as meaty. Obviously most of these compounds contain sulphur. She listed 78 compounds as having meaty-like aromas; 65 are heterocyclic sulphur compounds; seven are aliphatic sulphur compounds; and six are non-sulphur heterocyclics. Twenty-five of these compounds have been identified in meat, and most could be formed from the Maillard reaction.

Acetaldehydes formed by Strecker degradation of alanine and other aldehydes are important since these compounds can react with H_2S, ammonia and methane thiol also formed by Strecker degradation to yield dithiazines, thiols, sulphides and trithianes. A scheme for the formation of these sulphur compounds is given in Figure 9.5. These reactions can be extrapolated to the formation of many similar compounds as has been diagrammed in previous reviews of meat flavour.

Sulphur compounds from furan-like components. As outlined above, furan-like derivatives are very prominent products of Maillard reactions carried out in model systems, and the compounds emphasized form secondary reaction products with H_2S and ammonia to form meat flavours.

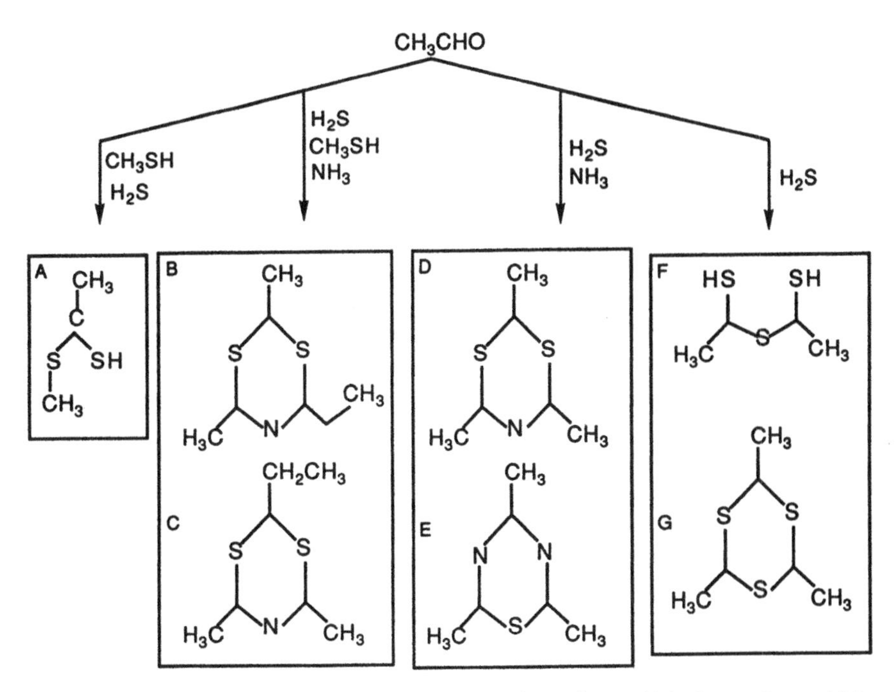

Figure 9.5 Sulphur compounds formed from reaction of acetaldehyde, methane thiol, ammonia and hydrogen sulphide. A = 1-methylthio-1-ethanethiol; B = 4-ethyl-2,6-dimethyl-dihydro-1,3,5-dithiazine; C = 2-ethyl-4,6-dimethyldihydro-1,3,5-dithiazine; D = 5,6-dihydro-2,4,6-trimethyl-1,3,5-dithiazine; E = 5,6-dihydro-2,4,6-trimethyl-1,3,5-thiadiazine; F = bis-(1-mercaptoethyl) sulphide; G = 2,4,6-trimethyl-1,3,5-trithiane.

Sugar degradation products like maltol, isomaltal, 4-hydroxy-5-methyl-3(2*H*)-furanone, 2,5-dimethyl-4-hydroxy-3(2*H*)-furanone and 2-hydroxy-3-methyl-2-cyclopentene-1-one (cyclotene) can exchange oxygen in the ring with nitrogen or sulphur.

Van den Ouweland and Peer (1975) contributed significantly to the knowledge of sulphur heterocyclic formation when they reacted H_2S with 4-hydroxy-5-methyl-3(2*H*)-furanone to form a number of 'meat-like' mercapto-substituted furan and thiophenone derivatives (Figures 9.6 and 9.7). They concluded that the reaction proceeds via a 2,4-diketone intermediate which then reacts with H_2S to form thiophenones (Figure 9.6). These workers strongly believe, however, that these compounds are not formed via a Maillard-type reaction in meat, but are derived from nucleic acid derivatives and ribose-5-phosphate. One of these compounds, 4-mercapto-5-methyl-tetrahydro-3-furanone, has a very strong meaty odour and was identified by Ching (1979) after heating freeze-dried defatted beef with triglycerides. It has a meaty or maggi odour and could be considered to be a meat flavour impact compound.

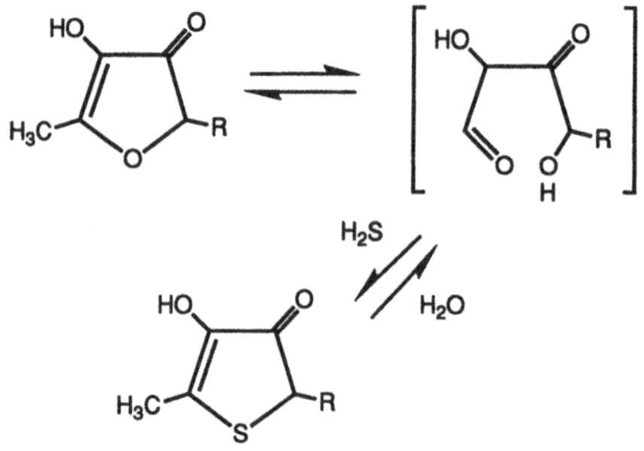

4-HYDROXY-3(2H)-FURANONES (R = H, CH_3)

Figure 9.6 Initial reaction between furanones and hydrogen sulphide to form *S*-heterocyclics. Adapted from van den Ouweland and Peer (1975).

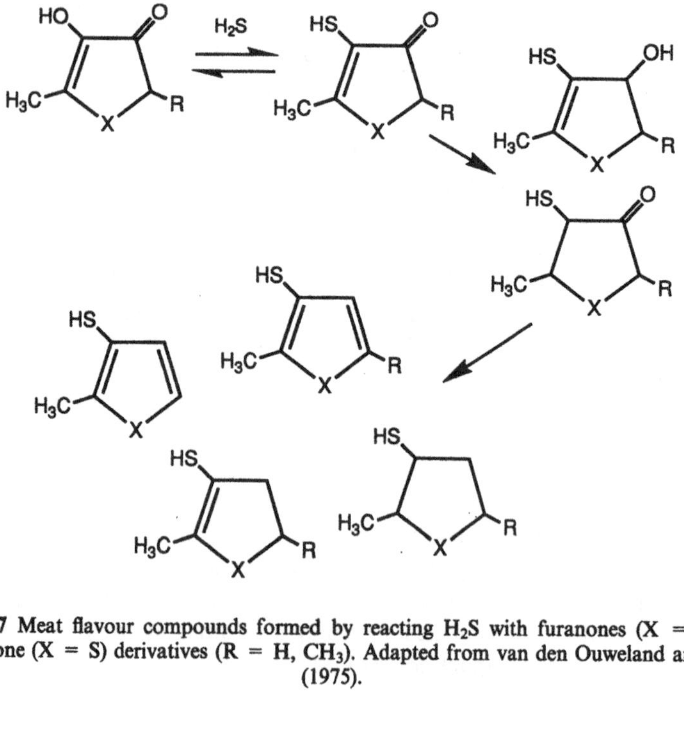

Figure 9.7 Meat flavour compounds formed by reacting H_2S with furanones (X = O) or thiophenone (X = S) derivatives (R = H, CH_3). Adapted from van den Ouweland and Peer (1975).

The importance of the reaction between furanones and sulphur-containing compounds in meat flavour was demonstrated by Bodrero *et al.* (1981), who used surface response methodology to demonstrate that this type of reaction gave the highest predictive score perceived for cooked beef aroma. 4-Hydroxy-2,5-dimethyl-3(2*H*)-furanone, a Maillard reaction product also formed by cyclization and by reacting L-rhamnose with piperidine acetate (Hodge and Osman, 1976), was reacted with cystine by Shu and Ho (1989) under various conditions and found to produce numerous sulphur-containing compounds with meaty aroma. The major ones were 3,5-dimethyl-1,2,4-trithiolanes, thiophenones and thiazoles. The amount of moisture in a reaction mixture of glycerol and water, the oxygen content and the pH of the reaction were found to be important parameters in the formation of these volatiles.

Shu *et al.* (1985) proposed a mechanism similar to that of van den Ouweland and Peer (1975) for the formation of sulphur derivatives, including 2,5-dimethyl-2,4-dihydroxy-3-(2*H*)-thiophenone, which they described as having a pot roast aroma and flavour. Seventy-five percent moisture resulted in maximum yield of 3,5-dimethyl-1,2,4-trithiolane and thiophenones at pH 4.7. Greater quantities of 3,5-dimethyl-1,2,4-trithiolane was formed anaerobically compared to oxygen environments.

Similar compounds that might possibly have their derivation from reaction of 4-hydroxy-5-methyl-3(2*H*)-furanone with H_2S have recently been identified by MacLeod and Ames (1986) as meat flavour character impact compounds. These compounds are 2-methyl-3-(methylthio)-furan and 2-methyl-3(methyldithio)-furan, which have odour thresholds in water of 0.05 and 0.01 ppm, respectively, and both have meaty aromas below 1 ppb. As described later, disulphides having similar structures are also considered to be meat flavour impact compounds.

Cyclotene (2-hydroxy-3-methylcyclopent-2-enone) formed from 5-hydroxy-5,6-dihydromaltol (Figure 9.3) and from hydroxyacetone condensation is also a precursor of volatile compounds having meaty aromas. Nishimura *et al.* (1980) described a reaction between ammonia, H_2S and cyclotene as having meaty odours and containing 1,2,4-trithiolane, 5-trithiane and 1,2,4,6-tetrathiepane, as well as 2-methylcyclopentanone and 3-methyl-cyclopentanone, which were described as having a roasted beef odour (Nishimura *et al.*, 1980). Tricyclic pyrazines and dihydro-cyclopentapyrazine are also formed by heating cyclotene and alanine (Rizzi, 1976). Other prominent intermediates from Amadori rearrangement products that can react further to form compounds with meat-like odours are furfural and isomaltol. Furfural, when heated with H_2S and ammonia, also formed sulphur-heterocyclics having meaty odour, probably being first degraded to furan aldehyde (Shibamoto, 1977). Isomaltol is also a prominent precursor of heterocyclic compounds related to cooked meat flavour (Tonsbeek *et al.*, 1968).

Mention must be made of the excellent studies by Werkhoff *et al.* (1989; 1990) of the formation of numerous sulphur-containing compounds with meaty odour by heating cystine, thiamine, ascorbic acid and MSG in a model system. Although many of the constituents identified undoubtedly are derived from thiamine degradation (MacLeod, 1986; Werkhoff *et al.*, 1990), Maillard-type reactions could be involved in the formation of some compounds, particularly compounds like 1-(2-methyl-2-thienyl-thio)-ethanethiol and 1-(2-methyl-3-furylthio)-ethanethiol which can be formed respectively from 2-methyl-3-furathiol and 2-methyl-3-thiophenethiol (Figure 9.8). Some of these compounds react with furans and thiophenes that could be derived from Maillard reaction, which are then substituted with sulphur in the 2 or 3 positions. Several of these sulphur-substituted compounds have characteristic meat flavour notes and are likely to have importance in meat flavour.

Farmer and Patterson (1991) identified five important disulphides having meaty odour from the headspace of freshly cooked (80°C) beef. Two were the same as those identified by Werkhoff *et al.* (1989). Two of these compounds, *bis* (2-methyl-3-furyl) disulphide and 2-furfuryl-2-methyl-3-furyl disulphide (Figure 9.9), both have strong meaty/roasted/burnt odours. *Bis* (2-methyl-3-furyl) disulphide, 2-methyl-3-furanthiol and 2-furfurylthiol have been identified as important beef odour constituents (Gasser and Grosch, 1988) and have a very low threshold of 2 parts in 10^{14} in water (Buttery *et al.*, 1984). This is one of the lower thresholds for flavour compounds so far identified. Grosch (1991) reported that the meat flavour impact compound, 2-methyl-3-furanthiol, previously identified in beef and pork could be synthesized by heating meat precursors, and that oxidation products from this compound (*bis*-(2-methyl-3-furyl) disulphide and 2-furfurylthiol) should also be considered meat flavour impact compounds. Preliminary results by Farmer and Patterson (1991) also indicate that the threshold of 2-furfuryl-2-methyl furyl-disulphide is very low. It is likely that other 3-methyl-3-furylthio compounds similar to those synthesized by Werkhoff *et al.* (1990) will also be found in meat volatiles at low levels.

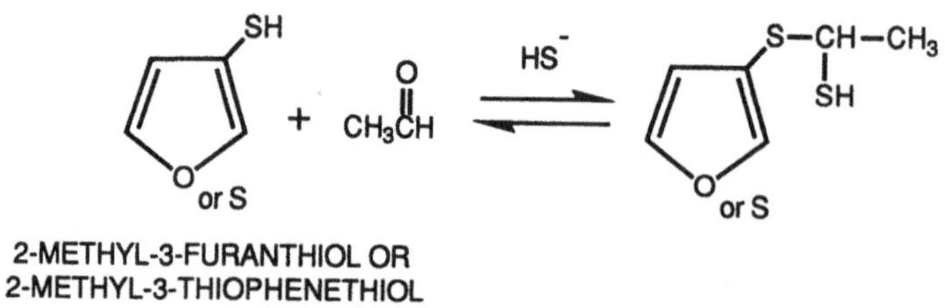

Figure 9.8 Synthesis of 1-[(2-methyl-3-thienyl)thio] ethanethiol and 1-[(2-methyl-2-furyl)thio] ethanethiol from acetaldehyde hydrogen sulphide and methyl furan derivatives.

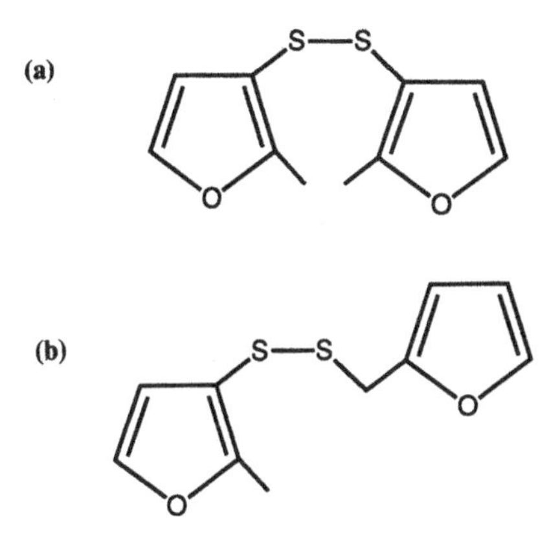

Figure 9.9 Meat flavour impact compounds identified by Werkhoff *et al.* (1989) and found in meat by Farmer and Patterson (1991). (a) bis-2-methyl-3-furyl disulphide (odour threshold 2 parts per 10^4; (b) 2-furfuryl-2-methyl-3-furyl disulphide.

9.3.4 Synthetic flavours from the Maillard reaction

The patent literature abounds with formulations for synthetic meat flavour. Most contain precursors that participate in Maillard reactions. Buckholz (1989) stated that several hundred patents had appeared in the literature based on non-enzymatic browning technology for meat flavour production, at least 50 patents of which include amino acids and sugars in their reaction mixtures. Most of the patents of for the formulation of meaty flavours utilize cysteine, cystine or methionine as a source for sulphur, but more recently thiamine has also been incorporated into reaction flavours (Buckholz, 1989).

Bailey and Um (1991) recently prepared a synthetic mixture of natural products that served both as an antioxidant and as a meat flavour enhancer during storage of cooked beef and pork. The antioxidant was prepared by heating 0.2 M glucose and 0.2 M histidine at 150°C for 2 h in 40% glycerol. This mixture was then used to disperse 0.5% cysteine, 0.5% thiamine, 1.0% glycine, 3.0% autolysed yeast and 2% beef fat, which was then heated at 125°C for 1 h. The resulting synthetic meat flavour, when added to beef roast prior to cooking, was found superior to all commercial samples tested for protecting meaty flavour of cooked roast beef during storage. Roast beef prepared with this flavour mixture was accepted at a high level by consumers during storage at 4°C.

The discovery of the elixir of meat flavour is just on the horizon, but this mixture may be very unstable and we will continue to the dependent upon Maillard reaction intermediates as essential ingredients for the production and stabilization of desirable meaty flavours. A résumé of these

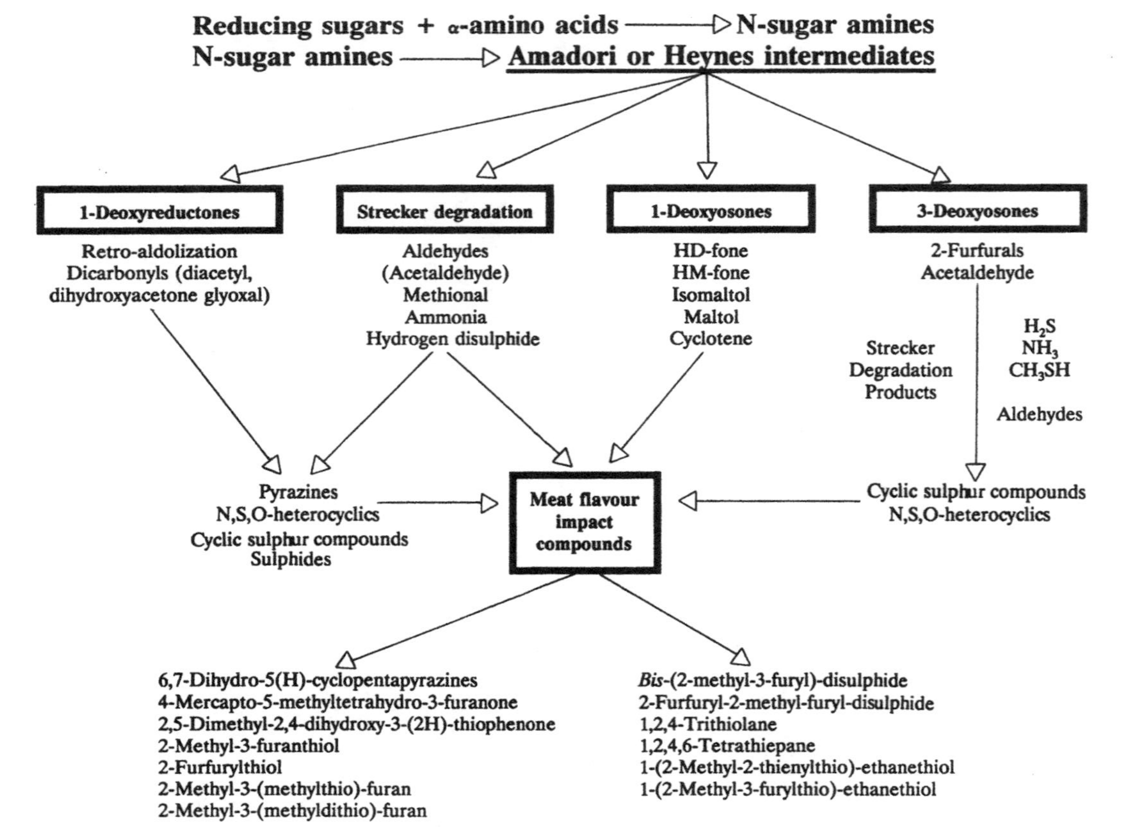

Figure 9.10 Formation of meat flavour impact compounds by Maillard reaction of sugars and amines.

activities in the formation of meat flavour impact compounds is given in Figure 9.10.

9.4 Summary

Reducing sugars can react with amino acids to produce *N*-glycosylamines or *N*-fructosylamines which rearrange to Amadori or Heynes intermediates that deaminate to form carbonyls such as 1-deoxyreductones, 1-deoxyosones and 3-deoxyosones that form many precursors for meaty flavour. Precursors for the 1-deoxyreductones can react with Strecker degradation products such as aldehydes; ammonia and hydrogen sulphide to form pyrazines; nitrogen–sulphur heterocyclics such as thiazoles and thiazolines; cyclic sulphur compounds such as dithianes, trithianes and trithiolanes; and mono-, di-, tri- and tetra-sulphides. Meat flavour precursors from the 1-deoxyosone such as HO-Fone (4-hydroxy-5-methyl-3(2*H*)-furanone), HM-Fone (4-hydroxy-2,5-dimethyl-3(2*H*)-furanone), isomaltol, maltol or cyclotene can react with Strecker degradation products to form many meat flavour impact compounds. These include: 4-mercapto-5-methyl-tetrahydro-3-furanone (maggi odour); 2,5-dimethyl-2,4-dihydroxy-3-(2*H*)-thiophenone (pot roast); 2-methyl-3-(methylthio)-furan (meaty); *bis*-(2-methyl-3-furyl)-disulphide (2 parts/10^4 parts water—meaty); 1-(2-methyl-2-thienylthio)-ethanethiol (meaty); 1-(2-methyl-3-furylthiol-ethanethiol (meaty); and some non-aromatic cyclic sulphur compounds formed by heating cyclotene and H_2S.

Important chemical structures for meaty-roasty flavours formed from Maillard reactions appear to be aromatic nitrogen derivatives such as pyrazines; *N,S*-heterocyclics such as thiazoles, 2-methyl-3-thio (or sulphide)-furans; 2-methyl-5-thio (or sulphide)-thiophenes; sulphur-substituted tetrahydrofuranones; and non-aromatic ring sulphur derivatives containing two or more sulphur moieties.

References

Anet, E.F.L.J. and Reynolds, T.M. (1957). Chemistry of non-enzymic browning. I. Reactions between amino acids, organic acids and sugars in freeze-dried apricots and peaches. *Austr. J. Chem.*, **10**, 182–192.

Bailey, M.E. (1983). The Maillard reaction and meat flavor. In *The Maillard Reaction in Foods and Nutrition*, ed. G.R. Waller and M.S. Feather. ACS Symposium Series 215, Washington, DC, pp. 169–184.

Bailey, M.E. and Einig, R.G. (1989). Reaction flavors of meat. In *Thermal Generation of Aromas*, ed. T.H. Parliment, R.J. McGorrin and C.-T. Ho. ACS Symposium Series 409, Washington, DC, pp. 421–432.

Bailey, M.E. and Um, K.-W. (1991). Maillard reaction products and lipid oxidation. Paper #119 Agric. and Food Chem. Div., Fourth Chemical Congress of North America, New York, 25–30 August.

Baines, D.A. and Mlotkiewicz, J.A. (1984). The chemistry of meat flavor. In *Recent Advances in Chemistry of Meat*, ed. A.J. Bailey. Royal Society Chemistry, London, pp. 119–164.

Baltes, W., Kunert-Kirchhoff, J. and Reese, G. (1989). Model reactions on generation of thermal aroma compounds. In *Thermal Generation of Aromas*, ed. T.H. Parliment, R.J. McGorrin and C.-T. Ho. ACS Symposium Series 409, pp. 143–155.

Barone, R. and Chanon, R. (1982). Computer application of non-interactive program of simulation of organic synthesis in Maillard's reaction: a proposition for new heterocyclic compounds for flavours. In *Chemistry of Heterocyclic Compounds in Flavours and Aromas*, ed. G. Vernin. Ellis Horwood Ltd, Chichester, pp. 249–261.

Bender, A.E., Wood, T. and Palgrove, J.A. (1958). Analyses of tissue constituents. Extract of fresh ox-muscle. *J. Sci. Food Agric.*, **9**, 812–817.

Bodrero, K.O., Pearson, A.M. and Magee, W.T. (1981). Evaluation of the contribution of flavor volatiles to the aroma of beef by surface response methodology. *J. Food Sci.*, **46**, 26–31.

Buckholz, L.L. Jr. (1989). Maillard technology as applied to meat and savory flavors. In *Thermal Generation of Aromas*, ed. T.H. Parliment, R.J. McGorrin and C.-T. Ho. ACS Symposium Series 409, Washington, DC, pp. 406–420.

Buttery, R.G., Haddon, W.F., Seifert, R.M. and Turnbaugh, J.G. (1984). Thiamine odor and *bis*-(2-methyl-3-furyl) disulfide. *J. Agric. Food Chem.*, **32**, 674–676.

Ching, J.C.Y. (1979). Volatile flavor compounds from beef and beef constituents. PhD thesis, University of Missouri, Columbia, MO.

Cramer, D. (1983). Chemical compounds implicated in lamb flavor. Overview. *Food Technol.*, **37**, 249–257.

Danehy, J.P. (1986). Maillard reactions: nonenzymatic browning in food systems with special reference to the development of flavor. *Adv. in Food Res.*, **30**, 77–138.

Einig, R. (1983). Interaction of meat aroma volatiles with soy proteins. PhD thesis, University of Missouri, Columbia, MO.

Eriksson, C. (1981). *Maillard Reactions in Food.* Pergamon Press, Oxford.

Eskin, N.A.M. (1990). *Biochemistry of Foods*. Academic Press, San Diego.

Farmer, L.J. and Patterson, R.L.S. (1991). Compounds contributing to meat flavour. *Food Chem.*, **40**, 201–205.

Feather, M.S. (1981). Amine-assisted sugar dehydration reactions. *Prog. Food Nutr. Sci.*, **5**, 37–45.

Feather, M.S. (1985). Some aspects of the chemistry of nonenzymatic browning (the Maillard reaction). In *Chemical Changes in Food During Processing*, ed. T. Richardson and J.W. Finley. AVI Publishing Co., Westport, CT, pp. 289–303.

Feather, M.S. and Harris, J.F. (1973). Dehydration reactions of carbohydrates. *Adv. Carbohydr. Chem.*, **28**, 161–224.

Feeney, R.E. and Whitaker, J.R. (1982). The Maillard reaction and its prevention. In *Food Protein Deterioration. Mechanisms and Functionality*, ed. J.P. Cherry. ACS Symposium Series 206, Washington, DC, pp. 201–229.

Finot, P.A., Aeschbacher, H.U., Hurrell, R.F. and Liardon, R. (1990). *The Maillard Reaction in Food Processing, Human Nutrition and Physiology*. Birkhäuser Verlag, Basel.

Flament, I., Kohler, M. and Aschiero, R. (1976). Sur láromé de viande de boeuff grillée. II. Dehydro-6,7-5H-cyclopenta[b] pyrazines, identification et mode de formation. *Helv. Chim. Acta.*, **59**, 2308–2313.

Fujimaki, M., Namiki, M. and Kato, N. (1986). *Amino Carbonyl Reactions in Food and Biological Systems.* Elsevier, Amsterdam.

Gasser, U. and Grosch, W. (1988). Identification of volatile flavour compounds with high aroma values from cooked beef. *Z. Lebensm. Unters-Forsch.*, **186**, 489–494.

Grosch, W. (1991). Studies on the formation of meat-like flavour compounds. Paper #70 Agric. and Food Chem. Div., Fourth Chemical Congress of North America, New York, 25–30 August.

Heath, H.B. and Reineccius, G.A. (1986). *Flavor Chemistry and Technology.* AVI Publishing Co., Westport, CT.

Hicks, K.B. and Feather, M.S. (1975). Studies on the mechanism of formation of 4-hydroxy-5-methyl-3(2H)-furanone, a component of beef flavor, from Amadori products. *J. Agric. Food Chem.*, **23**, 957–960.

Hicks, K.B., Harris, D.W., Feather, M.S. and Loeppky, R.N. (1974). Production of 4-hydroxy-5-methyl-3(2H)-furanone, a component of beef flavor, from a 1-aino-1-deoxy-D-fructuronic acid. *J. Agric. Food Chem.*, **22**, 724–725.

Ho, C.-T. (1980). Meat flavor: past, present and future. In *Proceedings of the Meat Industry Research Conference*. American Meat Institute Foundation, Arlington, VA, pp. 41–49.

Hodge, J.E. (1953). Chemistry of browning reactions in model systems. *J. Agric. Food Chem.*, **1**, 928–943.

Hodge, J.E. (1967). Origin of flavors in food. Nonenzymatic browning reactions. In *Symposium on Foods: Chemistry and Physiology of Flavors*, ed. H.W. Schultz, E.A. Day and L.M. Libbey. AVI Publishing Co., Westport, CT, pp. 465–491.

Hodge, J.E. and Osman, E.M. (1976). Carbohydrates. In *Principles of Food Science. Part I. Food Chemistry*, ed. O.R. Fennema. Marcel Dekker, New York, pp. 41–138.

Hodge, J.E., Mills, F.D. and Fisher, B.E. (1972). Compounds of browned flavor derived from sugar–amine reactions. *Cereal Science Today*, **17**, 34–40.

Hornstein, I. and Crowe, P.F. (1960). Flavor studies on beef and pork. *J. Agric. Food Chem.*, **8**, 494–498.

Hornstein, I. and Crowe, P.F. (1963). Meat flavor: lamb. *J. Agric. Food Chem.*, **11**, 147–149.

Hornstein, I. and Wasserman, A. (1987). Chemistry of meat flavor. In *The Science of Meat and Meat Products, 3rd Edition*, ed. J.F. Price and B.S. Schweigert. Food and Nutrition Press, Westport, CT, pp. 329–347.

Hurrell, R.F. (1982). Maillard reaction in flavour. In *Food Flavours. Part A. Introduction*, ed. I.D. Morton and A.J. MacLeod. Elsevier, Amsterdam, pp. 399–437.

Hurrell, R.F. and Carpenter, K.J. (1977). Maillard reactions in foods. In *Physical, Chemical and Biological Changes in Food Caused by Thermal Processing*, ed. T. Hoyem and O. Kvåle. Applied Science, London, pp. 168–184.

IFT Symposium (1983). 42nd Annual Meeting. *Food Technol.*, **37**, 225–268.

Katz, I. (1981). Recent progress in some aspects of meat flavor chemistry. In *Flavour Research: Recent Advances*, ed. R. Teranishi, R.A. Flath and H. Sugisawa. Marcel Dekker, New York, pp. 217–229.

Lawrie, R.A. (1982). The flavour of meat and meat analogues. *Food Flavor International*, **4**(2), 10–13, 15.

Ledl, F. (1990). Chemical pathways of the Maillard reaction. In *The Maillard Reaction in Food Processing, Human Nutrition and Physiology*, ed. P.A. Finot, H.U. Aeschbacher, R.F. Hurrell and R. Liardon. Birkhäuser Verlag, Basel, pp. 19–42.

Ledl, F., Beck, J., Sengl, M., Osiander, H., Estendorfer, S., Severin, T. and Huber, B. (1989). Chemical pathways of the Maillard reaction. In *The Maillard Reaction in Aging, Diabetes and Nutrition*, ed. J.W. Baynes and V.M. Monnier. Alan R. Liss, Inc., New York, pp. 23–42.

Mabrouk, A.F. (1979). Flavor of browning reaction products. In *Food Taste Chemistry*, ed. J.C. Boudreau. ACS Symposium Series 115, Washington, DC, pp. 205–245.

MacLeod, G. (1984). The flavour of meat. In *Proceedings of the Institute of Food Science and Technology (UK)*, Vol. 17, No. 4, pp. 184–197.

MacLeod, G. (1986). The scientific and technological basis of meat flavours. In *Developments in Food Flavors*, ed. G.G. Birch and M.G. Lindley. Elsevier, London, pp. 191–223.

MacLeod, G. and Ames, J.M. (1986). The effect of heat on beef aroma: comparisons of chemical composition and sensory properties. *Flavor and Fragrance Journal*, **1**, 91–104.

MacLeod, G. and Seyyedain-Ardebili, M. (1981). Natural and simulated meat flavors (with particular reference to beef). *CRC Crit. Rev. Food Sci. Nutr.*, **14**, 309–437.

Macy, R.L. Jr., Naumann, H.D. and Bailey, M.E. (1964a). Water-soluble flavor and odor precursors of meat. I. Qualitative study of certain amino acids, carbohydrates, non-amino compounds and phosphoric acid esters of beef, pork and lamb. *J. Food Sci.*, **29**, 131–141.

Macy, R.L. Jr., Naumann, H.D. and Bailey, M.E. (1964b). Water-soluble flavor and odor precursors of meat. II. Effects of heating on amino acids, nitrogen constituents and carbohydrates in lyophilized diffusates from aqueous extracts of beef, pork and lamb. *J. Food Sci.*, **29**, 142–148.

Maillard, L.-C. (1916). Synthesis des matieres humiques par action des acides amines sur les sucres reducteurs. *Ann. Chim. (Paris)*, **5**(9), 258–317.

Mauron, J. (1981). The Maillard reaction in food: a critical review from the nutritional standpoint. In *Maillard Reactions in Food*, ed. C. Eriksson. Pergamon Press, Oxford, pp. 5–35.

Moody, W.G. (1983). Beef flavor—a review. *Food Technol.*, **37**(5), 227–232.

Mottram, D.S. (1991). Meat. In *Volatile Compounds in Food and Beverages*, ed. H. Maarse. Marcel Dekker, New York, pp. 107–177.

Mottram, D.S., Croft, S.E. and Patterson, R.L.S. (1984). Volatile components of cured and uncured pork: the role of nitrite and the formation of nitrogen compounds. *J. Sci. Food Agric.*, **35**, 233–239.

Namiki, M. (1988). Chemistry of Maillard reactions: recent studies on the browning reaction mechanism and the development of antioxidants and mutagens. *Adv. Food Res.*, **32**, 115–184.

Namiki, M. and Hayashi, T. (1983). In *The Maillard Reaction in Foods and Nutrition*, ed. G.R. Waller and M.S. Feather. ACS Symposium Series 215, Washington, DC, pp. 21–46.

Nishimura, G., Mihara, S. and Shibamoto, T. (1980). Compounds produced by the reaction of 2-hydroxy-2-cyclopentene-1-one with ammonia and hydrogen sulfide. *J. Agric. Food Chem.*, **28**, 39–43.

Nursten, H.E. (1981). Recent developments in studies of the Maillard reaction. *Food Chemistry*, **6**, 263–277.

Nursten, H.E. (1986). Maillard browning reactions in dried foods. In *Concentration and Drying of Foods*, ed. D. MacCarthy. Elsevier, London, pp. 53–68.

O'Brien, J. and Morrissey, P.A. (1989). Nutritional and toxicological aspects of the Maillard reaction in foods. *CRC Crit. Rev. Food Sci. Nutr.*, **28**, 210–248.

Ohloff, G. and Flament, I. (1978). Heterocyclic constituents of meat aroma. *Heterocycles*, **11**, 663–695.

Parliment, T.H., McGorrin, R.J. and Ho, C.-T. (1989). *Thermal Generation of Aromas*. ACS Symposium Series 409, Washington, DC.

Paulsen, H. and Pflughaupt, K.-W. (1980). Glycosylamines. In *The Carbohydrates, Chemistry and Biochemistry*, 2nd Ed., Vol. 1B, ed. W. Pigman, D. Horton and J.D. Wander. Academic Press, New York, pp. 881–927.

Peer, H. A. and van den Ouweland, G.A.M. (1968). Synthesis of 4-hydroxy-5-methyl-3(2H)-furanone from D-ribose-5-phosphate. *Rec. Trav. Chim.*, **87**, 1017–1020.

Reynolds, T.M. (1963). Chemistry of nonenzymic browning. I. The reaction between aldoses and amines. *Adv. Food Res.*, **12**, 1–52.

Reynolds, T.M. (1965). Chemistry of nonenzymic browning. II. *Adv. Food Res.*, **14**, 167–283.

Rhee, K.S. (1989). Chemistry of meat flavor. In *Flavor Chemistry of Lipid Foods*, ed. D.B. Min and T.H. Smouse. American Oil Chemists' Society, Champaign, IL, pp. 166–189.

Rizzi, G.P. (1976). Non-enzymic transamination of unsaturated carbonyls: a general source of nitrogenous flavour compounds in foods. In *Phenolic, Sulfur and Nitrogen Compounds in Food Flavors*, ed. G. Charalambous and I. Katz. ACS Symposium Series 26, Washington, DC, pp. 122–132.

Schieberle, P. (1991). Studies on the formation of furanol. Paper #52 Agric. and Food Chem. Div., Fourth Chemical Congress of North America, New York, 25–30 August.

Severin, T.H. and Seilmeier, W. (1968). Studien zur Maillard reaktion. III. Unwandlung von geniose under einfluss von methylammoniumacetat. *Z. Lebensm. Unters-Forsch*, **137**, 4–6.

Shahidi, F. (1989). Flavor of cooked meats. In *Flavor Chemistry: Trends and Developments*, ed. R. Teranishi, R.G. Buttery and F. Shahidi. ACS Symposium Series 388, Washington, DC, pp. 188–201.

Shahidi, F., Rubin, L.J. and D'Souza, L.A. (1986). A review of the composition, techniques, analyses and sensory evaluation. *CRC Crit. Rev. Food Sci. Nutr.*, **24**(2), 141–243.

Shibamoto, T. (1977). Formation of sulfur- and nitrogen-containing compounds from the reaction of furfural with hydrogen sulfide and ammonia. *J. Agric. Food Chem.*, **25**, 206–208.

Shibamoto, T. (1980). Heterocyclic compounds found in cooked meats. *J. Agric. Food Chem.*, **28**, 237–243.

Shin-Lee, S.Y. (1988). Warmed-over flavor and its prevention by Maillard reaction products. PhD thesis, University of Missouri, Columbia, MO.

Shu, C.-K. and Ho, C.-T. (1989). Parameter effects on the thermal reaction of cystine and 2,5-dimethyl-4-hydroxy-3(2H)-furanone. In *Thermal Generation of Aromas*, ed. T.H. Parliment, R.J. McGorrin and C.-T. Ho. ACS Symposium Series 409, Washington, DC, pp. 229–241.

Shu, C.-K., Hagedorn, M.L., Mookherjee, B.D. and Ho, C.-T. (1985). Two novel 2-hydroxy-3(2H)-thiophenones from the reaction between cystine and 2,5-dimethyl-4-hydroxy-3(2H)-furanone. *J. Agric. Food Chem.*, **33**, 638–641.

Teranishi, R., Flath, R.A. and Sugisawa, H. (1971). *Flavour Research: Recent Advances*. Marcel Dekker, New York.

Tonsbeek, C.H.T., Plancken, A.J. and v.d. Weerdhof, T. (1968). Components contributing to beef flavor. Isolation of 4-hydroxy-5-methyl-3(2H)-furanone and its 2,5-dimethyl homolog from beef broth. *J. Agric. Food Chem.*, **16**, 1016–1021.

van den Ouweland, G.A.M. and Peer, H.G. (1975). Components contributing to beef flavor. Volatile compounds produced by the reaction of 4-hydroxy-5-methyl-3(2H)-furanone and its thio analog with hydrogen sulfide. *J. Agric. Food Chem.*, **23**, 501–505.

van den Ouweland, G.A.M., Demole, E.D. and Enggist, P. (1989). Processed meat flavor development and the Maillard reaction. In *Thermal Generation of Aromas*, ed. T.H. Parliment, R.J. McGorrin and C.-T. Ho. ACS Symposium Series 409, Washington, DC, pp. 433–441.

Vernin, G. and Parkanyi, C. (1982). Mechanisms of formation of heterocyclic compounds in Maillard and pyrolysis reactions. In *Chemistry of Heterocyclic Compounds in Flavours and Aromas*, ed. G. Vernin. Halsted Press, New York, pp. 151–207.

Waller, G.R. and Feather, M.S. (1983). *The Maillard Reaction in Food and Nutrition*. ACS Symposium Series 215, Washington, DC.

Wasserman, A.E. (1979). Chemical basis for meat flavor: a review. *J. Food Sci.*, **44**, 6–11.

Wasserman, A.E. and Gray, N. (1965). Meat flavor. I. Fractionation of water-soluble flavor precursors of beef. *J. Food Sci.*, **30**, 801–807.

Wasserman, A.E. and Spinelli, A.M. (1970). Sugar–amino acid interaction in the diffusate of water extract of beef and model systems. *J. Food Sci.*, **35**, 328–332.

Wasserman, A.E. and Spinelli, A.M. (1972). Effect of some water-soluble compounds on aroma of heated adipose tissue. *J. Agric. Food Chem.*, **20**, 171–174.

Weenan, H. (1991). Analysis and reactivity of glucosones. Paper #51 Agric. and Food Chem. Div., Fourth Chemical Congress of North America, New York, 25–30 August.

Werkhoff, P., Emberger, R., Güntert, M. and Köpsel, M. (1989). Isolation and characterization of volatile sulfur-containing meat flavor components in model systems. In *Thermal Generation of Aromas*, ed. T.H. Parliment, R.J. McGorrin and C.-T. Ho. ACS Symposium Series 409, Washington, DC, pp. 460–478.

Werkhoff, P., Brüning, J., Emberger, R., Güntert, M., Köpsel, M., Kuhn, W. and Surburg, H. (1990). Isolation and characterization of volatile sulfur-containing meat flavor components in model systems. *J. Agric. Food Chem.*, **38**, 777–791.

Williams, J.C. (1976). Chemical and non-enzymatic changes in intermediate moisture foods. In *Intermediate Moisture Foods*, ed. R. Davies, G.G. Birch and K.J. Parker. Applied Science, London, pp. 100–119.

Wood, T. (1961). The browning of ox-muscle extracts. *J. Sci. Food Agric.*, **12**, 61–69.

Wood, T. and Bender, A.E. (1957). Analysis of tissue constituents. Commercial ox-muscle extract. *Biochem. J.*, **67**, 366–373.

Yaylayan, V. and Sporns, P. (1987). Novel mechanisms for the decomposition of 1-(amino acid)-1-deoxy-D-fructoses (Amadori compounds): a mass spectrometric approach. *Food Chem.*, **26**, 283–305.

10 The flavour of cured meat

N. RAMARATHNAM and L.J. RUBIN

10.1 Introduction

The origin of the use of nitrate/nitrite in the curing of meat is lost in history, but it is strongly believed that the preservation of meat with salt preceded the intentional use of nitrate by many centuries. It appears that meat preservation was first practised in the saline deserts of Hither Asia and in coastal areas (Binkert and Kolari, 1975). Salt was in common use in ancient Palestine as early as 1600 BC because of its availability from the salt-rich Dead Sea. The technology of sea-salt production was also known by at least 1200 BC by the Chinese, who early made salt from drilled wells (Jensen, 1953, 1954). Salt from the sea, desert, and as found as an efflorescence on the walls of caves and stables was used by ancient peoples in the curing of meat. These salts contained nitrates as impurities, and though saltpetre or 'nitre' was gathered in ancient China and India long before the Christian era, the reddening effect of nitrates on meat was not mentioned until late Roman times.

The curing of meat as practised today is based upon the ancient art practised through aeons of time, and perhaps to a far greater extent upon the scientific principles developed since about the turn of the century. The permission to conduct the first and subsequent series of experiments with the direct use of nitrite in meat curing, under federal inspection, was given in 1923 by the Bureau of Animal Industry of the United States Department of Agriculture (Kerr *et al.*, 1926). On the basis of the results obtained in those experiments, the use of sodium nitrite in meat-curing was formally regulated by the USDA in 1925, and this authorization by the Bureau of Animal Industry restricted the amount of sodium nitrate used in the cured-meat product to 200 parts per million (mg/kg). Present day meat-curing practice involves the addition of sodium nitrite and salt along with other additives such as sugar, certain reducing agents, phosphates and, where appropriate, seasonings, to impart characteristic properties to the end product. It is estimated that over 70% of pork is cured with nitrite. In the curing pickle, the function of salt, which constitutes the bulk of the mixture, is to contribute to taste, act as a preservative, and enhance the functional properties of meat protein; sugar enhances the flavour of the cured product; the reducing agents such as ascorbates and erythorbates accelerate the development of the characteristic pink colour of cured meat; phosphates help in retaining the moisture of the cured product, and therefore play an important role in contributing to the mouthfeel and juiciness of the processed product; and seasonings and spices impart additional aroma and taste.

10.2 Advantages of nitrite in the meat-curing process

Nitrite is a unique and critical ingredient in the meat-curing system. It imparts the characteristic pink colour to the cured meat (Eakes *et al.*, 1975; Giddings, 1977) and provides oxidative stability to meat by preventing lipid oxidation (Pearson *et al.*, 1977; Fooladi *et al.*, 1979; MacDonald *et al.*, 1980; Shahidi *et al.*, 1987a; Yun *et al.*, 1987). This effect is complex, but it is believed to be associated with bringing forth the cured-meat flavour and the prevention of the warmed-over flavour (WOF) in meat (Mottram and Rhodes, 1974; Skjelkvale and Tjaberg, 1974; Rubin and Shahidi, 1988). Nitrite has an antimicrobial effect which is important in preventing the outgrowth of *Clostridium botulinum* and the formation of deadly toxin (Hauschild *et al.*, 1982; Pierson and Smooth, 1982; Wood *et al.*, 1986), particularly under conditions of product mishandling. To date, no other single chemical additive is known that can perform all the functions of sodium nitrite for use as a curing agent for meat.

10.3 Antioxidant role of nitrite in cured meats

The overall acceptance of meat products depends to a major extent on their flavour quality. Raw meat has little odour and only a blood-like taste, and cooking develops its flavour (Crocker, 1948. Bender and Ballance, 1961). When meat is cooked under very mild conditions, like heating in water, it develops a characteristic desirable flavour attributable to the animal species. The final flavour spectrum of each species depends to a major extent on the nature and amount of the non-volatile flavour precursors such as the amino acids, amines, sugars, and fatty acids present in the raw meat.

Lipids in meat make an important contribution toward the overall flavour and mouthfeel of the cooked meat. They serve as a reservoir of fat-soluble compounds that volatilize upon heating to form aroma compounds, and can themselves undergo degradation and autoxidation to produce a wide range of carbonyl compounds. When meat is cured using nitrite, the flavour of the resultant meat, through desirable, is not the same as the flavour of uncured cooked meat. When volatile constituents from nitrite-treated and untreated ham, beef or chicken, were passed through a solution of 2,4-dinitrophenylhydrazine, the effluent stream in all cases had a characteristic cured-ham aroma (Cross and Ziegler, 1965; Minor *et al.*, 1965). This observation by Cross and Ziegler indicated that 'cured-meat' flavour, irrespective of the meat source, was comprised of essentially the same constituents as were generated by a combination of several reactions in addition to the suppression of lipid oxidation.

While the flavour of freshly cooked meat is generally described as 'meaty', this flavour deteriorates upon storage for an extended time. Storage of cooked meat, or prolonged storage of raw meat prior to cooking, gives rise to 'old, stale, oxidized, rancid, or warmed-over' flavour, caused by the oxidation of unsaturated fatty acids in meat. Nitrite has been shown to retard lipid oxidation and development of warmed-over flavour (WOF) in cooked meat and meat products. Younathan and Watts (1959) studied lipid oxidation in cured and uncured pork and found the highest thiobarbituric acid (TBA) values in uncured samples regardless of storage period. Sato and Hegarty (1971) were able to eliminate WOF in cooked ground beef by adding nitrite at a level of 2000 mg/kg of beef, and inhibit it at a level of 50 mg/kg, as indicated by TBA values. The antioxidant effect of nitrite in the meat-curing process, using TBA values and sensory scores, was also demonstrated by other workers (Hadden *et al.*, 1975; Love and Pearson, 1976; MacDonald *et al.*, 1980). In a recent study, Yun *et al.* (1987) demonstrated that the flavour acceptability of cooked pork decreased as the TBA number or hexanal content increased. While the TBA values of the control (uncured cooked meat) stored at 4°C increased from 4.2 (0 week) to 10.8 (4 weeks), the values for meat cured with 150 ppm nitrite remained constant at 0.1 throughout the 4 weeks of storage. Sensory evaluation of the nitrite-cured and uncured meat samples after storage for 24 h at 4°C indicated differences in their acceptability scores, the nitrite-treated samples being preferred with a preset score of 5 to the untreated meat with an average score of 3.8.

Initial studies on the oxidation of unsaturated lipids in meat products implicated the haem proteins to be the major pro-oxidants (Tappel, 1952; Watts, 1954; Maier and Tappel, 1959; Younathan and Watts, 1959). In recent years, however, data have been presented to demonstrate that the non-haem iron released during cooking, from the bound haem pigments, accelerated lipid oxidation in cooked meat (Sato and Hegarty, 1971; Love and Pearson, 1976). The exact mechanism of action of nitrite as an antioxidant in the elimination of WOF in cooked cured meats is not yet thoroughly understood. It has, however, been suggested by Pearson *et al.* (1977) that nitrite may either inhibit the action of natural pro-oxidants in the muscle or stabilize the lipid components of the membranes.

10.4 Chemistry of cured-meat flavour

Although nitrite is closely associated with cured-meat flavour, the chemistry behind the formation and composition of this unique flavour is not clearly understood (Gray and Pearson, 1984). There have been only a few reports focussing attention mainly on the composition of cured-meat flavour (Bailey and Swain, 1973; MacDougall *et al.*, 1975). Some of the recent reviews on meat flavour chemistry have given special emphasis to the discussion of the nature of cured-meat flavour (Gray *et al.*, 1981; Rhee, 1989; Shahidi, 1989).

Nearly 1000 compounds have been so far identified in the volatiles of pork, beef, chicken and lamb. The general chemical composition of volatiles in uncured and cured pork, as surveyed by Shahidi *et al.* (1986), is listed in Table 10.1. While the total number of carbonyl compounds and hydrocarbons identified in uncured pork (118) is higher compared with that in cured pork (45), the number of sulphur compounds identified in each of them is the same, i.e. 31. Hornstein and Crowe (1960) were among the first to report that the fat, and more specifically the carbonyl compounds derived from fat oxidation, contributed to differences in flavour among species. In their study, initiated to develop satisfactory methods for the isolation, separation and identification of volatile compounds in

Table 10.1 Composition of volatile components found in uncured and cured pork

Class	Uncured pork	Cured pork
Aldehydes	35	29
Alcohols	24	9
Carboxylic acids	5	20
Esters	20	9
Ethers	6	–
Furans	29	5
Hydrocarbons	45	4
Ketones	38	12
Lactones	2	–
Oxazoles/oxazolines	4	–
Phenols	9	1
Pyrazines	36	1
Pyridines	5	1
Pyrroles	9	–
Thiazoles/thiazolines	5	–
Thiophenes	11	–
Other nitrogen compounds	6	3
Other sulphur compounds	20	31
Halogenated compounds	4	1
Miscellaneous	1	11
Total	314	137

Table 10.2 Volatile constituents identified in nitrite-treated ham

Aldehydes
Methanal (formaldehyde)[b,c]
Ethanal (acetaldehyde)[a,b,c,d]
Propanal[a,b,c]
β-Methylthiopropanal (methional)[d]
2-Methylpropanal[b,c]
Propenal (acrolein)[d]
Butanal[a,b,c]
2-Methylbutanal[a]
3-Methylbutanal[a,b,d]
Pentanal[a,b,c,d]
Hexanal[a,b,c]
2-Hexenal[c,d]
Heptanal[c,d]
2-Heptenal[c]
2,4-Heptadienal[c]
Octanal[c,d]
2-Octenal[c]
Nonanal[c,d]
2-Nonenal[c,d]
2,4-Nonadienal[c,d]
Decanal[c,d]
2-Decenal[c]
2,4-Decadienal[c,d]
Undecanal[c,d]
2-Undecenal[c]
2,4-Undecadienal[c]
Dodecanal[c,d]
2-Dodecenal[c,d]
2,4-Dodecadienal[c]

Carboxylic acids
Methanoic (formic)[b]
Ethanoic (acetic)[b,d]
Propanoic[b]
Butanoic[b,d]
4-Methylpentanoic[b]
Hexanoic[d]
Octanoic[c,d]
Decanoic[c]
Dodecanoic[c]
Tetradecanoic[c]
Hexadecanoic[c]
Hexadecenoic[c]
Octadecanoic[c]
Octadecenoic[c]
Octadecadienoic[c]
Octadecatrienoic[c]
Eicosanoic[c]
Eicosenoic[c]
Eicosadienoic[c]
Eicosatrienoic[c]

Alcohols
Methanol[c]
Ethanol[b]
Propanol[c]
3-Methylbutanol[d]
Hexanol[c,d]
4-Hexanol[d]
Heptanol[c]
Octanol[c,d]
2-Octanol[c]

Esters
Ethylmethanoate[c]
Methylethanoate[c]
Methylpropanoate[c]
Ethylbutanoate[c]
Methylhexanoate[c]
Ethylhexanoate[c]
Methyloctanoate[c]
Methyldecanoate[c]
Methyldodecanoate[c]

Furans
Methylfuran[d]
Pentylfuran[d]
Heptylfuran[d]
2-Acetylfuran[d]
Furyl alcohol[d]

Hydrocarbons
Heptane
Pentadecane (and isomers)[d]
1-Pentadecene[d]
Hexadecane (and isomers)[d]

Ketones
2-Propanone (acetone)[a,b,d]
2-Butanone[b,c]
2,3-Butanedione[b,c,d]
2-Pentanone[c]
2,3-Pentanedione[d]
2,6-Hexanedione[c]
2-Octanone[d]
2-Nonanone[d]
2-Undecanone[d]
2-Dodecanone[d]
2-Tridecanone[d]
Pentadecanone[d]

Phenols
Dimethylphenol[d]

Pyridines
2-Methylpyridine[d]

Pyrazines
2-Methylpyrazine[d]

Other nitrogen compounds
Ammonia[b]
Methylamine[b]
N-ethylpyrrolidone[d]

Sulphur compounds
Hydrogen sulphide[b]
Methylmercaptan[d,e]
Ethylmercaptan[e]
n-Propylmercaptan[e]
n-Butylmercaptan[e]
3-Methylbutylmercaptan[e]
4-Methyl-4-mercaptan-pentane-2-one[e]
Methyl ethylsulphide[e]
Ethylenesulphide[e]
Propylenesulphide[e]
Ethyl isobutylsulphide[e]
2,2′-bis(Ethylthio)-propane
Thiacyclohexane[d]
Ethylthioacetate[e]
n-Butylthioacetate[e]
n-Butylthiopropionate[e]
iso-Butylthiobutanoate[e]
Methylthioacetate[e]
Dimethyldisulphide[d,e]
Ethyl methyldisulphide[d,e]
Diethyldisulphide[e]
1,3-Thioxalane[e]
Dimethyltrisulphide[d,e]
Methyl ethyltrisulphide[e]
Diethyltrisulphide[e]
3,5-Dimethyl-1,2,4-trithiolane[d]
3,6-Dimethyl-1,2,4-trithiolane[d]
Thiophene[e]
2-Methylthiophene[e]
2-Formylthiophene[e]
2,4,6-Triethyl perhydro-1,3,5-dithiazine (thialdine)[d]

Halogenated compounds
Dichlorobenzene[d]

Miscellaneous compounds
Benzonitrile[f]
Phenylacetonitrile[f]
Nonanenitrile[f]
Decanenitrile[f]
Undecanenitrile[f]
Dodecanenitrile[f]
Tridecanenitrile[f]
Pentylnitrate[f]
Hexylnitrate[f]
Heptylnitrate[f]
Octylnitrate[f]

[a]Cross and Ziegler (1965); [b]Ockerman *et al.* (1964); [c]Lillard and Ayres (1969); [d]Bailey and Swain (1973); [e]Golovnya *et al.* (1982); [f]Mottram (1984).

dry-cured country-style hams, Ockerman *et al.* (1964) tentatively identified numerous aldehydes, ketones, acids, bases and sulphur compounds. Cross and Ziegler (1965) concluded from gas chromatographic examination that the composition of volatiles of cured and uncured ham was qualitatively quite similar, but there were distinct quantitative differences. Pentanal and hexanal were present in appreciable quantities in the uncured product, but were barely detectable in the volatiles of the cured meat. It was suggested that the absence of these aldehydes in the cured meat was responsible for the flavour differences between cured and uncured ham, and that it was brought about by the inhibition of the autoxidation of unsaturated lipids in the presence of nitrite. It was also demonstrated by Cross and Ziegler (1965) that when volatiles from uncured and cured ham, beef and chicken were passed through a solution of 2,4-dinitrophenylhydrazine, the effluents from all samples had aromas similar to cured ham, as mentioned previously. This observation led them to postulate that carbonyl components were not essential for cured meat flavour, and that the cured-ham flavour represented the basic meat flavour derived from precursors other than the triglycerides. They further suggested that the differences in the cooked meat flavour depended on the spectrum of carbonyls generated during oxidation of lipids, which in themselves differed in composition depending on species. This preliminary observation on volatiles from uncured and cured meat emphasizing the importance of carbonyls in causing species-specific differences did not receive the attention it deserved.

Attempts to identify the flavour compounds in country-cured hams, isolated by steam-distillation under reduced pressure, showed the presence of various carbonyls, alcohols and esters (Lillard and Ayres, 1969). Piotrowski *et al.* (1970) studied various fractions of ham to isolate and identify the cured-ham aroma and to follow changes occurring in pork during curing, cooking and smoking. It was found that components of precursors of cured and smoky aroma could be extracted from hams with a mixture of chloroform–methanol (2:1 v/v). Further examination of this extract did not show the presence of any single compound that had 'meaty' or 'cured-ham' aroma. A list of individual constituents identified so far in the volatiles of cured ham is given in Table 10.2. In contrast with the large number of papers published on the chemistry of uncured-meat flavour, the information available on the chemistry of cured-meat flavour has been contributed by a small group of researchers listed in Table 10.2. Most of the compounds given in this table have, however, been identified and reported previously in other cooked meats.

A large number of components in the volatiles of meat have been isolated and identified in the past two decades, and exhaustive review papers on meat flavour have been published (Herz and Chang, 1970; Bailey and Swain, 1973; Dwivedi, 1975; Chang and Peterson, 1977; Wasserman, 1979; Gray *et al.*, 1981; MacLeod and Seyyedain-Ardebili, 1981;

Ramaswamy and Richards, 1982; Moody, 1983; Baines and Mlotkiewicz, 1984; Shahidi *et al.*, 1986; Rhee, 1989). Although over 700 components in beef and half that number in pork, chicken and lamb have been characterized, the search for individual character-impact components possessing notes specific for beef, pork, chicken or lamb has remained largely unsuccessful. Numerous reports have indicated that carbonyl components make a significant contribution to the flavour of uncured meat (Hornstein and Crowe, 1960; Jacobsen and Koehler, 1963; Sanderson *et al.*, 1966; Langer *et al.*, 1970). However, no attempt has been made to prevent the formation of such carbonyl compounds by using antioxidants such as nitrite or other specific reagents, and to study the organoleptic properties of the resulting aroma mixture in depth. Thus, the nature of cured-meat flavour, which is made apparent by suppression of lipid oxidation by nitrite and which is the basic flavour of cooked meat (Rubin and Shahidi, 1988), remains a mystery so far.

In our attempts to unravel the complex nature of cured-meat flavour, we have taken a stepwise approach. We have kept the cooking conditions mild, so as to limit the formation of heterocyclic nitrogen and sulphur compounds, and also have used a rather modest-sized sample of 250–400 g of ground meat. As the first step, volatile components from uncured and cured pork isolated by two classical procedures, steam distillation followed by solvent extraction and continuous steam distillation–extraction (SDE), were compared (Ramarathnam *et al.*, 1991a). In this study we deliberately focussed our attention on characterization of carbonyls and hydrocarbons that usually make up the bulk of the meat-flavour spectrum. The aroma concentrates isolated by the two methods were analysed by gas chromatography, and the individual components were further identified and quantified using GC–MS with hexanal and decanal as internal standards.

Typical gas chromatograms of the uncured and cured pork aroma concentrates extracted by the conventional steam distillation method and SDE method using the modified Likens–Nickerson flavour-extraction apparatus are given in Figures 10.1 and 10.2. It was observed that the aroma concentrates from cured meat isolated by either technique had fewer constituents than the uncured-meat sample. It was also found that many of the volatile constituents present in the aroma concentrate of uncured pork in the retention time region of 15–40 min were either absent or present in much lower concentrations in the cured-meat aroma concentrate. This result is attributable to the suppression of lipid oxidation due to the presence of nitrite in cured meat. It was observed that the relative concentrations of the individual components in the aroma concentrates isolated by the SDE method were higher than those present in the aroma concentrates prepared by the conventional steam-distillation method. This could be due to the partial loss of volatiles during the

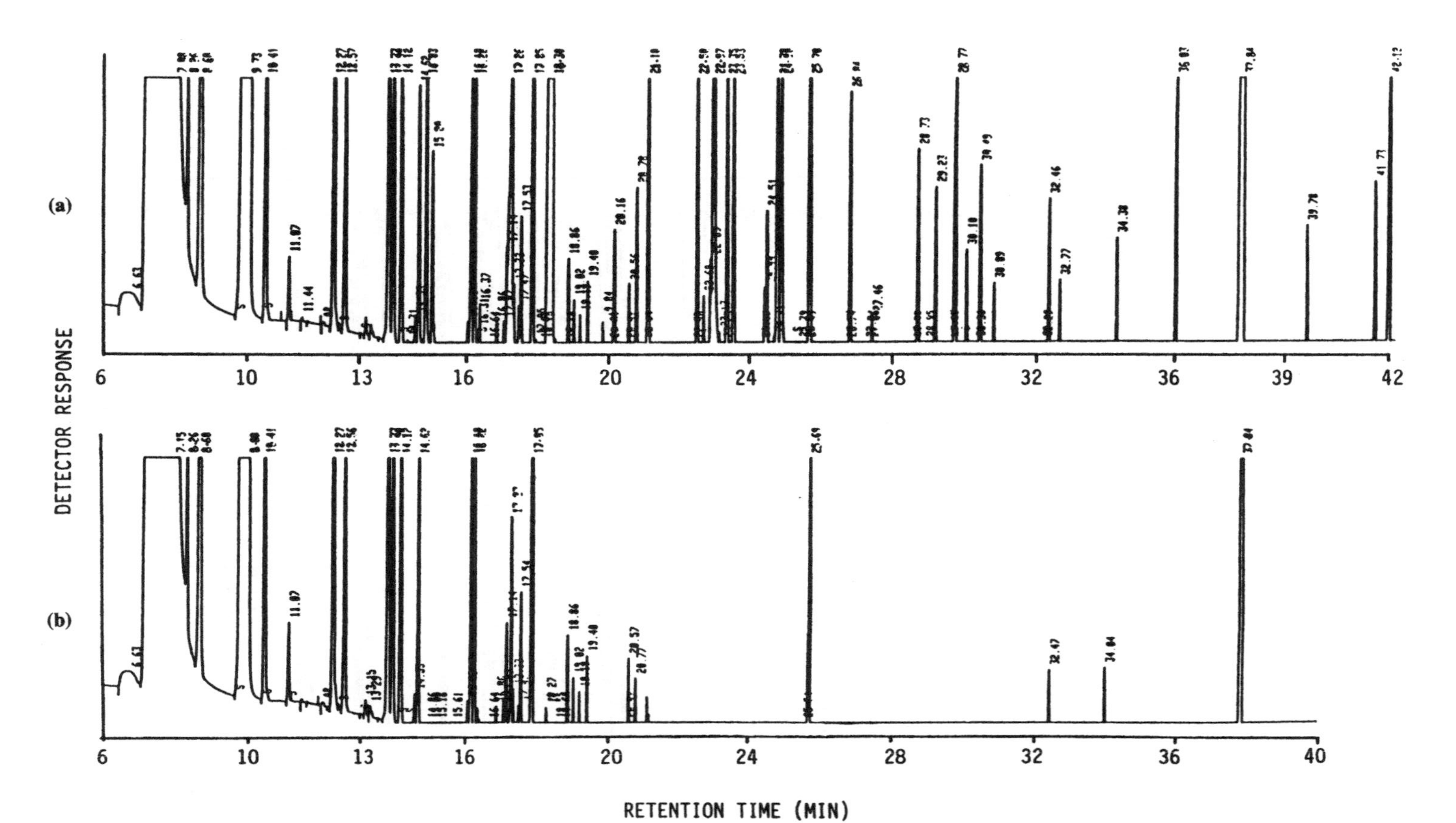

Figure 10.1 Typical gas chromatograms of (a) uncured-pork-flavour and (b) cured-pork-flavour concentrates isolated by steam distillation.

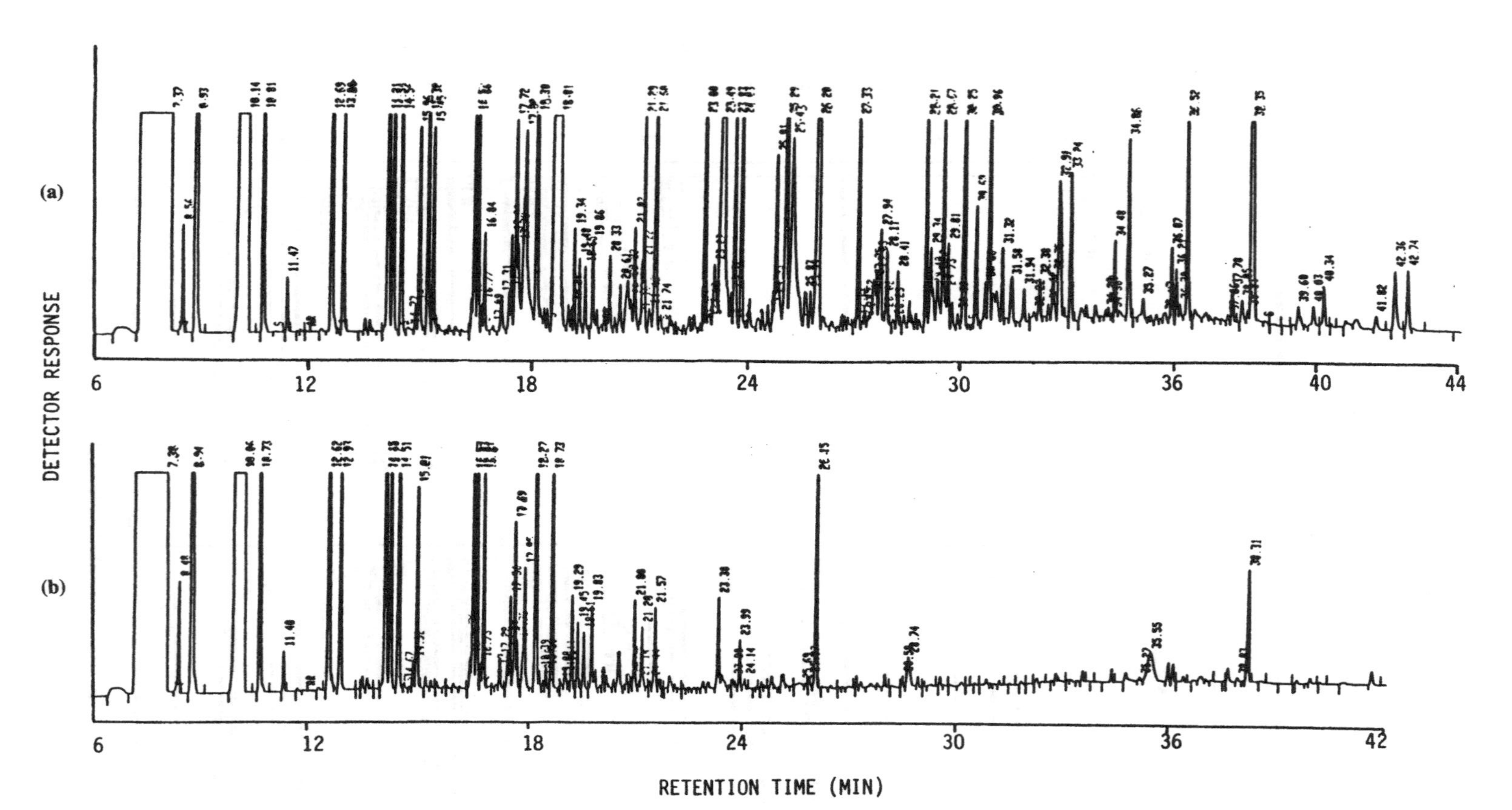

Figure 10.2 Typical gas chromatograms of (a) uncured-pork-flavour and (b) cured-pork-flavour concentrates isolated by the SDE method.

extraction and concentration steps of the latter method, which involves twice the volume of extraction solvent as the SDE method. Also, the SDE method, being a closed system, was more effective in preventing the loss of the volatiles.

The separated constituents in uncured and cured pork are reported in Table 10.3. In all 50 hydrocarbons, 37 carbonyls, 6 acids, and 2 alcohols were identified. Table 10.3 lists these components and also shows in which of the samples prepared by the two isolation methods the components were identified. It was observed that the aroma concentrates of uncured and cured pork isolated by the SDE method were resolved into 77 and 72 components, while those extracted by the conventional steam distillation contained only 59 and 51 components, respectively. This shows that the Likens–Nickerson flavour extraction apparatus, being a closed system, was more effective in extracting aroma components than the conventional steam-distillation method. Of the components identified, hexanal was found to be present in uncured meat at a concentration of 12.66 ± 0.08 mg/kg, while in the cured meat it was found to be present as a minor component to the extent of 0.030 ± 0.004 mg/kg, which amounts to only 0.24% of that present in uncured meat. Shahidi *et al.* (1987b) found the hexanal content of cured pork to be 2% of the value observed for uncured pork, which is quantitatively similar to the values obtained by us. The difference in magnitude is probably due to differences in the extraction technique and sample size.

As the next step, using the SDE method as the choice of extraction technique, we have prepared aroma concentrates from uncured and cured beef and chicken, and by use of GC–MS compared the qualitative and quantitative differences in them. We have compared the data for carbonyls in beef and chicken with those of pork we reported earlier (Ramarathnam *et al.*, 1991a), and have made an attempt to identify the components responsible for species differences (Ramarathnam *et al.*, 1991b). Analysis of the aroma concentrates isolated from uncured and cured beef and chicken using GC–MS showed that the aroma concentrates isolated from uncured and cured beef had 59 and 40 components, respectively (Figures 10.3a and b), while those of chicken were resolved into 48 and 36 components (Figures 10.4a and b). Of the separated constituents, 31 hydrocarbons, 26 carbonyls, three alcohols, and two acids were identified in uncured and cured beef (Table 10.4). The corresponding figures for chicken (Table 10.5) were 29 hydrocarbons, 26 carbonyls, and two alcohols. Our earlier work on pork volatiles showed that aroma concentrates from cooked uncured and cured pork resolved into 77 and 72 components, respectively (Ramarathnam *et al.*, 1991a). The differences in the total number of components and individual carbonyls identified among the three species could be attributed to the differences in their fat content, and also to a great extent to the differences in their fatty acid

Table 10.3 Components of the aroma concentrates of uncured and cured pork

Peak no.	RT (min)	Component	Detected in[a]				Content[b] (mg/kg)	
			A	B	A⁺	B⁺	A	B
1	2.47	2-Methylhexane	+	+	+	+	1.20 ± 0.18	1.01 ± 0.06
2	2.56	3-Methylhexane	+	+	+	+	0.69 ± 0.04	0.59 ± 0.01
3	2.70	2,2-Dimethylhexane	+	+	+	+	0.33 ± 0.04	0.28 ± 0.03
4	2.88	3-Hexanone	+	+	–	–	0.42 ± 0.06	tr
5	3.23	Unidentified	–	+	–	+	–	tr
6	3.41	2,4-Dimethylhexane	+	+	+	+	0.93 ± 0.08	0.68 ± 0.15
7	3.55	4-Methyl-2-pentanone	–	+	–	–	–	tr
8	3.68	2,3-Dimethylhexane	–	+	–	+	–	tr
9	3.76	3,3-Dimethylhexane	–	+	–	+	–	0.03 ± 0.01
10	3.90	4-Methylheptane	–	+	+	+	–	0.09 ± 0.01
11	4.02	2,5-Dimethylhexane	+	+	+	+	0.23 ± 0.09	0.17 ± 0.03
12	4.16	3-Methylheptane	+	+	+	+	0.21 ± 0.06	0.09 ± 0.01
13	4.35	2,2,5-Trimethylhexane	+	+	+	+	1.27 ± 0.09	1.08 ± 0.06
14	4.44	2,2,4-Trimethylhexane	–	+	–	+	–	0.09 ± 0.06
15	4.69	Hexanal	+	+	+	+	12.66 ± 0.08	0.03
16	4.76	Unidentified	–	–	–	+	–	tr
17	4.96	2,3,5-Trimethylhexane	+	+	+	+	0.12 ± 0.02	0.10 ± 0.02
18	5.10	2,4-Dimethylheptane	–	+	+	+	–	0.07 ± 0.01
19	5.26	2,6-Dimethylheptane	+	+	+	+	tr	0.07 ± 0.02
20	5.35	2,5-Dimethylheptane	+	+	+	+	0.17 ± 0.02	0.15 ± 0.03
21	5.63	1,2,4-Trimethylcyclohexane	–	+	–	+	–	0.03 ± 0.01
22	5.76	2-Hexenal	+	+	+	+	tr	tr
23	5.92	3-Methyl-4-heptanone	–	+	–	+	–	tr
24	6.02	1,3-Dimethylbenzene	–	+	–	+	–	tr
25	6.09	2,5-Dimethyloctane	–	+	–	+	–	0.04 ± 0.02
26	6.17	4-Ethyl-2,2-dimethylhexane	–	+	–	+	–	0.12 ± 0.02
27	6.20	2,2,3-Trimethylhexane	+	–	+	–	0.23 ± 0.11	–
28	6.30	2,2,4-Trimethylheptane	+	+	+	+	0.12 ± 0.01	0.10 ± 0.03
29	6.46	1,2-Dimethylbenzene	–	+	–	–	–	0.04 ± 0.03
30	6.50	2-Heptanone	+	–	+	–	0.20 ± 0.06	–
31	6.55	3,3,5-Trimethylheptane	–	+	+	+	–	0.05 ± 0.01
32	6.66	3-Methyl-2-nonene	–	+	–	+	–	0.04 ± 0.01
33	6.74	3-Methylhexanal	+	–	+	–	0.65 ± 0.14	–
34	6.89	3,5-Dimethyloctane	–	+	–	+	–	tr
35	7.74	2,4,6-Trimethyloctane	–	+	–	+	–	tr
36	7.81	2-Heptenal	+	–	+	–	0.34 ± 0.04	–
37	7.83	Benzaldehyde	+	+	+	+	0.11 ± 0.01	0.04 ± 0.05
38	8.12	3-Methyloctane	+	–	–	–	0.11 ± 0.01	–
39	8.18	Unidentified	+	–	–	–	1.77 ± 0.015	–
40	8.33	2,3-Octanedione	+	–	–	–	0.88 ± 0.09	–
41	8.36	1,3,5-Trimethylbenzene	–	+	–	–	–	tr
42	8.40	1-Nonen-3-ol	+	–	+	–	0.75 ± 0.05	–
43	8.46	3,6-Dimethyloctane	–	+	–	–	–	tr
44	8.56	Unidentified	–	+	–	+	–	0.04 ± 0.01
45	8.66	3-Ethoxy-2-methyl-1-propene	+	–	+	–	0.66 ± 0.01	–
46	8.85	2,3,4-Trimethyloctane	–	+	–	–	–	tr
47	9.00	D-Limonene	–	+	–	–	–	0.02
48	9.13	3-Ethyl-2-methyl-1,3-hexadiene	+	–	+	–	0.14 ± 0.01	–
49	9.40	4,4,5-Trimethyl-2-hexene	+	–	–	–	0.13 ± 0.02	–
50	9.47	5-Methylundecane	–	+	–	–	–	tr
51	9.57	5,5-Dimethyl-2-hexene	–	+	–	–	–	tr
52	9.65	(*E*)-2-octenal	+	–	+	–	0.99 ± 0.01	–
53	9.86	2-Octen-1-ol	+	–	+	–	0.69 ± 0.15	–
54	10.20	3,7-Dimethylnonane	+	+	–	–	tr	tr
55	10.40	Methylcyclohexane	+	+	+	+	2.10 ± 0.33	0.33 ± 0.08
56	11.30	2-Nonenal	+	–	+	–	0.39 ± 0.05	–
57	11.34	4-Ethylbenzaldehyde	+	–	+	–	tr	–
58	11.45	5-Undeca-3(*Z*),5-diyne	+	–	–	–	tr	–
59	11.60	5-Undeca-3(*E*),5-diyne	+	–	–	–	tr	–

Table 10.3 *continued*

Peak no.	RT (min)	Component	Detected in[a]				Content[b] (mg/kg)	
			A	**B**	**A⁺**	**B⁺**	**A**	**B**
60	11.68	Naphthalene	+	+	–	–	0.12 ± 0.03	0.04 ± 0.01
61	11.81	Dodecane	+	+	+	–	0.28 ± 0.06	tr
62	11.96	Decanal	+	+	+	–	tr	tr
63	12.13	2,4-Nonadienal	+	–	+	–	tr	–
64	12.40	Unidentified	+	–	–	–	tr	–
65	12.87	2-Undecenal	+	–	+	–	0.39 ± 0.07	–
66	13.13	Unidentified	+	–	–	–	tr	–
67	13.27	2-Undecanone	–	+	–	–	–	0.02
68	13.32	4,6-Dimethylundecane	–	+	–	–	–	tr
69	13.40	Tridecane	+	+	+	–	0.49 ± 0.04	tr
70	13.49	Undecanal	+	–	–	–	tr	–
71	13.70	(*E,E*)-2,4-decadienal	+	–	+	–	0.69 ± 0.16	–
72	13.75	(*E,Z*)-2,4-decadienal	+	–	+	–	0.41 ± 0.15	–
73	14.11	Unidentified	+	–	+	–	tr	–
74	14.22	5-Tridecanone	–	+	–	–	–	tr
75	14.35	2-Dodecenal	+	–	+	–	0.43 ± 0.08	–
76	14.76	Tetradecane	+	–	–	–	0.14 ± 0.04	–
77	14.92	Dodecanal	+	–	–	–	tr	–
78	15.10	2,4-Undecadienal	+	–	–	–	tr	–
79	15.70	4-Pentylbenzaldehyde	+	–	+	–	tr	–
80	15.87	1-Pentadecene	+	–	–	–	tr	–
81	16.07	Pentadecane	+	+	+	+	0.19 ± 0.05	0.08 ± 0.07
82	16.29	Tridecanal	+	+	+	+	0.25 ± 0.05	0.09 ± 0.01
83	16.97	Dodecanoic acid	–	+	–	–	–	tr
84	17.31	Unidentified	–	+	–	–	–	0.14 ± 0.01
85	17.36	Hexadecane	+	+	+	+	tr	0.04 ± 0.01
86	17.49	3-Tridecen-1-yne	+	–	–	–	tr	–
87	17.55	Tetradecanal	+	+	+	+	0.40 ± 0.14	0.03 ± 0.01
88	18.51	Heptadecane	+	+	–	–	0.13 ± 0.02	0.11 ± 0.03
89	18.55	2-Pentadecanone	+	+	–	+	tr	0.06 ± 0.02
90	18.75	Hexadecanal	+	+	+	+	0.65 ± 0.05	0.06 ± 0.02
91	19.37	Tridecanoic acid	–	–	+	+	–	tr
92	19.48	Unidentified	+	–	–	–	0.12 ± 0.02	–
93	19.64	1,14-Tetradecanediol	+	+	+	+	tr	0.05 ± 0.01
94	19.88	17-Octadecenal	+	–	+	+	tr	–
95	20.01	16-Octadecenal	+	+	+	+	8.34 ± 0.35	2.20 ± 1.26
96	20.56	Unidentified	+	–	–	–	0.17 ± 0.03	–
97	20.78	Pentadecanitrile	+	+	–	+	0.21 ± 0.05	0.12 ± 0.04
98	20.99	15-Octadecenal	+	+	+	+	0.70 ± 0.04	0.14 ± 0.04
99	21.52	Hexedecanoic acid	+	+	+	+	0.97 ± 0.07	0.14 ± 0.02
100	21.81	9-Octadecenal	+	+	+	+	0.81 ± 0.06	0.14 ± 0.02
101	21.86	5-Octadecenal	–	+	+	+	–	0.05 ± 0.01
102	22.05	Octacanal	+	+	+	+	1.19 ± 0.11	0.19 ± 0.09
103	23.17	9,12-Octadecadienoic acid	+	+	–	–	tr	0.13 ± 0.03
104	23.22	9-Octadecenoic acid	+	+	+	+	0.16 ± 0.05	tr
105	23.37	Octadecanoic acid	–	+	–	+	–	tr

[a]Qualitative information only.
+, detected; –, not detected.
A and **A⁺** are uncured meat flavour constituents isolated by SDE and steam distillation methods.
B and **B⁺** are cured meat flavour constituents isolated by SDE and steam distillation methods.
[b]Concentration of constituents in uncured (**A**) and cured meat (**B**), isolated by the SDE method.
Reported values are mean ±S.D., $n = 3$.
tr, trace amount (< 0.01 mg/kg).

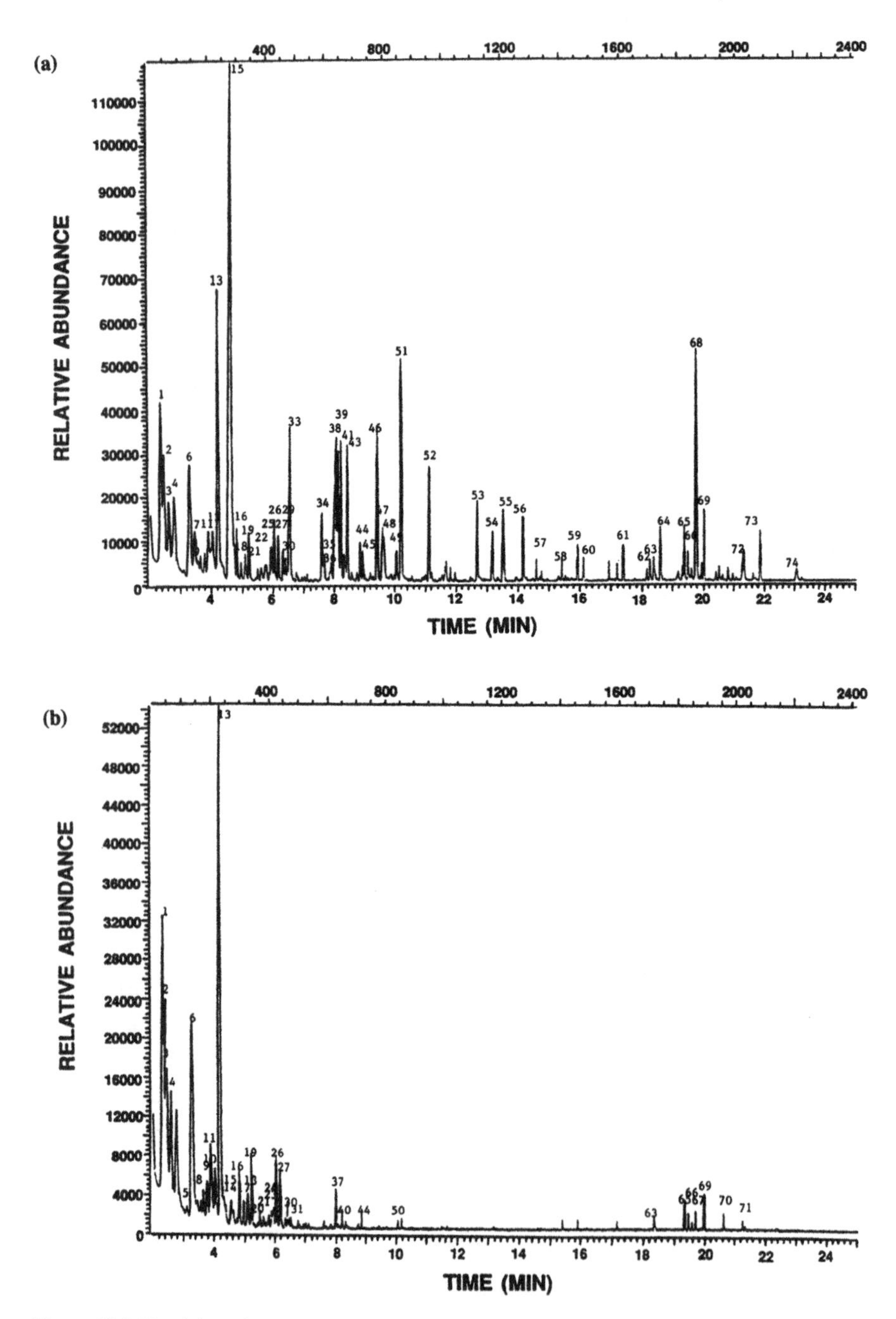

Figure 10.3 Total ion chromatograms (TIC) of (a) uncured-beef-flavour and (b) cured-beef-flavour concentrates isolated by the SDE method.

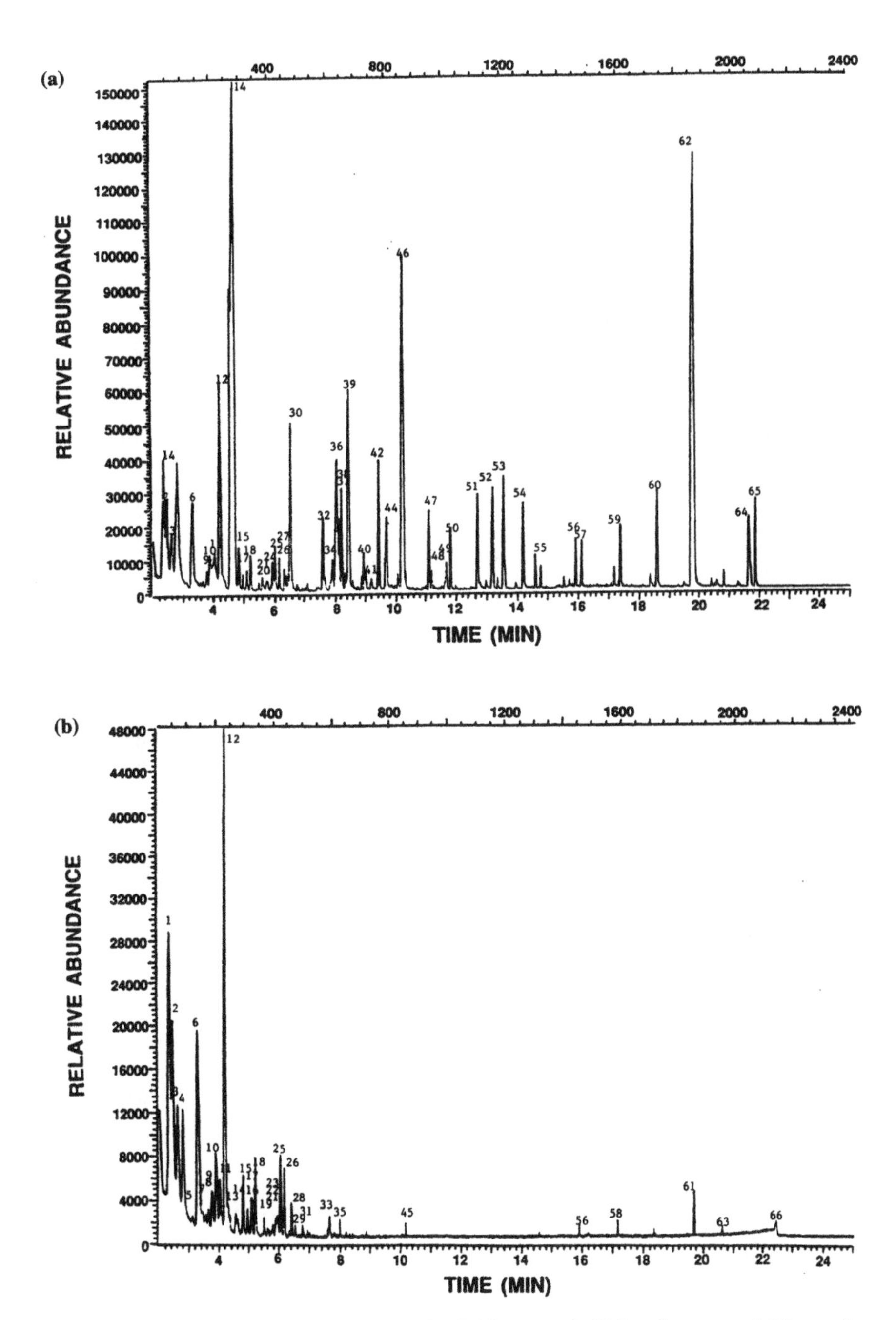

Figure 10.4 Total ion chromatograms (TIC) of (a) uncured-chicken-flavour and (b) cured-chicken-flavour concentrates isolated by the SDE method.

Table 10.4 Components of the aroma concentrates of uncured and cured beef, isolated by the SDE method

Peak no.	RT (min)	Component	Content (mg/kg)	
			Uncured	Cured
1	2.35	2-Methylhexane	1.82 ± 0.12	1.58 ± 0.09
2	2.45	3-Methylhexane	1.11 ± 0.08	0.94 ± 0.04
3	2.60	2,2-Dimethylhexane	0.71 ± 0.07	0.60 ± 0.03
4	2.79	3-Hexanone	1.08 ± 0.09	0.57 ± 0.02
5	3.12	Unidentified	–	0.04 ± 0.02
6	3.30	2,4-Dimethylhexane	1.46 ± 0.12	1.22 ± 0.11
7	3.47	2-Hexanone	0.38 ± 0.06	–
8	3.57	4-Methyl-2-pentanone	–	0.03 ± 0.01
9	3.65	3,3-Dimethylhexane	0.12 ± 0.04	0.10 ± 0.02
10	3.77	4-Methylheptane	0.13 ± 0.05	0.12 ± 0.02
11	3.89	2,5-Dimethylhexane	0.42 ± 0.09	0.04 ± 0.02
12	4.03	3-Methylheptane	0.36 ± 0.08	0.14 ± 0.04
13	4.21	2,2,5-Trimethylhexane	2.27 ± 0.12	1.91 ± 0.09
14	4.55	2,2,4-Trimethylhexane	–	0.12 ± 0.04
15	4.65	Hexanal	8.15 ± 0.17	0.05 ± 0.02
16	4.81	2,3,5-Trimethylhexane	0.27 ± 0.08	0.19 ± 0.08
17	4.96	2,3,4-Trimethylhexane	0.11 ± 0.05	0.07 ± 0.02
18	5.09	2,6-Dimethylheptane	0.20 ± 0.11	0.12 ± 0.05
19	5.21	2,5-Dimethylheptane	0.37 ± 0.15	0.24 ± 0.09
20	5.50	1,2,4-Trimethylcyclohexane	–	0.05 ± 0.02
21	5.61	3,3-Dimethylhexanal	0.11 ± 0.04	0.03 ± 0.02
22	5.75	Unidentified	0.10 ± 0.01	–
23	5.80	3-Methyl-4-heptanone	–	0.04 ± 0.01
24	5.87	1,3-Dimethylbenzene	–	0.04 ± 0.02
25	5.95	2,5-Dimethyloctane	0.30 ± 0.05	0.10 ± 0.04
26	6.04	4-Ethyl-2,2-dimethylhexane	0.33 ± 0.11	0.19 ± 0.07
27	6.17	2,2,4-Trimethylhexane	0.28 ± 0.12	0.15 ± 0.07
28	6.33	1,2-Dimethylbenzene	–	0.03 ± 0.01
29	6.34	2-Heptanone	0.21 ± 0.07	–
30	6.41	3,3,5-Trimethylheptane	0.10 ± 0.02	0.07 ± 0.02
31	6.47	Unidentified	–	0.03 ± 0.01
32	6.53	Unidentified	–	0.04 ± 0.01
33	6.56	3-Methylhexanal	1.09 ± 0.07	–
34	7.60	(*E*)-2-heptenal	0.45 ± 0.04	–
35	7.67	Benzaldehyde	0.20 ± 0.02	–
36	7.90	3-Methyloctane	0.14 ± 0.05	–
37	8.00	1,3,5-Trimethylbenzene	–	0.11 ± 0.02
38	8.10	1-Hepten-3-ol	1.73 ± 0.15	–
39	8.16	2,3-Octanedione	0.65 ± 0.11	–
40	8.21	3,6-Dimethyloctane	–	0.04 ± 0.02
41	8.24	Unidentified	0.75 ± 0.06	–
42	8.35	3-Ethoxy-2-methyl-1-propene	0.11 ± 0.04	–
43	8.46	Octanal	0.69 ± 0.07	–
44	8.87	D-Limonene	0.18 ± 0.05	0.04 ± 0.02
45	8.95	3-Ethyl-2-methyl-1,3-hexadiene	0.15 ± 0.11	–
46	9.45	(*E*)-2-octenal	1.07 ± 0.15	–
47	9.62	2-Octen-1-ol	0.38 ± 0.09	–
48	9.66	Unidentified	0.23 ± 0.05	–
49	10.07	3,7-Dimethylnonane	0.17 ± 0.04	–
50	10.18	Unidentified	–	0.03 ± 0.01
51	10.23	Nonanal	1.44 ± 0.07	–

Table 10.4 *continued*

Peak no.	RT (min)	Component	Content (mg/kg)	
			Uncured	Cured
52	11.13	2-Nonenal	0.68 ± 0.11	–
53	12.71	2-Undecenal	0.44 ± 0.05	–
54	13.19	Tridecane	0.31 ± 0.03	–
55	13.53	(*E,E*)-2,4-decadienal	0.42 ± 0.08	–
56	14.18	2-Dodecenal	0.35 ± 0.06	–
57	14.60	Tetradecane	0.10 ± 0.02	–
58	15.43	Unidentified	0.10 ± 0.05	–
59	15.92	Pentadecane	0.17 ± 0.08	–
60	16.12	Tridecanal	0.12 ± 0.04	–
61	17.38	Tetradecanal	0.17 ± 0.03	–
62	18.25	Unidentified	0.12 ± 0.02	–
63	18.37	2-Pentadecanone	0.19 ± 0.08	0.05 ± 0.02
64	18.60	Hexadecanal	0.28 ± 0.07	–
65	19.38	1,14-Tetradecanediol	0.27 ± 0.09	0.07 ± 0.01
66	19.50	Octadecane	0.15 ± 0.09	0.05 ± 0.02
67	19.73	17-Octadecenal	–	0.04 ± 0.01
68	19.79	16-Octadecenal	1.81 ± 0.14	–
69	20.02	Unidentified	0.37 ± 0.15	0.09 ± 0.02
70	20.65	Pentadecanitrile	–	0.04 ± 0.01
71	21.27	15-Octadecenal	–	0.03 ± 0.02
72	21.35	Hexadecanoic acid	0.32 ± 0.04	–
73	21.87	Octadecanal	0.26 ± 0.11	–
74	23.06	9-Octadecenoic acid	0.15 ± 0.03	–

–, not detected.
Reported values are mean ±S.D., $n = 3$.

compositions. Pork used in our investigation had a fat content of 10.4 ± 0.1% (Ramarathnam *et al.*, 1991a), while beef and chicken had fat contents of 6.5 ± 0.2% and 2.4 ± 0.3%, respectively. It is also well known that the composition of polyunsaturated fatty acids (PUFA) differs widely among the three species (Fogerty *et al.*, 1990).

The separated constituents in beef and chicken volatiles are reported in Tables 10.4 and 10.5 respectively. Among the carbonyl components identified in beef and chicken, a distinct difference was observed in the hexanal content of uncured and cured meat. Hexanal content in uncured beef was 8.15 ± 0.17 mg/kg (peak 15, Table 10.4), whereas the content of this lipid oxidation product in uncured chicken was 9.84 ± 0.17 mg/kg (peak 14, Table 10.5). The corresponding values for cured beef and chicken were 0.05 ± 0.02 mg/kg and 0.11 ± 0.04 mg/kg, respectively. A comparison of differences in the content of carbonyl components in the three species—beef, chicken and pork—is summarized in Table 10.6. The concentration of 3-hexanone in uncured and cured beef was 1.08 ± 0.09 mg/kg and 0.57 ± 0.02 mg/kg, respectively, while the corresponding levels in cooked uncured and cured chicken were 5.78 ± 0.13 mg/kg and 1.45 ± 0.07 mg/

Table 10.5 Components of the aroma concentrates of uncured and cured chicken, isolated by the SDE method

Peak no.	RT (min)	Component	Content (mg/kg)	
			Uncured	Cured
1	2.36	2-Methylhexane	4.27 ± 0.28	3.03 ± 0.04
2	2.46	3-Methylhexane	2.55 ± 0.06	1.80 ± 0.11
3	2.61	2,2-Dimethylhexane	1.57 ± 0.08	1.17 ± 0.14
4	2.80	3-Hexanone	5.78 ± 0.13	1.45 ± 0.07
5	3.13	Unidentified	–	0.09 ± 0.02
6	3.30	2,4-Dimethylhexane	3.76 ± 0.12	2.78 ± 0.09
7	3.58	4-Methyl-2-pentanone	–	0.06 ± 0.02
8	3.66	3,3-Dimethylhexane	–	0.11 ± 0.05
9	3.78	4-Methylheptane	0.38 ± 0.11	0.29 ± 0.04
10	3.89	2,5-Dimethylhexane	0.94 ± 0.08	0.70 ± 0.06
11	4.04	3-Methylheptane	0.92 ± 0.12	0.48 ± 0.08
12	4.22	2,2,5-Trimethylhexane	5.81 ± 0.16	4.37 ± 0.09
13	4.55	2,2,4-Trimethylhexane	–	0.20 ± 0.02
14	4.60	Hexanal	9.84 ± 0.17	0.11 ± 0.04
15	4.82	2,3,5-Trimethylhexane	0.67 ± 0.05	0.42 ± 0.04
16	4.95	2,3,4-Trimethylhexane	–	0.17 ± 0.05
17	5.10	2,6-Dimethylheptane	0.46 ± 0.07	0.29 ± 0.03
18	5.22	2,5-Dimethylheptane	0.78 ± 0.08	0.63 ± 0.06
19	5.50	1,2,4-Trimethylcyclohexane	–	0.11 ± 0.02
20	5.62	2-Hexenal	0.40 ± 0.04	–
21	5.79	3-Methyl-4-heptanone	0.44 ± 0.05	0.09 ± 0.04
22	5.88	1,3-Dimethylbenzene	–	0.11 ± 0.03
23	5.92	Unidentified	–	0.11 ± 0.02
24	5.96	2,5-Dimethyloctane	0.86 ± 0.07	0.13 ± 0.05
25	6.04	2,2,3-Trimethylhexane	0.83 ± 0.06	0.51 ± 0.07
26	6.17	2,2,4-Trimethylheptane	0.65 ± 0.05	0.40 ± 0.03
27	6.34	2-Heptanone	0.54 ± 0.05	–
28	6.41	3,3,5-Trimethylheptane	–	0.20 ± 0.02
29	6.53	Unidentified	–	0.09 ± 0.02
30	6.58	3-Methylhexanal	4.22 ± 0.04	–
31	6.76	3,5-Dimethyloctane	–	0.11 ± 0.05
32	7.60	(*E*)-2-heptenal	1.78 ± 0.05	–
33	7.66	Benzaldehyde	–	0.12 ± 0.04
34	7.93	3-Methyloctane	0.96 ± 0.09	–
35	8.00	Unidentified	–	0.11 ± 0.04
36	8.10	1-Hepten-3-ol	4.91 ± 0.05	–
37	8.15	2,3-Octanedione	1.23 ± 0.01	–
38	8.23	Unidentified	1.98 ± 0.12	–
39	8.49	Octanal	5.08 ± 0.14	–
40	8.96	3-Ethyl-2-methyl-1,3-hexadiene	0.64 ± 0.06	–
41	9.20	4,4,5-Trimethyl-2-hexene	0.37 ± 0.03	–
42	9.46	(*E*)-2-octenal	3.07 ± 0.11	–
43	9.65	2-Octen-1-ol	0.69 ± 0.07	–
44	9.72	Unidentified	1.60 ± 0.09	–
45	10.18	Unidentified	–	0.08 ± 0.02
46	10.28	Nonanal	11.59 ± 0.12	–
47	11.13	2-Nonenal	1.46 ± 0.09	–
48	11.21	4-Ethylbenzaldehyde	0.36 ± 0.04	–
49	11.69	Dodecane	0.45 ± 0.05	–
50	11.83	Decanal	1.05 ± 0.06	–
51	12.72	2-Undecenal	1.94 ± 0.09	–

Table 10.5 *continued*

Peak no.	RT (min)	Component	Content (mg/kg)	
			Uncured	Cured
52	13.20	Tridecane	2.21 ± 0.11	–
53	13.56	(*E,E*)-2,4-decadienal	2.48 ± 0.15	–
54	14.20	2-Dodecenal	1.90 ± 0.09	–
55	14.60	Tetradecane	0.52 ± 0.07	–
56	15.92	Pentadecane	0.82 ± 0.09	0.07 ± 0.01
57	16.12	Tridecanal	0.85 ± 0.08	–
58	17.17	Hexadecane	–	0.09 ± 0.03
59	17.40	Tetradecanal	1.14 ± 0.08	–
60	18.61	Hexadecanal	2.13 ± 0.05	–
61	19.73	17-Octadecenal	–	0.23 ± 0.03
62	19.89	16-Octadecenal	5.86 ± 0.18	–
63	20.65	Pentadecanitrile	–	0.06 ± 0.04
64	21.67	9-Octadecenal	1.85 ± 0.09	–
65	21.90	Octadecanal	1.88 ± 0.04	–
66	22.46	Unidentified	–	0.18 ± 0.04

–, not detected.
Reported values are mean ±S.D., $n = 3$.

kg, respectively. This component was found in uncured pork at a level of 0.42 ± 0.06 mg/kg, while in cured pork it was present only in small traces. The absence of 4-methyl-2-pentanone in the uncured meat of all the three species and its presence in small amounts, 0.03 ± 0.01 mg/kg in beef and 0.06 ± 0.02 mg/kg in chicken, and traces in pork, indicate that this component may be one of the typical constituents of the 'cured-meat' volatiles. Whether it is a constituent of the spectrum of volatiles which forms the 'cured-meat' flavour remains to be established experimentally. The fact that 4-methyl-2-pentanone is absent in uncured meat and present in small amounts in the cured product indicates that this compound is not derived from lipid oxidation. It may be formed as a result of a Maillard reaction.

3,3-Dimethylhexanal was present in uncured and cured beef at a concentration of 0.01 ± 0.04 mg/kg and 0.03 ± 0.02 mg/kg, respectively, and was not detected in chicken or pork. 3-Methyl-4-heptanone was present in chicken at a fairly high level of 0.44 ± 0.05 mg/kg, while it was absent in both pork and beef. It may be an important component of uncured cooked-chicken flavour. Octanal was not detected in cured beef, cured chicken, and in cured or uncured pork, while in uncured chicken it was present as a major component (5.08 ± 0.14 mg/kg), and in uncured beef it was detected to the extent of 0.69 ± 0.07 mg/kg.

Striking differences were also observed between the contents of nonanal and 16-octadecenal. Nonanal was absent in uncured pork, but was present in uncured chicken at the very high concentration of 11.59 ± 0.12 mg/kg, and in uncured beef the level of this compound was only 1.44 ± 0.07 mg/

Table 10.6 Carbonyls in the aroma concentrates of uncured and cured beef, chicken and pork, isolated by the SDE method

Component	Content (mg/kg)					
	Beef		Chicken		Pork	
	Uncured	Cured	Uncured	Cured	Uncured	Cured
3-Hexanone	1.08 ± 0.09	0.57 ± 0.02	5.78 ± 0.13	1.45 ± 0.07	0.42 ± 0.06	tr
2-Hexanone	0.38 ± 0.06	–	–	–	–	–
4-Methyl-2-pentanone	–	0.03 ± 0.01	–	0.06 ± 0.02	–	tr
Hexanal	8.15 ± 0.17	0.05 ± 0.02	9.84 ± 0.17	0.11 ± 0.04	12.66 ± 0.08	0.03 ± 0.05
2-Hexenal	–	–	0.40 ± 0.04	–	tr	tr
3,3-Dimethylhexanal	0.11 ± 0.04	0.03 ± 0.02	–	–	–	–
3-Methyl-4-heptanone	–	0.04 ± 0.01	0.44 ± 0.05	0.09 ± 0.04	–	tr
2-Heptanone	0.21 ± 0.07	–	0.54 ± 0.05	–	0.20 ± 0.06	–
3-Methylhexanal	1.09 ± 0.07	–	4.22 ± 0.04	–	0.65 ± 0.14	–
(*E*)-2-heptenal	0.45 ± 0.04	–	1.78 ± 0.05	–	0.34 ± 0.04	–
Benzaldehyde	–	–	–	–	0.11 ± 0.01	0.04 ± 0.05
2,3-Octanedione	–	–	–	–	0.88 ± 0.09	–
Octanal	0.69 ± 0.07	–	5.08 ± 0.14	–	–	–
(*E*)-2-octenal	1.07 ± 0.15	–	3.07 ± 0.11	–	0.99 ± 0.10	–
Nonanal	1.44 ± 0.07	–	11.59 ± 0.12	–	–	–
2-Nonenal	0.68 ± 0.11	–	1.46 ± 0.09	–	0.39 ± 0.05	–
4-Ethylbenzaldehyde	–	–	0.36 ± 0.04	–	tr	–
Decanal	–	–	1.05 ± 0.06	–	tr	tr

2-Undecenal	0.44 ± 0.05	–	1.94 ± 0.09	–	0.39 ± 0.07	–
(*E,E*)-2,4-decadienal	0.42 ± 0.08	–	2.48 ± 0.15	–	0.69 ± 0.16	–
(*E,Z*)-2,4-decadienal	–	–	–	–	0.41 ± 0.15	–
2-Dodecenal	0.35 ± 0.06	–	1.90 ± 0.09	–	0.43 ± 0.08	–
Tridecanal	0.12 ± 0.04	–	0.85 ± 0.08	–	0.25 ± 0.05	0.09 ± 0.01
Tetradecanal	0.17 ± 0.03	–	1.14 ± 0.08	–	0.40 ± 0.14	0.03 ± 0.01
2-Pentadecanone	0.19 ± 0.08	0.05 ± 0.02	–	–	tr	0.06 ± 0.02
Hexadecanal	0.28 ± 0.07	–	2.13 ± 0.05	–	0.65 ± 0.05	0.06 ± 0.02
17-Octadecenal	–	0.04 ± 0.01	–	0.23 ± 0.03	tr	–
16-Octadecenal	1.81 ± 0.14	–	5.86 ± 0.18	–	8.34 ± 0.35	2.20 ± 1.26
15-Octadecenal	–	0.03 ± 0.02	–	–	0.70 ± 0.04	0.14 ± 0.04
9-Octadecenal	–	–	1.85 ± 0.09	–	0.81 ± 0.06	0.14 ± 0.02
Octadecanal	0.26 ± 0.11	–	1.88 ± 0.04	–	1.19 ± 0.11	0.19 ± 0.09

–, not detected.
tr, trace amount (< 0.01 mg/kg).
Reported values are mean ±S.D., $n = 3$.

kg. 16-Octadecenal was present in uncured chicken at a concentration of 5.86 ± 0.18 mg/kg. This component was present in uncured beef at a concentration of only 1.81 ± 0.14 mg/kg, while in uncured pork the concentration of 16-octadecenal was found to be 8.34 ± 0.35 mg/kg. Nonanal was absent in the cured meat of all the three species, while 16-octadecenal was detected in cured pork at a concentration of 2.20 ± 1.26 mg/kg.

Thus, the presence and absence of certain carbonyls, or the differences in their concentration in the volatiles among the three species, can be a major contributory factor to the differences in the aroma nuances observed in them. Carbonyl compounds, which are formed due to the oxidation of unsaturated lipids and during the non-enzymatic amino-carbonyl reactions, have been implicated as significant contributors to the flavour of uncured meat, but not of cured meat. Since the aroma concentrates of cured pork, beef and chicken are similar and the concentration of the individual carbonyls, with the exception of 3-hexanone and 16-octadecenal, is less than 1 mg/kg, it is therefore evident that the 'cured-meat flavour' or the 'basic flavour of cooked meat', which is devoid of any lipid-oxidation product, should originate from non-triglyceride precursors. Removal of carbonyls by the use of carbonyl-specific reagents should result essentially in a simplified basic-meat flavour mixture (Cross and Ziegler, 1965; Minor *et al.*, 1965). Although the nature of such a mixture seems to be much simpler than that of uncured meat, the elucidation of the compounds which are responsible for the cured-meat flavour is not easy. This simplified mixture still has a second major group of volatiles, the hydrocarbons, that make practically no contribution to the 'meaty note' detectable in cured meat. Minute traces of aroma-effective heterocyclic components having very low flavour–threshold values can present enormous difficulties in their isolation and identification (MacLeod and Ames, 1986). The TIC profiles have clearly shown that the flavour spectrum of cured meat is indeed simple (Figures 10.3b and 10.4b). The components identified however, do not show the presence of sulphur and nitrogenous substances. Suitable modifications to the existing isolation and analytical techniques should be helpful in overcoming this problem. Preliminary experiments on the isolation of volatiles from the three meat species using the purge-and-trap technique, which is milder than the SDE method, have been successful in identifying certain heterocyclic compounds (data not shown). It is also believed that the heterocyclic compounds could be preferentially extracted from cooked meat by using supercritical carbon dioxide at relatively low temperatures and in a completely inert atmosphere. Work is currently being planned in this direction.

Among the hydrocarbons identified, cured and uncured chicken had the highest concentration of low-boiling homologues of branched hexane, heptane, and octane (Table 10.5), while the levels of such components in cured and uncured beef (Table 10.4) were only slightly higher than those of pork (Table 10.3). Hydrocarbons of specific interest, that were absent in the uncured meat of all the three species but present in the cured meat, are 2,2,4-trimethylhexane which was present in cured beef to the extent of 0.12 ± 0.04 mg/kg (peak 14, Table 10.4), 0.20 ± 0.02 mg/kg in cured chicken (peak 13, Table 10.5), and 0.09 ± 0.06 mg/kg in cured pork (peak 14, Table 10.3); 1,2,4-trimethylcyclohexane detected in cured beef to the extent of 0.05 ± 0.02 mg/kg (peak 20, Table 10.4), 0.11 ± 0.02 mg/kg in cured chicken (peak 19, Table 10.5), and 0.03 ± 0.01 mg/kg in cured pork (peak 21, Table 10.3); 1,3-dimethylbenzene, present in cured beef to the extent of 0.04 ± 0.02 mg/kg (peak 24, Table 10.4), 0.11 ± 0.03 mg/kg in cured chicken (peak 22, Table 10.5), and in small traces in cured pork (peak 24, Table 10.3). D-Limonene (peak 44, Table 10.4) was detected both in uncured beef (0.18 ± 0.05 mg/kg) and cured beef (0.04 ± 0.02 mg/kg), and was absent in chicken. This compound was also absent in uncured pork, while in cured pork it was present to the extent of 0.02 mg/kg (peak 47, Table 10.3). Hydrocarbons are formed due to the breakdown of unsaturated fatty acids during autoxidation of lipids. The differences in the concentration of most of the hydrocarbons detected in pork, chicken and beef can be attributed to the differences in the contents of total fat and unsaturated fatty acids.

In the first phase of this major study we have deliberately concentrated on the carbonyl spectrum of both uncured and cured meat from the three important species on this continent—beef, pork and chicken. The quantitative information on carbonyls and hydrocarbons which we provide here using the SDE technique has not been hitherto reported. We have kept the cooking conditions mild, which in itself would limit the formation of heterocyclic sulphur and nitrogen compounds, and the sample rather modest in size. This simplified system threw the emphasis on the carbonyls and much new data are presented here. We now plan to proceed to the isolation of heterocyclic meat-flavour components using the purge-and-trap and supercritical fluid extraction techniques. Cooking conditions will be kept mild, e.g. heating in water as opposed to roasting, to keep the system in the first instance as simple as possible.

10.5 Conclusion

In the study of meat flavour volatiles, much attention has been focussed on the characterization of the key components responsible for the flavour of meat from different species. Though higher in fat content, the number of volatiles detected in pork is far fewer than in beef, in which more than 700 components have been detected in the past two decades (Shahidi *et al.*, 1986). This could be due to the extensive investigations carried out on beef mainly because of its commercial importance and consumer preference (Baines and Mlotkiewicz, 1984). Nevertheless, the current literature available on meat flavour does not provide a clear path to the formulation of essences that could impart 'meaty' or 'cured-meat' type flavour notes.

Our approach to this problem, at this stage, is a fundamental one. The first step, that of providing quantitative information for carbonyls and hydrocarbons present in the three main species of meat consumed in most parts of the world, has now been completed. Of the various components identified in the present investigation in the three meat species, 4-methyl-2-pentanone, 2,2,4-trimethylhexane, 1,2,4-trimethylcyclohexane, and 1,3-dimethylbenzene could be contributing either directly as individual constituents or indirectly as synergists in the formation of the 'cured-meat' aroma. Though these components were detected in small amounts in the cured-meat flavour concentrates of all three species, they were however, absent in the cooked uncured meat. 16-Octadecenal, benzaldehyde, 2,3-octanedione, and (*E,Z*)-2,4-decadienal may be responsible for the species-specific flavour notes in pork, while 2-hexanone and 3,3-dimethylhexanal have been uniquely identified in beef. The characteristic 'chicken-like' flavour perhaps includes a complex mixture of 3-hexanone, 2-hexenal, 3-methyl-3-heptanone, 3-methylhexanal, (*E*)-2-heptenal, octanal, (*E*)-2-octenal, nonanal, 16-octadecenal, 4-ethyl benzaldehyde and decanal. The preparation of a 'nature-identical' chicken flavour would involve a sophisticated methodology, both in the sensory evaluation and in the formulation.

In our future work formation of such carbonyls, and perhaps also of hydrocarbons, in cured-meat aroma concentrates will be avoided. The minor components such as those belonging to the heterocyclic family, which may be important in the spectrum of volatiles producing the cured-meat flavour, will be isolated in greater yields so that they can be detected and quantified by the instrumentation currently available. Work is in progress in this direction and the results will be published in due course.

References

Bailey, M.E. and Swain, J.W. (1973). Influence of nitrite on meat flavour. In *Proceedings of the Meat Industry Research Conference*, American Meat Science Association, Chicago, pp. 29–45.

Baines, D.A. and Mlotkiewicz, J.A. (1984). The chemistry of meat flavour. In *Recent Advances in the Chemistry of Meat*, ed. A.J. Bailey. Royal Soc. Chem. London, pp. 119–164.

Bender, A.E. and Ballance, P.E. (1961). A preliminary examination of the flavour of meat extract. *J. Sci. Food Agric.*, **12**, 683–687.

Binkerd, E.F. and Kolari, O.E. (1975). The history and use of nitrate and nitrite in the curing of meat. *Food Cosmet. Toxicol.*, **13**, 655–661.

Chang, S.S. and Peterson, R.J. (1977). Symposium: the basis of quality in muscle foods. Recent developments in the flavour of meat. *J. Food Sci.*, **42**, 298–305.

Crocker, E.C. (1948). The flavour of meat. *Food Res.*, **13**, 179–183.

Cross, C.K. and Ziegler, P. (1965). A comparison of the volatile fractions from cured and uncured meats. *J. Food Sci.*, **30**, 610–614.

Dwivedi, B.K. (1975). Meat flavour. *CRC Crit. Rev. Food Technol.*, **5**, 487–535.

Eakes, B.D., Blumer, T.N. and Monroe, R.J. (1975). Effect of nitrate and nitrite on colour and flavour of country-style hams. *J. Food Sci.*, **40**, 973–976.

Fogerty, A.C., Whitfield, F.B., Svoronos, D. and Ford, G.L. (1990). Changes in the composition of the fatty acids and aldehydes of meat lipids after heating. *Int. J. Food Sci. Technol.*, **25**, 304–312.

Fooladi, M.H., Pearson, A.M., Coleman, T.H. and Merkel, R.A. (1979). The role of nitrite in preventing development of warmed-over flavour. *Food Chem.*, **4**, 283–292.

Giddings, C.G. (1977). The basis of colour in muscle foods. *CRC Crit. Rev. Food Sci. Nutr.*, **9**, 81–114.

Golovnya, R.V., Garbuzov, V.G., Grigor'eva, I. Ya, Zharich, S.L. and Bol'shakov, A.S. (1982). Gas chromatographic analysis of volatile sulphur-containing compounds of salted and salted-boiled pork. *Nahrung*, **26**, 89–96.

Gray, J.I. and Pearson, A.M. (1984). Cured meat flavour. In *Advances in Food Research*, ed. C.O. Chichester, E.M. Mrak and B.S. Scheweigart. Vol. 29. Academic Press, New York, pp. 1–86.

Hadden, J.P., Ockerman, H.W., Cahill, V.R., Parrett, N.A. and Borton, R.J. (1975). Influence of sodium nitrite on the chemical and organoleptic properties of comminuted pork. *J. Food Sci.*, **40**, 626–630.

Hauschild, A.H.W., Hilsheimer, R., Jarvis, G. and Raymond, D.P. (1982). Contribution of nitrite to the control of *Clostridium botulinum* in liver sausage. *J. Food Prot.*, **45**, 500–506.

Herz, K.O. and Chang, S.S. (1970). Meat flavor. *Adv. Food Res.*, **18**, 1–83.

Hornstein, I. and Crowe, P.F. (1960). Flavour studies on beef and pork. *J. Agric. Food Chem.*, **8**, 494–498.

Jacobsen, M. and Koehler, H.E. (1963). Components of the flavour of lamb. *J. Agric. Food Chem.*, **11**, 336–339.

Jensen, L.B. (1953). Early preparations of foods. In *Man's Foods*. The Garrard Press, Champaign, U.S.A., pp. 159–170.

Jensen, L.B. (1954). Introduction and history. In *Microbiology of Meats*. The Garrard Press, Champaign, U.S.A., pp. 1–11.

Kerr, R.H., Marsh, C.T.N., Shroeder, W.F. and Boyer, E.A. (1926). The use of sodium nitrite in curing of meat. *J. Agric. Res.*, **33**, 541–551.

Langer, H.J., Heckel, V. and Malek, E. (1970). Aroma substances in ripening dry sausage. *Fleischwirtschaft*, **50**, 1193–1199.

Lillard, D.A. and Ayres, J.C. (1969). Flavour compounds in country-cured hams. *Food Technol.*, **23**, 251–254.

Love, J.D. and Pearson, A.M. (1976). Metmyoglobin and nonheme iron as prooxidants in egg-yolk phospolipid dispersions and cooked meat. *J. Agric. Food Chem.*, **24**, 494–498.

MacDonald, B., Gray, J.I. and Gibbins, L.N. (1980). Role of nitrite in cured meat flavour: antioxidant role of nitrite. *J. Food Sci.*, **45**, 893–897.

MacDougall, D.B., Mottram, D.S. and Rhodes, D.N. (1975). Contribution of nitrite and nitrate to the colour and flavour of cured meats. *J. Sci. Food Agric.*, **26**, 1743–1754.

MacLeod, G. and Ames, J.M. (1986). Capillary gas chromatography–mass spectrometric analysis of cooked ground beef aroma. *J. Food Sci.*, **51**, 1427–1434.

MacLeod, G. and Seyyedain-Ardebili, M. (1981). Natural and simulated meat flavours (with particular reference to beef). *CRC Crit. Rev. Food Sci. Nutr.*, **14**, 309–437.

Maier, V.P. and Tappel, A.L. (1959). Studies of unsaturated fatty acid oxidation catalyzed by hematin compounds. *J. Am. Oil Chem. Soc.*, **36**, 12–15.

Minor, L.J., Pearson, A.M., Dawson, L.E. and Schweigert, B.S. (1965). Chicken flavour: the identification of some chemical components and the importance of sulphur compounds in the cooked volatile fraction. *J. Food Sci.*, **30**, 686–696.

Moody, W.G. (1983). Beef flavour—a review. *Food Technol.*, **37**, 227–232, 238.

Mottram, D.S. (1984). Organic nitrates and nitriles in the volatiles of cooked cured pork. *J. Agric. Food Chem.*, **32**, 343–345.

Mottram, D.S. and Rhodes, D.N. (1974). Nitrite and the flavour of cured meat. In *Proceedings of the International Symposium of Nitrite Meat Products*, Zeist, Pudoc, Wageningen, Holland, p. 161.

Ockerman, H.W., Blumer, T.W. and Craig, H.B. (1964). Volatile chemical compounds in dry-cured hams. *J. Food Sci.*, **29**, 123–129.

Pearson, A.M., Love, J.D. and Shorland, F.B. (1977). Warmed-over flavour in meat, poultry and fish. *Adv. Food Res.*, **23**, 1–74.

Pierson, M.D. and Smoot, L.A. (1982). Nitrite, nitrite alternatives, and the control of *Clostridium botulinum* in cured meats. *CRC Crit. Rev. Food Sci. Nutr.*, **17**, 141–187.

Piotrowski, E.G., Zaika, L.L. and Wasserman, A.E. (1970). Studies on aroma of cured ham. *J. Food Sci.*, **35**, 321–325.

Ramarathnam, N., Rubin, L.J. and Diosady, L.L. (1991a). Studies on meat flavour. 1. Qualitative and quantitative differences in uncured and cured pork. *J. Agric. Food Chem.*, **39**, 344–350.

Ramarathnam, N., Rubin, L.J. and Diosady, L.L. (1991b). Studies on meat flavour. 2. A quantitative investigation of the volatile carbonyls and hydrocarbons in uncured and cured beef and chicken. *J. Agric. Food Chem.* **39**, 1839–1847.

Ramaswamy, H.S. and Richards, J.F. (1982). Flavour of poultry meat—a review. *Can. Inst. Food Sci. Technol. J.*, **15**, 7–18.

Rhee, K.S. (1989). Chemistry of meat flavour. In *Flavour Chemistry of Lipid Foods*, ed. D.B. Min and T.H. Smouse. American Chemical Society, Champaign, pp. 166–189.

Rubin, L.J. and Shahidi, F. (1988). Lipid oxidation and the flavour of meat products. In *Proceedings of the 34th International Congress of Meat Science Technology*, Brisbane, Australia. pp. 295–301.

Sanderson, A., Pearson, A.M. and Schweigert, B.S. (1966). Effect of cooking procedure on flavour components of beef—carbonyl compounds. *J. Agric. Food Chem.*, **14**, 245–247.

Sato, K. and Hegarty, G.R. (1971). Warmed-over flavour in cooked meats. *J. Food Sci.*, **36**, 1098–1102.

Shahidi, F. (1989). Flavour of cooked meats. In *Flavour Chemistry—Trends and Developments*, ed. R. Teranishi, R.G. Buttery and F. Shahidi. American Chemical Society, Washington, DC, pp. 188–201.

Shahidi, F., Rubin, L.J. and D'Souza, L.A. (1986). Meat flavour volatiles: a review of the composition, techniques of analysis, and sensory evaluation. *CRC Crit. Rev. Food Sci. Nutr.*, **24**, 141–243.

Shahidi, F., Rubin, L.J. and Wood, D.F. (1987a). Control of lipid oxidation in cooked ground pork with antioxidants and dinitrosyl ferrohemochrome. *J. Food Sci.*, **52**, 564–567.

Shahidi, F., Yun, J., Rubin, L.J. and Wood, D.F. (1987b). The hexanal content as an indicator of oxidative stability and flavour acceptability in cooked ground pork. *Can. Inst. Food Sci. Technol. J.*, **20**, 104–106.

Skjelvale, R. and Tjaberg, T.B. (1974). Comparison of salami sausage produced with and without addition of sodium nitrite and sodium nitrate. *J. Food Sci.*, **39**, 520–524.

Tappel, A.L. (1952). Linoleate oxidation catalyzed by hog muscle and adipose tissue extract. *Food Res.*, **17**, 550–559.

Wasserman, A.E. (1979). Symposium on meat flavour. Chemical basis for meat flavour: a review. *J. Food Sci.*, **44**, 6–11.

Watts, B.M. (1954). Oxidative rancidity and discolouration in meat. *Adv. Food Res.*, **5**, 1–52.

Wood, D.S., Collins-Thompson, D.L., Usborne, W.R. and Picard, B. (1986). An evaluation of antibotulinal activity in nitrite-free curing systems containing dinitrosyl ferrohemochrome. *J. Food Prot.*, **49**, 691–695.

Younathan, M.T. and Watts, B.M. (1959). Relationship of meat pigments to lipid oxidation. *Food Res.*, **24**, 728–734.

Yun, J., Shahidi, F., Rubin, L.J. and Diosady, L.L. (1987). Oxidative stability and flavour acceptability of nitrite-free meat-curing systems. *Can. Inst. Food Sci. Technol. J.*, **20**, 246–251.

11 Contribution of smoke flavourings to processed meats

C.M. HOLLENBECK

11.1 Introduction

The flavouring, colouring and preservation of foods, primarily meats, by applying wood smoke were probably discovered accidentally, and over a long period of time. From what we can learn, the smoking of foods has existed almost as long as cooking.

Wood smoke is composed of numerous chemicals and the possibilities of their reactions with food components are almost infinite. However, over the years the components of smoke have been divided into major classes, with each class considered the primary source for given functions in the treatment of foods. For example, the acidic compounds, primarily aliphatic acids, contribute to flavour, the skin formation and peeling properties of skinless frankfurters. The phenolic compounds give flavour, bacteriostatic activity and antioxidation, while the carbonyls react with proteins to give the food a smoke colour. These are only rough generalizations because there are many overlapping functions between the so-called classes.

During pyrolysis of wood to produce smoke, there are many variables which complicate any generalization as to the source of the important components of wood smoke. However, it is generally assumed that cellulose and hemicellulose form the carbonyl and acid fractions and lignin forms the phenolic fractions.

A quick look at the structure of these wood components, and some of the literature on their breakdown products, seems to verify these generalizations (Maga, 1988). The generally proposed chemical structures of the three major components of wood are shown in Figure 11.1. Some of the proposed products and pathways for their formation during pyrolysis are shown in Table 11.1, and Figures 11.2, 11.3 and 11.4.

11.2 Pyrolysis of cellulose

Although, as mentioned before, the products of pyrolysis from the cellulose component of wood are numerous and variable, depending on cellu-

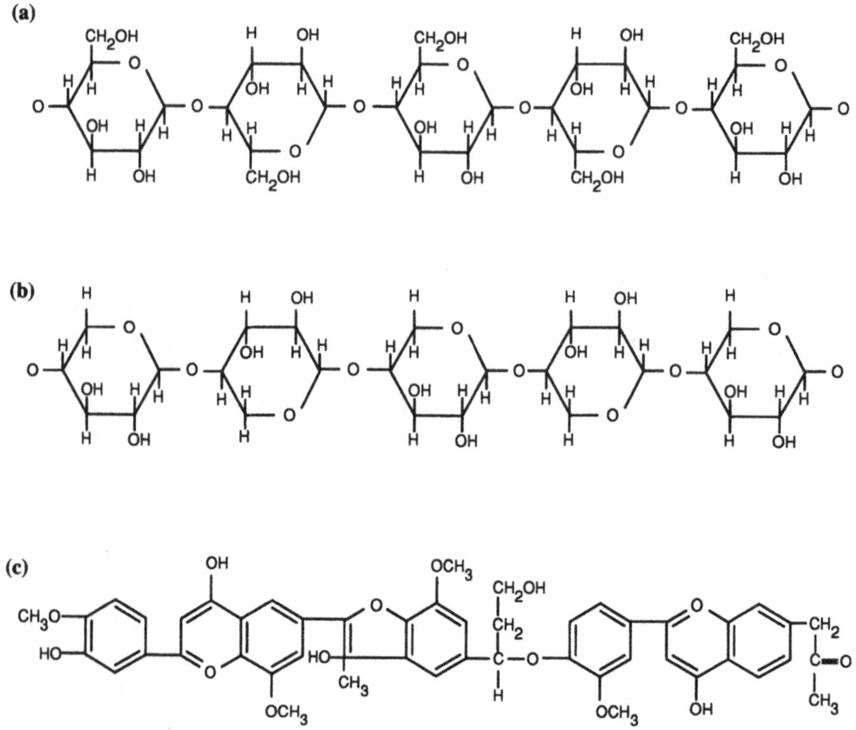

Figure 11.1 Structural inter-relationships of the three major components of wood: (a) cellulose; (b) hemicellulose; (c) lignin.

Table 11.1 Pyrolysis products (600°C) of cellulose

Compound	Relative %
Acetaldehyde	2.3
Furan	1.6
Acetone/propionaldehyde	1.5
Propanal	3.2
Methanol	2.1
2,3-Butanedione	2.0
1-Hydroxy-2-propanone	2.1
Glyoxal	2.2
Acetic acid	6.7
2-Furaldehyde	1.1
Formic acid	0.9
5-Methyl-2-furaldehyde	0.7
2-Furfuryl alcohol	0.5
Carbon dioxide	12.0
Water	18.0
Char	15.0
Tar	28.0

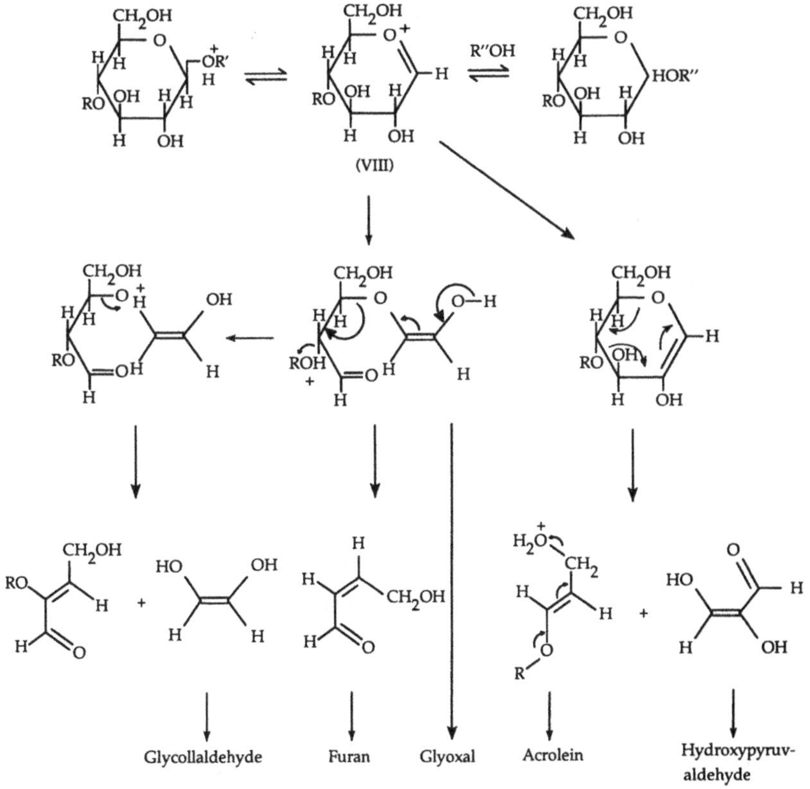

(VIII)

Glycollaldehyde

Furan

Glyoxal

Acrolein

Hydroxypyruv-aldehyde

Figure 11.2 Mechanism of the formation of carbonyl compounds.

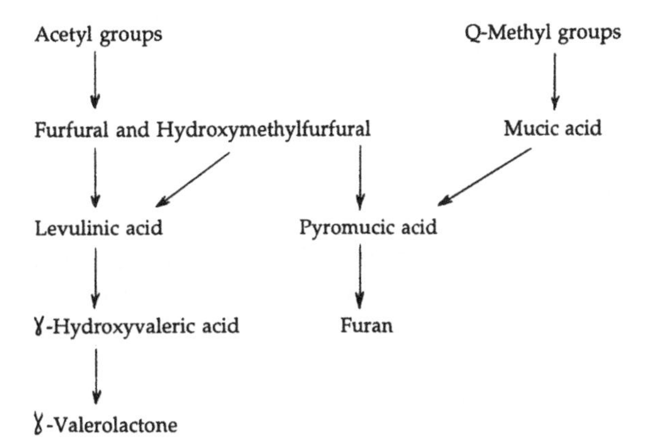

Figure 11.3 Thermal degradation products from hemicellulose.

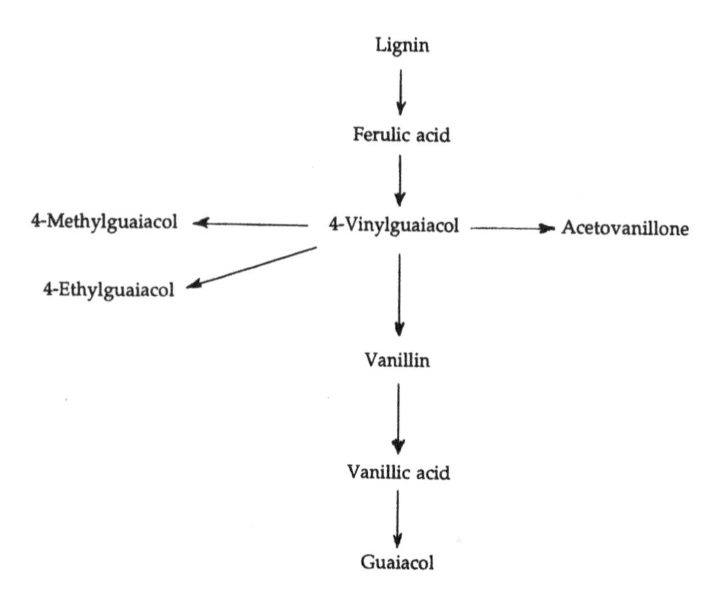

Figure 11.4 Thermal degradation products of ferulic acid.

losic source and conditions of pyrolysis, a listing of some of the major products, reported by Shafizadeh (1984) is shown in Table 11.1. The important compounds from a smoke flavouring standpoint are the aliphatic acids and the aldehydes.

Byrne *et al.* (1966) proposed a series of pathways for the breakdown of cellulose to form the lower molecular weight aldehydes (Figure 11.2), which we have found to be very important in the smoke colour formation in processed meats and other foods.

11.3 Pyrolysis of hemicellulose

Fengel and Wegener (1984) (Figure 11.3) proposed some pathways for the degradation of hemicellulose. The major products formed from the pentosic structure of hemicellulose are furan derivatives and various lactones. We, and others, have found many of these compounds among components of smoke flavourings. Obviously, they contribute to the overall flavour and chemical properties of smoke, but their contributions are not as easily identified as some of the derivations of cellulose and lignin.

11.4 Pyrolysis of lignin

The pyrolysis products from lignin (the proposed pathways by Gilbert and Knowles (1975) in Figure 11.4) are primarily phenolic. A quick look at the proposed lignin molecular structure shows many possibilities on how it could be thermally fractured to form many phenolic derivatives. These phenolic compounds are the major sources of the 'smoky' flavour of smoke and smoked foods.

11.5 Smoke colour formation in processed meats

As indicated earlier, the major smoke colour-forming reaction is of the Maillard type between the aldehydes of smoke and the amino groups on the proteins of the meat. We have found that certain 'active' aldehydes in smoke, like hydroxyacetaldehyde (glycoaldehyde), account for most of the smoke brown colour formed in the smoke–meat reaction. Namiki and Hayashi (1983) (Figure 11.5) showed a possible pathway to the formation of a pyrazine polymer (colour-basic) in the sugar–amino compound reaction by way of a hydroacetaldehyde intermediate. We have found hydroxyacetaldehyde (glycolaldehyde) to be a very active browning agent

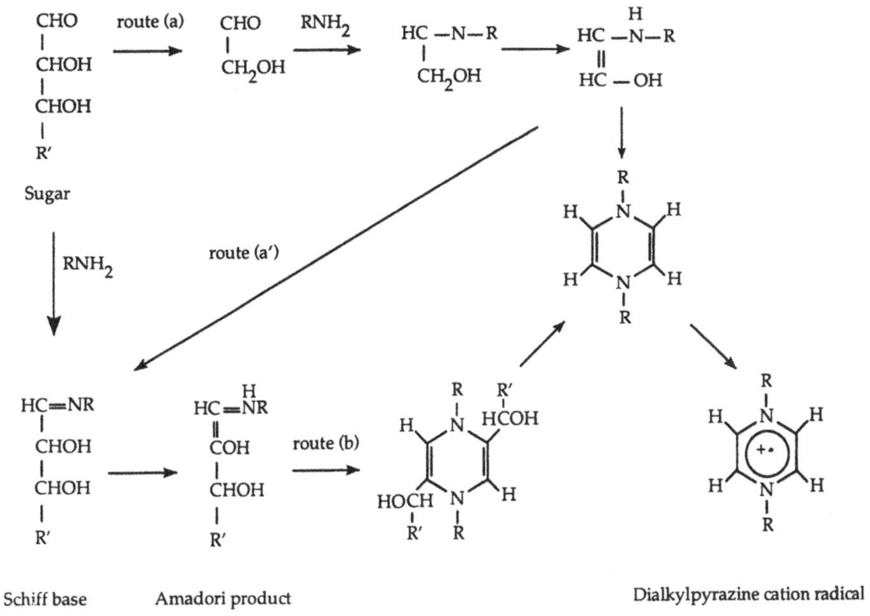

Figure 11.5 Two posssible pathways for radical formation in the reaction of sugars with amino compounds.

Table 11.2 Colour forming potential of selected carbonyls

Colour intensity	Browning index (pure carbonyls)[1]	Glycine paper (pure carbonyls)[2]	Wiener dips (pure carbonyls)[2]
Darkest	Glycoaldehyde (18.9)	Glyoxal	Glycoaldehyde
↓	Pyruvaldehyde (14.8)	Pyruvaldehyde	Glyoxal
	Glyoxal (6.4)	Glycoaldehyde	Formaldehyde
	Diacetyl (5.3)	Diacetyl	Pyruvaldehyde
	Furfural (4.9)	Acetol	Acetol
	Acetol (0)	Furfural	3-Hydroxy-2-butanone
Lightest	Formaldehyde (0)	3-Hydroxy-2-butanone	Furfural

[1]10% solutions
[2]glycoaldehyde 0–3% solution; all others 5%

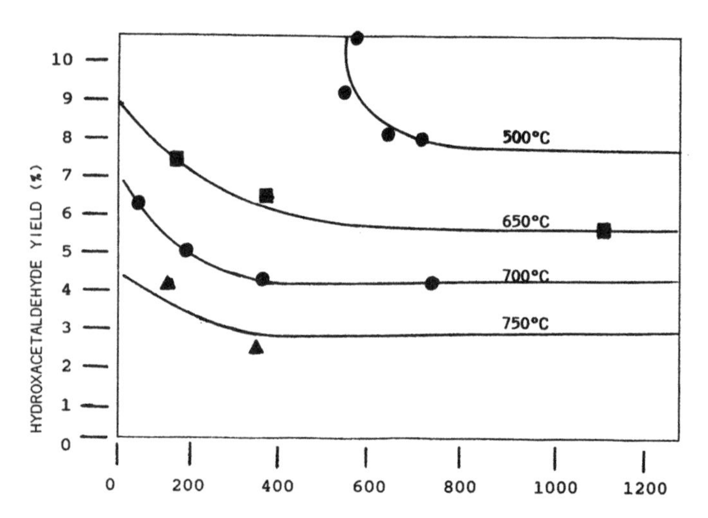

Figure 11.6 Fast pyrolysis of wood (hydroxyacetaldehyde versus residence time).

in meats and other foods. In fact, in the production of liquid smoke flavourings, we like to keep the content of hydroxyacetaldehyde, and similar compounds like glyoxal and pyruvicaldehyde, as high as possible to ensure good browning potential. In Table 11.2 Underwood (private communication) shows the relative browning activity of various aldehydes measured in the different ways.

Since these active aldehydes are very reactive, it is very important in the production of liquid smoke-flavourings to quench the pyrolysis products as rapidly as possible after generation to slow their reaction with other smoke components. There has been a definite trend in liquid smoke-production technology to shorter and shorter dwell times in the hot pyrolysis zone. A very rapid pyrolysis process has been developed (Underwood, 1989) in which the dwell time is a fraction of a second compared to several seconds in the conventional process. As shown in Figure 11.6, this greatly increases the yield of hydroxyacetaldehyde, especially in the 500° to 600°C range of temperatures for pyrolysis.

11.6 Smoke flavour in processed meats

The total smoky flavour in processed meats is obviously a combination of unreacted smoke components and reacted smoke-meat components. The phenolic compounds are probably the major unreacted smoke components

Table 11.3 Flavours of smoke phenolics

Phenolic compound	Flavour response
o-, *m*- and *p*-Xylene	Petroleum-like
Phenylethylalcohol	Mild, warm rose-honey
1,2-Dimethoxybenzene	Burnt and woody
1,2-Dimethoxybenzene-4-methylbenzene	Wood, vanilla-like
1,2-Dimethoxybenzene-4-ethylbenzene	Sweet, woody, vanilla-like
Benzaldehyde	Oil of almond
Acetophenone	Sweet floral
m- and *p*-Methylacetophenone	Grassy, sweet floral
Phenol	Pungent characteristic
o-, *m*- and *p*-Cresol	Pungent characteristic
2,3- and 2,4-Xylenol	Pungent characteristic
2,6-, 3,4- and 3,5-Xylenol	Cresolic
2-Ethyl-5-methylphenol	Cresolic
2,3,5-Trimethylphenol	Cresolic
3-Ethyl-5-methylphenol	Cresolic
Guaiacol	Sweet smoky and somewhat pungent
3 (or 6)-Methylguaiacol	Weak phenolic
4-Methylguaiacol	Sweet smoky
4-Ethylguaiacol	Sweet smoky
4-Vinylguaiacol	Sweet smoky
4-Allyguaiacol	Woody
Syringol	Smoky
4-Methylsyringol	Mild heavy burnt
4-Ethylsyringol	Mild heavy burnt
4-Propylsyringol	Mild heavy burnt
4-Propenylsyringol	Mild heavy burnt
Pyrocatechol	Heavy sweet burnt
3-Methylpyrocatechol	Heavy sweet burnt
4-Methylpyrocatechol	Heavy sweet burnt

Table 11.4 Partition ratios of some phenolic smoke components at 28°C

Compound	$K_p = \frac{C_{oil}}{C_{water}}$
Syringol	3.0
Phenol	5.6
Guaiacol	9.2
4-Methyl syringol	11.0
m-Cresol	16.0
p-Cresol	16.0
o-Cresol	22.0
4-Methyl guaiacol	23.0
4-Ethyl phenol	42.0
4-Ethyl guaiacol	60.0
Eugenol	97.0
4-Propyl guaiacol	143.00

that contribute smokiness to the flavour. However, the unreacted acids, esters, lactones and carbonyls probably contribute to the overall flavour. Kim (1974) evaluated the flavour of individual smoke phenolics. His observations are shown in Table 11.3.

Based on Kim's observations, the predominant flavour phenols in smoke, the quaiacols, syringols and pyrocatechols, give meat most of its smoky flavour. These phenols are also more soluble in oil (fat) than in water as shown in Table 11.4 (Doerr and Fiddler, 1970). Furthermore, these phenols and their solubility in fat probably account for the observation that the fat portion in the processed meat usually has a more pleasant smokiness than the lean portion.

11.7 Fractionation of smoke flavourings

The above observation on the relative smokiness flavour of the fat and lean portions in processed meats led this author to develop an oil-based smoke flavouring (CharOil). This product is prepared by extracting aqueous smoke solution with a vegetable oil. The smoke components in CharOil are mainly phenolic, and it is strictly a flavouring agent with no colour-forming properties. It is used widely as a flavouring agent in processed meats, especially poultry meat products. As will be shown later, CharOil is also an excellent antioxidant for smoke-flavoured fatty foods.

As a means of concentrating the oil-soluble smoke phenols and making them soluble in pumping pickles for cured meats, Underwood and Wendorff (1981) extracted CharOil with polysorbate. They found that the phenolic compounds transferred mainly to the polysorbate layer and

Table 11.5 The removal of phenols from liquid smoke by adsorption on a resin column

Volume of CharSol C-10 solution (gal.)	Phenols (mg/ml)	Carbonyls (%)	Browning index	Brix
CharSol C-10 Feed	17.0	12.4	9.9	25.9
10	1.3	11.3	9.6	19.2
20	2.4	11.1	10.1	21.0
30	4.8	11.5	10.6	22.6
40	6.3	11.5	10.4	23.2
50	7.9	12.2	10.3	23.4
65	12.5	11.8	9.5	24.4
75	14.3	NA	NA	24.6
85	15.3	11.3	9.5	24.6
95	17.0	NA	NA	25.4
105	16.8	12.6	9.3	26.0

NA = not analysed

became water dispensable. Their product, Aro-Smoke P-50, is used widely as a smoke-flavouring agent in bacon, and other smoked meats, and is applied as a component of pumping pickle.

Among the many varieties of processed meats, there are several types that like a deep smoke colour but do not need a strong smoke flavour. With this need in mind, Underwood (1990) developed a high colour-forming, low-flavour smoke flavouring by removing phenols in a resin adsorption column. The data in Table 11.5 show the removal of phenols without appreciably affecting the levels of colour-forming carbonyls by passing a smoke solution (CharSol C-10) through a non-ionic resin adsorption column. The resulting smoke flavouring, CharSol LFB, is finding many uses in products needing dark surface colour.

11.8 Miscellaneous contributions of smoke flavouring

In addition to above-described contributions of flavour and colour, smoke flavourings also contribute to the foregoing quality (shelf-life) of processed meats. As mentioned in the foregoing, smoke flavourings are effective antibacterial and anti-fungal agents. Wendorff (1981) measured the bacteriostatic and fungistatic activities of some smoke flavourings against some of the common food bacteria and fungi. No doubt, the aliphatic acids and the phenols of smoke contribute to the inhibition of growth of both bacteria and fungi. These data are shown in Tables 11.6 and 11.7.

Furthermore, smoke solutions, especially the oil-based CharOil and its derivative Aro-Smoke P-50, are strong antioxidants. Some data developed by Wendorff (1981) with pork fat are shown in Table 11.8. It is widely known and easily shown that smoke-flavoured processed meats do not

Table 11.6 Antibacterial properties of smoke flavourings

Smoke flavouring	% inhibition			
	Escherichia coli	*Staphylococcus aureus*	*Pseudomonas aeruginosa*	*Lactobacillus viridescens*
CharSol C-6 (0.25% v/v, pH 2.4)	33	72	52	99
CharSol C-6 (0.25% v/v, pH 5.0)	11	31	51	97
CharSol C-6 (0.25% v/v, pH 7.0)	3	25	54	15
6.5% Acetic acid (0.25% v/v)	25	52	29	99
CharDex (0.10% w/v)	0	77	46	21
Aro-Smoke P-50 (0.25% w/v)	20	60	62	10
CharOil (0.25% v/v)	0	55	52	85

Table 11.7 Antifungal properties of smoke flavourings

Smoke flavouring	Zones of inhibition (mm^2)		
	Penicillium sp.	*Aspergillus niger*	*Aspergillus flavus*
CharSol C-6 (pH 2.4)	21	18	14
CharSol C-6 (pH 5.0)	19	16	13
CharSol C-6 (pH 7.0)	17	12	11
6.5% Acetic acid	16	12	13
CharDex (30% w/v)	12	9	8
Aro-Smoke P-50 (15% w/v)	9	9	8

Diameter of assay disc was 7 mm

Table 11.8 Antioxidant properties of natural smoke flavourings

Antioxidant	Peroxide value (meq/kg fat)				
	Initial	1 week	2 weeks	4 weeks	26 weeks
Untreated control	0.8	3.5	8.5	18.4	43.1
CharOil (0.4%)	0.8	1.2	2.0	2.4	3.7
CharOil (0.2%)	0.8	1.3	2.1	3.0	4.6
Aro-Smoke P-50 (0.04%)	0.8	1.2	2.0	2.2	3.4
Aro-Smoke P-50 (0.02%)	0.8	1.4	2.1	2.9	4.9
BHA (0.02%)	0.8	1.4	2.1	3.3	12.5
BHT (0.02%)	0.8	1.3	1.9	2.8	4.8
Propyl gallate (0.01%)	0.8	1.4	2.1	3.0	5.8

BHA = butylated hydroxyanisole; BHT = butylated hydroxytoluene develop rancidity as rapidly as their counterparts of cured, unsmoked meats.

11.9 Summary

Many of the pyrolytic breakdown compounds from cellulose, hemicellulose and lignin in smoke flavourings contribute to the flavour, colour and colloidal properties of smoke-flavoured processed meats. The reactive aldehydes from cellulose degradation are the primary colour-forming reactants, and the phenolic compounds from lignin are the major smoke flavour bases.

Extracting some of the phenolic compounds from liquid smoke-flavouring with vegetable oil produces a 'ham fat'-like flavouring. Transferring these phenolic substances from the oil to polysorbate renders these flavours water dispersible and easily injectable into the meat in the curing brine. Removing a major portion of the phenolic compounds from smoke flavouring by resin adsorption produces a smoke adjunct with excellent colour-forming properties with much less 'smoky' flavour. Smoke flavourings contribute bacteriostatic and antioxidant activity to processed meats which extend the shelf-life of the meat products.

References

Byrne, G.A., Gardener, D. and Holmes, F.H. (1966). The pyrolysis of cellulose and the action of flame retardants. *J. Appl. Chem.*, **16**, 81–87.

Doerr, R.C. and Fiddler, W. (1970). Partition ratios of some wood smoke phenols in two oil:water systems. *Agr. Food Chem.*, **18**, 937–939.

Gilbert, J. and Knowles, M.E. (1975). The chemistry of smoked foods. *J. Food Technol.*, **10**, 245–251.

Fengel, D. and Wegener, G. (1984). *Wood: Chemistry Ultra-structure, Reactions.* Walter de Gruyter, Berlin, chapter 12.

Kim, K., Kurata, T. and Fujimaki, M. (1974). Identification of flavor constituents in carbonyl, non-carbonyl, neutral and basic fractions of aqueous smoke condensates. *Agr. Biol. Chem.*, **38**(1), 53–63.

Maga, J.A. (1988). *Smoke in Food Processing.* CRC Press Inc., Boca Raton, Florida.

Namiki, M. and Hayashi, T. (1983). In *The Maillard Reaction in Foods and Nutrition*, ed. G.R. Waller and M.S. Feather. American Chemical Society Symposia Series, Washington, DC, p. 45.

Shafizadeh, F. (1984). In *The Chemistry of Solid Wood*, ed. R. Rowell. American Chemical Society, Washington, DC, chapter 13.

Underwood, G. (1990). Process making liquid smoke compositions and resin treated liquid smoke compositions. US Patent 4,959, 232.

Underwood, G. and Graham, R.G. (1989). Method of using fast pyrolysis liquids as liquid smoke. US Patent 4,876,108.

Underwood, G.L. and Wendorff, W.L. (1981). Smoke flavored hydrophilic liquid concentrate and process of producing same. US Patent 4,250,199.

Wendorff, W.L. (1981). Antioxidant and bacteriostatic properties of liquid smoke. In *Proceedings of Smoke Symposium*, Red Arrow Products Co., Manitowoc, WI, pp. 73–87.

12 Some aspects of the chemistry of meat flavour

D.S. MOTTRAM

12.1 Introduction

The flavours associated with cooked meats have proved particularly difficult to characterize, both for the sensory analyst and the flavour chemist. Meat flavour is influenced by compounds contributing to the sense of taste as well as those stimulating the olfactory organ. Other sensations such as mouthfeel and juiciness will also affect the overall flavour sensation. However it is the volatile compounds of cooked meat that determine the aroma attributes and contribute most to the characteristic flavours of meat. It has been one of the most researched of food flavours, with over 1000 volatile compounds having been isolated. A survey of the volatiles found in meat shows a much larger number from beef (880) than the other meats (361 from pork; 271 from lamb/mutton; 468 from chicken), but this is reflected in the much larger number of publications for beef (70) compared with pork (11), sheep meat (12) or poultry (20) (Mottram, 1991).

Meat flavour is thermally derived, since uncooked meat has little or no aroma and only a blood-like taste. Much research has been aimed at understanding the chemistry of meat aroma and the nature of the reactions occurring during cooking which are responsible for the formation of aroma compounds. The major precursors of meat flavour can be divided into two categories: water-soluble components (amino acids, peptides, carbohydrates, nucleotides, thiamine, etc.) and lipids. The main reactions during cooking which result in aroma volatiles, are the Maillard reaction between amino acids and reducing sugars, and the thermal degradation of lipids.

12.2 Meat flavour precursors

Most of the early work on meat flavour, during the 1950s and 1960s, was concerned with identifying those components of meat which, on heating, gave the characteristic flavour (Kramlich and Pearson, 1960; Hornstein and Crowe, 1960; Macey *et al.*, 1964; Wasserman and Gray, 1965). It was concluded that meat flavour precursors are low molecular weight water-soluble components and that the high molecular weight fibrillar and sarcoplasmic proteins are unimportant. The main flavour precursors were suggested to be free sugars, sugar phosphates, nucleotide-bound sugars, free amino acids, peptides, nucleotides and other nitrogenous components such as thiamine. A number of studies examined the changes which occurred in the quantities of these water-soluble compounds on heating. Depletions in the quantities of carbohydrates and amino acids were observed, the most significant losses occurring for cysteine and ribose. Subsequent studies of the aromas produced on heating mixtures of amino acids and sugars, confirmed the important role played by cysteine in meat flavour formation and led to the classic patent of Morton *et al.* (1960) which involved the formation of a meat-like flavour by heating a mixture of cysteine and ribose. Most subsequent patent proposals for 'reaction product' meat flavourings have involved sulphur, usually as cysteine or other sulphur-containing amino acids or hydrogen sulphide (MacLeod and Seyyedain-Ardebili, 1981; MacLeod, 1986).

The role of the lipid fractions of meat, both adipose tissue and fat contained within the lean, has been the subject of some debate which still continues. Hornstein and Crowe (1960, 1963) found that aqueous extracts of beef, pork and lamb had similar aromas when heated, while heating the fats yielded species-characteristic aromas. It was suggested that lipid provides volatile compounds which give the characteristic flavours of the different species, and the lean is responsible for a basic meaty flavour common to all species. It was also recognized that autoxidation of fat could produce undesirable flavour compounds and lead to rancidity. Fat can also serve as a solvent for aroma compounds, obtained either from extraneous sources or as part of the flavour-forming reactions, and it can therefore influence the release of flavour from the meat (Wasserman and Spinelli, 1972). However, it is now realized that this is an oversimplification of the role of fat in meat flavour and that lipid-derived volatiles have an important part to play in desirable meat aroma both as aroma compounds and as intermediates to other compounds.

12.3 Reactions leading to meat aroma

A wide range of temperature conditions exist during normal cooking of meat; the centre of a rare steak may only reach 50°C, the centre of roast meat may attain 70–80°C, while the outside of grilled or roast meat will be subjected to much higher temperatures and localized dehydration of the surface will occur. On the other hand, in stewing the meat remains at a temperature of 100°C, in the presence of excess water, for several hours. It is not surprising, therefore, that a wide range of different flavour sensations is perceived in cooked meats; some meats may be relatively bland while others may have strong meaty notes and others will be distinctly roasted.

The primary reactions occurring on heating which can lead to meat flavour include pyrolysis of amino acids and peptides, caramelization of carbohydrates, degradation of ribonucleotides, thiamine degradation, interaction of sugars with amino acids or peptides, and thermal degradation of lipid (MacLeod and Seyyedain-Ardebili, 1981; Mottram, 1991). This complicated array of reactions is made even more complex by the whole host of secondary reactions which can occur between the products of the initial reactions giving rise to the vast number of volatile compounds which contribute to meat flavour. The thermal decomposition of amino acids and peptides and the caramelization of sugars normally require temperatures over 150°C before aroma compounds are formed (Feather and Harris, 1973; Wasserman, 1979). Such temperatures are higher than those which are normally encountered during the cooking of meat, except for small areas of the surface which may dehydrate during grilling or roasting.

12.3.1 Maillard reaction

The Maillard reaction between reducing sugars and amino compounds does not require the very high temperatures associated with sugar caramelization and protein pyrolysis and readily produces aroma compounds at the temperatures associated with the cooking of food. It is, therefore, one of the most important routes to flavour compounds in cooked foods. It occurs much more readily at low moisture levels; hence in meat, flavour compounds produced by the Maillard reaction tend to be associated with the areas of the cooking meat which have been dehydrated by the heat source (van den Ouweland *et al.*, 1978).

The Maillard reaction in relation to flavour has been discussed in some detail in an earlier chapter, and has also been the subject of a number of reviews (Mauron, 1981; Hurrell, 1982; Vernin and Parkanyi, 1982). The essential features of the early stages of the reaction, showing the formation of Amadori products and dehydration and deamination to give furfurals, furanones and dicarbonyls compounds, are shown in Figure 12.1. These compounds may themselves make some contribution to meat flavour, but they are more important as reactants for other aroma volatiles.

An important reaction associated with the Maillard reaction is Strecker degradation which involves the oxidative deamination and decarboxylation of an α-amino acid in the presence of a dicarbonyl compound (Figure 12.2). This leads to the formation of an aldehyde, containing one fewer carbon atom than the original amino acid, and an α-aminoketone. The Strecker degradation of cysteine yields the expected Strecker aldehyde, mercaptoacetaldehyde and an α-amino ketone, but hydrogen sulphide, ammmonia and acetaldehyde are also formed from the breakdown of the

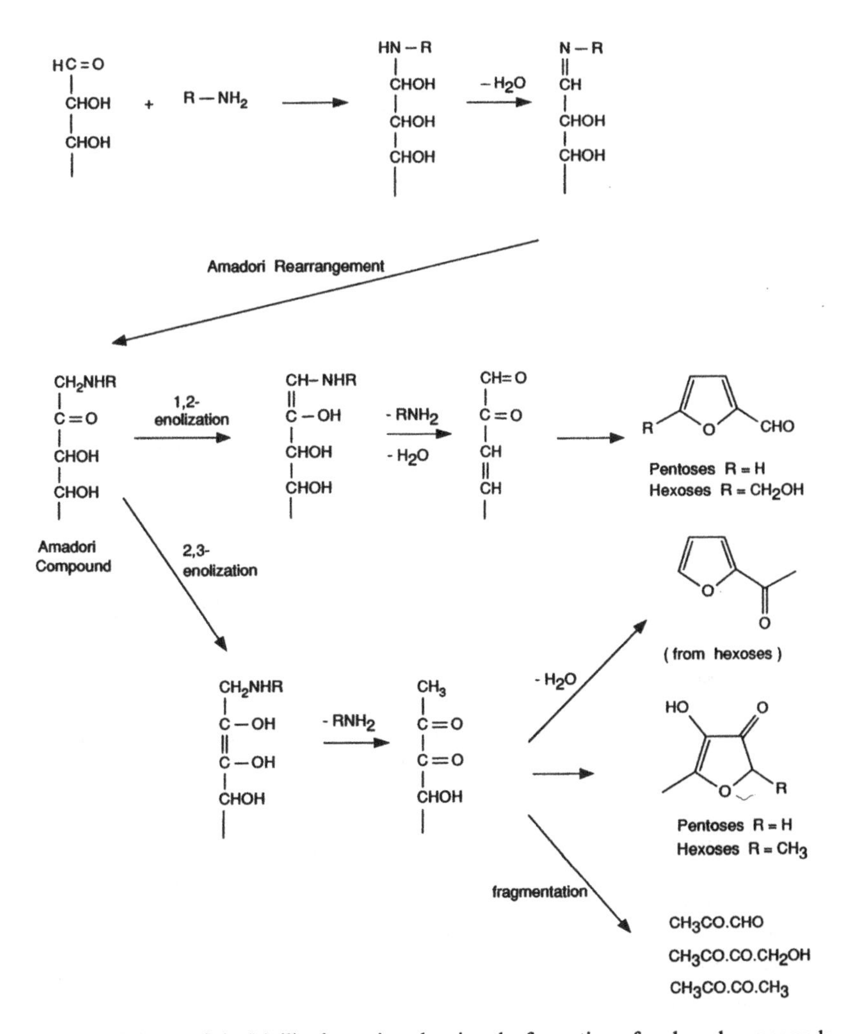

Figure 12.1 Scheme of the Maillard reaction showing the formation of carbonyl compounds.

intermediate mercaptoiminenol. These compounds are important as reactive intermediates for the formation of many highly odoriferous compounds which play important roles in meat flavour, and this emphasizes the importance of cysteine in the development of meat flavour. The Strecker degradation of methionine is another source of sulphur-containing intermediates; in this case methional, methanethiol and 2-propenal are produced (Schutte, 1974).

All these Maillard products are capable of further reaction, and the subsequent stages of the Maillard reaction involve the interaction of fur-

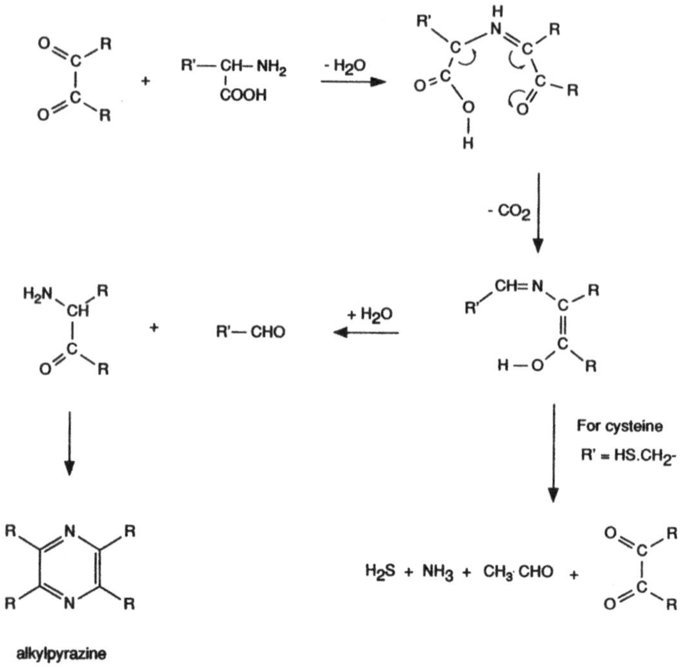

Figure 12.2 The Strecker degradation of α-amino acids showing the formation of hydrogen sulphide and ammonia in the Strecker degradation of cysteine.

furals, furanones and dicarbonyls with other reactive compounds such as amines, amino acids, hydrogen sulphide, thiols, ammonia, acetaldehyde and other aldehydes. These additional reactions lead to many important classes of flavour compounds including heterocyclics such as pyrazines, oxazoles, thiophenes, thiazoles and other heterocyclic sulphur compounds (Vernin and Parkanyi, 1982).

12.3.2 Lipid degradation

In addition to the subcutaneous fat and other fat depot tissues, triglycerides are present within the muscle in amounts which depend on the particular muscle and the age of the animal (Lawrie, 1985). All tissues also contain structural phospholipids. During cooking of meat, thermal degradation of lipids results in the formation of many volatile compounds, indeed more than half the volatiles reported in meat are lipid derived. One of the main routes to aroma volatiles during meat cooking is the thermally induced oxidation of the acyl chains of the lipids. Autoxidation of unsaturated fatty acid chains is also responsible for the undesirable

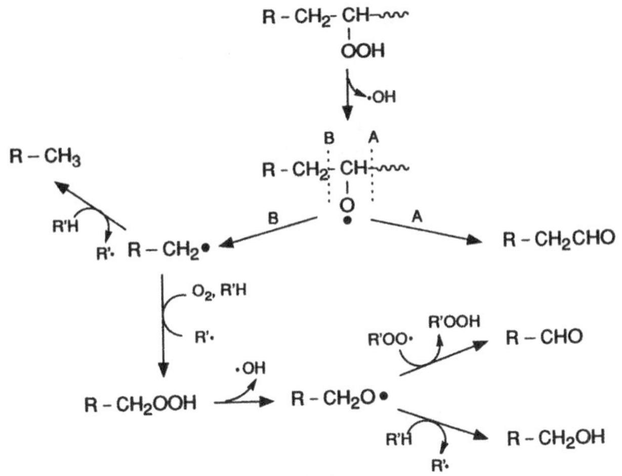

Figure 12.3 Breakdown of simple lipid hydroperoxides to give volatile products.

flavours associated with rancidity which develops during the storage of fatty foods. The reactions by which volatile aroma compounds are formed from lipids follow the same general routes for both thermal oxidation and rancid oxidation, although subtle changes in the mechanisms give rise to different profiles of volatiles in the two systems.

The oxidative breakdown of the unsaturated alkyl chains of lipids involves a free radical mechanism and the formation of intermediate hydroperoxides (Frankel, 1980). Decomposition of these hydroperoxides involves further free radical mechanisms and the formation of non-radical products including volatile aroma compounds (Forss, 1972; Grosch, 1982). The degradation of hydroperoxides (Figure 12.3) initially involves homolysis to give an alkoxy radical and a hydroxy radical; this is followed by cleavage of the fatty acid chain adjacent to the alkoxy radical. The nature of the volatile product from a particular hydroperoxide depends on the composition of the alkyl chain and the position where cleavage of the chain takes place (A or B). If the alkyl group is saturated and cleavage takes place at A, a saturated aldehyde results, while cleavage at B gives an alkyl radical which can give an alkane or, alternatively, can react with oxygen to give a hydroperoxide. The latter, just like the hydroperoxide of the lipid, breaks down to give an alkoxy radical which then gives a stable non-radical product such as an alcohol or aldehyde. One or more double bonds in the alkyl chain will give analogous compounds containing the double bonds, but the final range of products is more complex because further oxidation of the unsaturated chain can occur.

Other compounds formed in these and related reactions include ketones, furans and aromatic hydrocarbons, aldehydes and ketones. Free fatty acids can be formed from thermal hydrolysis of lipids while the gamma and delta lactones which are found in meat arise from the lactonization of hydroxy fatty acids (Watanabe and Sato, 1970).

In boiled and lightly grilled or roasted meat, lipid degradation products have been found to dominate the volatile extracts (Mottram *et al.*, 1982; Mottram, 1985). However, the contribution of many of these compounds to the overall flavour of the cooked meat may be small because their odour threshold values are relatively high. Those compounds which have sufficiently low odour threshold values for them to contribute directly to meat aroma are aldehydes, unsaturated alcohols and ketones and lactones.

12.4 Compounds contributing to meat flavour

Among the many aroma characteristics that can be differentiated in meat, those which are most clearly recognized are: fatty, species-related, roast, boiled-meat and the characteristic 'meaty' aroma which is associated with all meats regardless of species.

The characteristic flavour of the different meat species is generally believed to be derived from the lipid sources, although it has been suggested that interaction of lipid with other meat components may also be involved (Wasserman and Spinelli, 1972). Aldehydes, as major lipid degradation products, are probably involved in certain species' characteristics. The higher proportion of unsaturated fatty acids in the triglycerides of pork and chicken gives more unsaturated volatile aldehydes in these meats and these compounds may be important in determining the specific aromas of these species (Noleau and Toulemonde, 1987).

Sheep meat has been found to contain a number of methyl-branched saturated fatty acids which have not been reported in other meats. These acids have been associated with the characteristic flavour of mutton which results in low consumer acceptance of sheep meat in many countries, and which the Chinese described as 'soo' flavour (Wong *et al.*, 1975). Two acids, 4-methyloctanoic and 4-methylnonanoic, are considered to be primarily responsible for this flavour. The lipids of sheep contain significant quantities of methyl-branched fatty acids, unlike other species, and these are known to arise from the metabolic process occurring in the rumen of sheep. Branched acids with methyl substituents at even-numbered carbon atoms result from fatty acid synthesis utilizing methyl malonate (arising from propionate metabolism) instead of malonate in the chain lengthening (Christie, 1978). Fatty flavours, of course, originate from the lipid, and aldehydes (e.g. 2,4-decadienal), ketones and lactones will contribute to the fatty aromas associated with cooked meat.

Roast flavours appear to be associated with heterocyclic compounds (e.g. pyrazines and thiazoles) formed in the later stages of the Maillard reaction. Many different alkyl pyrazines have been found in meat volatiles as well as two classes of interesting bicyclic compounds, 6,7-dihydro-5(*H*)-cyclopentapyrazines and pyrrolo[1,2*a*]pyrazines (Flament *et al.*, 1976, 1977). This latter class of compounds has not been reported in any other food. Alkyl substituted thiazoles in general, have lower odour thresholds than pyrazines (Petit and Hruza, 1974), although they are found at lower concentrations in meat. Both classes of compounds increase markedly with the increasing severity of the heat treatment and, in well-done grilled meat, pyrazines are the dominant class of volatiles (Mottram, 1985).

One notable feature of the volatiles from cooked meat is the preponderance of sulphur-containing compounds. The majority occur at low concentrations, but their very low odour thresholds make them important contributors to the aromas of cooked meat. A comparison of boiled and roast beef shows that many more aliphatic thiols, sulphides and disulphides have been reported in the boiled meat (Table 12.1). A number of heterocyclic compounds with two or three sulphur atoms in 5- and 6-membered rings (e.g. trithiolanes, trithianes) have also been found in boiled beef, and thiophenes are much more prevalent in boiled beef than in the roast meat. Many of these sulphur compounds have low odour thresholds with sulphurous, onion-like and, sometimes, meaty aromas (Fors, 1983), and they probably contribute to the overall flavour by providing sulphurous notes which form part of the aroma of boiled meat.

Isolation of compounds with the characteristic 'meaty' aroma associated with all meats has been the subject of much research. These desirable characteristics of meat flavour have also been sought in the production of simulated meat flavourings which are of considerable importance in convenience and processed savoury foods. Furans (and also thiophenes) with

Table 12.1 Numbers of volatile sulphur compounds identified in cooked beef

Sulphur compounds	Number of compounds in cooked beef	
	Boiled	Grilled, roasted
Aliphatic thiols	28	–
Aliphatic sulphides	20	3
Aliphatic polysulphides	15	6
Thiophenes	26	13
Cyclic (two or three sulphurs)	15	1
Thiazoles	14	17
Thiazolines	4	1
Other	10	3

a thiol group in the 3-position appear to have meat-like aromas, as do the disulphides formed by oxidation of furan and thiophene thiols. A number of such compounds have been found in the volatiles from heated model systems containing hydrogen sulphide or cysteine, and pentoses or other sources of carbonyl compounds, and some are reported to have meaty aromas. 2-Methyl-3-furanthiol was shown to be one of the products formed in the reaction of hydrogen sulphide with 4-hydroxy-5-methyl-3(2*H*)-furanone, which is a dehydration product of ribose (van den Ouweland and Peer, 1975). The meaty characteristics of the reaction mixture led to patents dealing with a number of related compounds with potential as meat flavourings (MacLeod, 1986). 2-Methyl-3-furanthiol, and the analogous thiophenethiol, have been found in the volatiles from the reaction of cysteine and ribose (Farmer *et al.*, 1989; Farmer and Mottram, 1990a), from heated thiamine (van der Linde, *et al.*, 1979) and from a model meat flavour system containing cysteine, thiamine, glutamate and ascorbic acid (Werkhoff *et al.*, 1990). These thiols are readily oxidized to the corresponding disulphides, which have also been found in some of these model systems. A number of mixed sulphides and disulphides from furan and thiophene thiols and furylmethanethiol were found in the cysteine–ribose and the thiamine–cysteine–glutamate–ascorbate reaction mixtures, and several of these compounds were reported to have meaty aromas.

Although compounds with these structures had been quoted in patents relating to meat aroma, it is only recently that such compounds have been reported in meat itself (Figure 12.4). MacLeod and Ames (1986) identified 2-methyl-3-(methylthio)furan as a character impact compound in cooked beef. It has been reported to have a low odour threshold value (0.05 μg/kg) and a meaty aroma at levels below 1 μg/kg. Gasser and Grosch (1988)

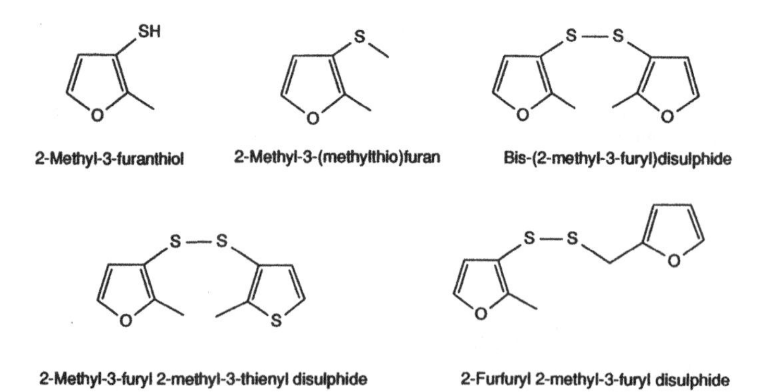

Figure 12.4 Some derivatives of 2-methyl-3-furanthiol found in meat, which contribute to meat-like aromas.

identified 2-methyl-3-furanthiol and the corresponding disulphide, bis-(2-methyl-3-furyl) disulphide, as major contributors to the meaty aroma of cooked beef. The odour threshold value of the disulphide has been reported as 0.00002 μg/kg, one of the lowest known threshold values (Buttery *et al.*, 1984). More recently, several other disulphides with furyl and thienyl groups have been reported in the volatiles of heated beef muscle, and evaluation of their aroma on elution from the GC column indicated meaty characteristics (Farmer and Patterson, 1991).

From the work on model systems and from attempts to synthesize structures with meat-like aromas, it has become clear that the furanthiols and thiophenethiols and their disulphides have particularly meaty aroma characteristics. The importance of these types of compound in meat flavour has now been confirmed by their detection in meat itself.

12.5 Pathways for the formation of some meat aroma volatiles

The Maillard reaction produces a number of pentose and hexose degradation products containing carbonyl groups, such as 2-oxopropanal, 2,3-butanedione, hydroxypropanone, 3-hydroxy-2-butanone, 2-furfural, 5-methyl-2-furfural, 5-hydroxymethyl-2-furfural and 5-methyl-4-hydroxy-3(2*H*)-furanone and its dimethyl homologue. Strecker degradation of amino acids by these dicarbonyl compounds yields aldehydes, while Strecker degradation of cysteine also produces hydrogen sulphide and ammonia. These compounds provide the main reactants for the formation of the large number of different heterocyclic compounds which are found in meat volatiles.

An important route to alkyl pyrazines is from α-aminoketones, which are formed in Strecker degradation or from the reaction of α-dicarbonyls with ammonia. Condensation of two aminoketone molecules yields a dihydropyrazine which oxidizes to the pyrazine (Figure 12.2).

Thiazoles and oxazoles may be formed by related mechanisms, involving the formation of an intermediate imine from the reaction of an aldehyde (e.g. a Strecker aldehyde) with ammonia (Figure 12.5). Reaction of the imine with a dicarbonyl produces an oxazoline which may undergo oxidation to the corresponding oxazole. If hydrogen sulphide is present it may react with the dicarbonyl to give an α-mercaptoketone which then yields thiazolines and thiazoles from reaction with the imine.

Many of the important aroma volatiles in meat contain sulphur. Hydrogen sulphide, formed from cysteine by Strecker degradation or hydrolysis, appears to be the essential intermediate for these compounds. Some of the reactions involving hydrogen sulphide in the formation of sulphur-containing heterocyclics are shown in Figure 12.6. Formylthiophenes and furylmethanethiol can be formed from furfurals, while the

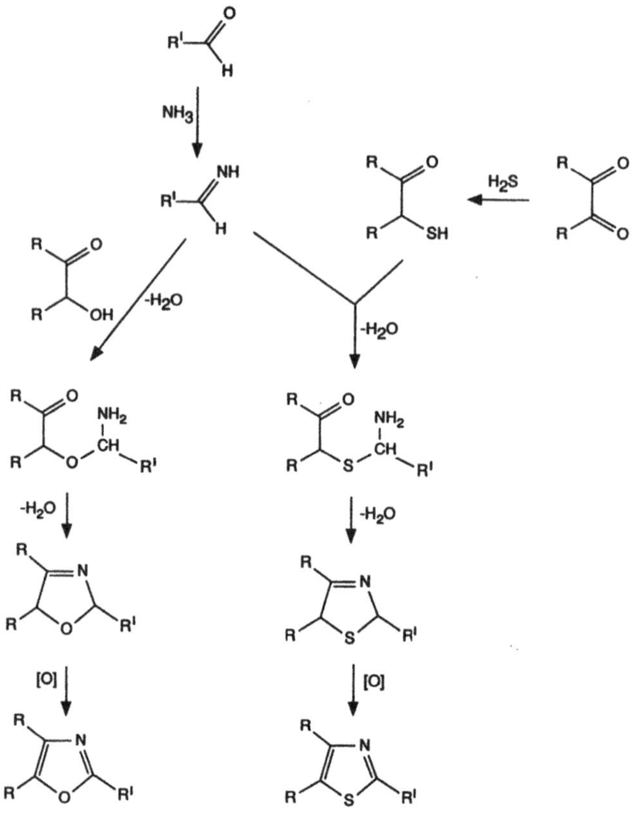

Figure 12.5 Formation of thiazoles, thiazolines, oxazoles and oxazolines by the reaction of aliphatic aldehydes with dicarbonyls, hydrogen sulphide and ammonia (Vernin and Parkanyi, 1982).

reaction of short-chain aldehydes, such as acetaldehyde, with hydrogen sulphide has been shown to yield a number of different sulphur compounds including trithiolanes and trithianes (Boelens *et al.*, 1974). One of the most important reactions for the formation of meaty aromas is the reaction of 4-hydroxy-5-methyl-3(2*H*)-furanone with hydrogen sulphide to give 2-methyl-3-furanthiol (van den Ouweland and Peer, 1975). Other products of the reaction include the thiophene analogue, even though it has not been reported in meat, and disulphides containing a 2-methyl-3-thienyl group have recently been found in cooked beef (Farmer and Patterson, 1991). The furanthiols and thiophenethiols have meaty aromas and they are readily oxidized to disulphides which are also meaty at very low concentrations. Although this reaction may only occur to a small extent,

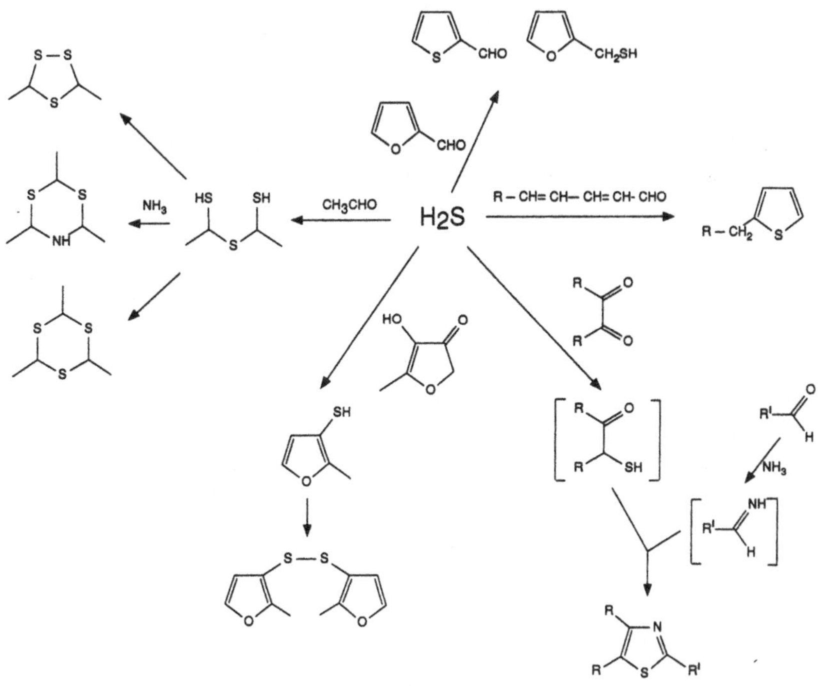

Figure 12.6 Some classes of volatile aroma compounds formed by the reaction of hydrogen sulphide with carbonyl compounds.

the exceedingly low odour threshold values of the compounds and their meaty character gives the reaction major importance for meat flavour.

Compounds produced in other reactions may also react with products of the Maillard reaction, especially hydrogen sulphide and ammonia. For example, aldehydes and other carbonyls formed during lipid oxidation could readily participate in some of the reactions described above.

12.6 Interaction of lipid with the Maillard reaction

In an examination of the contribution that lipids make to the development of aroma during the heating of meat, the phospholipids were shown to be particularly important (Mottram and Edwards, 1983). Sensory panels and consumer studies had failed to show any relationship between the meat flavour of lean meat and the amount of fat on the carcass, apparently confirming the early work on meat flavour in which meatiness was associated with the water-soluble flavour precursors and species characteristics with the lipid. However, the volatiles of cooked meat are often dominated by lipid-derived volatiles and such volatiles would be expected to have some effect on meaty flavour. The contribution of lipid to meat aroma was investigated by evaluating the change in aroma which occurred when lipids were extracted from meat prior to cooking. When inter- and intramuscular triglycerides were removed from lean muscle using hexane, the aroma after cooking could not be differentiated from the untreated material in sensory triangle tests; both preparations were judged to be meaty. However, when a more polar solvent (chloroform–methanol) was used to extract all the lipids—phospholipids as well as triglycerides—a very marked difference in aroma resulted; the meaty aroma was replaced by a roast, biscuit-like aroma. Comparison of the aroma volatiles from these meat preparations showed that the control and the material extracted with hexane had similar profiles, dominated by aliphatic aldehydes and alcohols, while removal of phospholipids as well as triglycerides gave a very different profile; the lipid oxidation products were lost, but there was a large increase in the amounts of alkyl pyrazines. This implied that in normal meat, lipids or their degradation products inhibit the formation of heterocyclic compounds by participating in the Maillard reaction.

12.6.1 Model systems

These results, showing a reduction in amounts of volatile Maillard products in defatted cooked meat, led to investigations of the effect of phospholipids on the volatile products of heated mixtures of amino acids and sugars (Whitfield *et al.*, 1988; Salter *et al.*, 1988; Farmer *et al.*, 1989; Farmer and Mottram, 1990b). Several amino acids were used, including cysteine, and ribose was chosen as the reducing sugar because of its recognized role in the formation of meat flavour, and concentrations were selected to approximate their relative concentrations in muscle. Reactions were carried out in aqueous solution buffered at pH 5.6 with phosphate under pressure in sealed glass tubes at 140°C. The effects of several phospholipid preparations, including egg-yolk phospholipids and triglycerides extracted from beef, on the volatiles from the reaction, were examined. In the absence of lipid the reaction mixtures yielded complex mixtures of volatiles including furfurals, furanones, alkylpyrazines and pyrroles. The volatiles from reactions involving cysteine were dominated by sulphur-containing heterocyclics, particularly thiophenes, thienothiophenes, dithiolanones, dithianones, trithiolanes and trithianes, together with 2-methyl-3-furanthiol, 2-furanmethanethiol and 2-methyl-3-thiophenethiol.

In the presence of phospholipids a reduction in the amounts of many of these volatiles was observed, confirming the observations in meat that phospholipids exert a quenching effect on the quantities of heterocyclic compounds formed in the Maillard reaction. As would be expected the inclusion of phospholipids in the reaction mixtures produced many volatiles derived from lipid degradation, such as hydrocarbons, alkylfurans, saturated and unsaturated alcohols, aldehydes and ketones. In addition, the reaction mixtures contained several compounds derived from the interaction of the lipid or its degradation products with Maillard reaction intermediates. In reaction mixtures containing cysteine, ribose and phospholipid, the most abundant volatile compounds were 2-pentylpyridine, 2-pentylthiophene, 2-hexylthiophene and 2-pentyl-2*H*-thiapyran. Smaller amounts of other 2-alkylthiophenes with n-alkyl substituents between C_4 and C_8 were found together with 2-(1-hexenyl)thiophene, 1-heptanethiol and 1-octanethiol. All these compounds are probably formed by the reaction of lipid breakdown products with hydrogen sulphide or ammonia derived from cysteine. Figure 12.7 shows a scheme for the formation of 2-pentylpyridine, 2-hexylthiophene and 2-pentyl-2*H*-thiapyran from 2,4-decadienal which is one of the major oxidation products of polyunsaturated fatty acids.

The particular role of the phospholipids, compared with triglycerides, was also examined (Farmer and Mottram, 1990b). The structural phos-

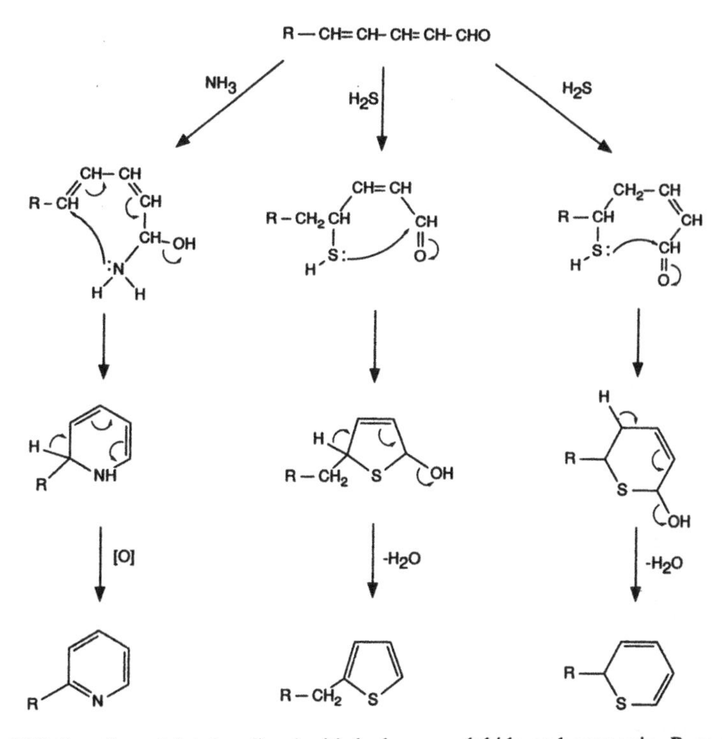

Figure 12.7 Reaction of 2,4-decadienal with hydrogen sulphide and ammonia. R = C_5H_{11}. (Farmer and Mottram, 1990b.)

pholipids contain a high proportion of polyunsaturated fatty acids, particularly those with three or more double bonds such as arachidonic acid (20:4), and these would be expected to break down during heating to give products which could react with Maillard products. The triglycerides in meat contain only a very small proportion of polyunsaturated fatty acids and this may explain the observations on defatted meat, which suggested that phospholipids rather than triglycerides were important for meat flavour. Reaction mixtures of cysteine and ribose were prepared with different lipids: triglycerides (BTG) and phospholipids (BPL) extracted from beef, and commercial egg phosphatidylcholine (PC) and phosphatidylethanolamine (PE). The aroma of the reaction mixture without any lipid was described as 'sulphurous, rubbery' but there was a distinct underlying meaty aroma. Addition of the beef triglyceride did not affect the aroma; however when beef phospholipids were used, the meaty aroma was more intense and the sulphurous notes less pronounced. Similarly, the addition of PC or PE gave mixtures with increased meatiness, with the mixture containing PE exhibiting the most meaty character.

The effect of these different lipids on the formation of selected volatiles was also examined (Table 12.2). The lipid preparations differed in the way they influenced the profile of volatiles from the reaction mixtures. All the phospholipids produced 2-pentylpyridine, 2-pentylthiapyran and the 2-alkylthiophenes, but the quantities differed markedly. Only trace amounts of these compounds were found in the triglyceride-containing system. Many Maillard reaction products showed marked reductions on addition of lipid, although not all the volatiles were affected to the same extent, and the different lipids did not all behave in the same way. In general BTG showed some effect on the Maillard volatiles, but this was not as marked as the phospholipid preparations.

Table 12.2 Relative concentration of some selected heterocyclic compounds formed from the reaction between cysteine and ribose with different lipids. (Mottram and Salter, 1989; Farmer and Mottram, 1990b)

Compound	No lipid	BTG	BPL	PC	PE
2-Mercapto-3-pentanone	1	0.72	0.49	0.53	0.53
3-Mercapto-2-pentanone	1	0.77	0.47	0.50	0.49
2-Furylmethanethiol	1	0.67	0.63	0.62	0.72
2-Methyl-3-furanthiol	1	0.40	0.15	0.27	0.24
2-Thiophenethiol	1	0.32	0.03	0.46	0.14
2-Methyl-3-thiophenethiol	1	0.08	0.01	0.20	0.10
2-Pentylpyridine	0	0.09	1	18.6	1.5
2-Pentylthiophene	0	0	1	23.9	10.9
2-Hexylthiophene	0	0.15	1	6.6	2.4
2-Pentyl-2*H*-thiapyran	0	0.01	1	11.0	4.0

These results clearly demonstrated the important role that lipids, especially phospholipids, play in these Maillard reactions which are the basis of meat flavour formation. The early stages of the Maillard reaction give rise to many reactive intermediates, of which dicarbonyls, furfurals, furanones, Strecker aldehydes, ammonia and hydrogen sulphide are the most important. These intermediates provide the reactants for most of the important classes of meat aroma volatiles. The relative amounts of the different volatiles produced from these intermediates must depend on the concentrations produced by the early Maillard reaction, and the relative rates of the different reactions. The addition of lipid (especially unsaturated lipids) and lipid degradation products (such as aldehydes, ketones and alcohols) to this mixture of Maillard intermediates provides competing reactions which produce other volatiles and affect the relative proportions of compounds produced by the other reactions. Hydrogen sulphide and ammonia are extremely important intermediates in many of the aroma forming reactions in meat, and their interaction with lipid degradation products is a clear example of how lipid will influence the relative proportions of heterocyclic Maillard reaction products.

12.6.2 Meat volatiles from lipid–Maillard interaction

Examination of the volatiles identified in meat shows a number of compounds which could be formed from the interaction of lipid with the Maillard reaction (Table 12.3). These include a number of alkylpyridines in roast lamb fat, including 2-pentylpyridine which has also been found in all the other main species of meat. The reaction of ammonia or amino acids with 2,4-dienals is a likely route to this type of compound (Figure 12.7). Related reactions involving hydrogen sulphide instead of ammonia provide pathways to the 2-alkylthiophenes which also have been found in roast or pressure cooked beef and fried chicken. Several thiazoles with long n-alkyl substituents in the 2-position have been reported in roast beef, fried chicken and fried bacon by the research group at Rutgers University. Some other fried foods, in particular potatoes, have also been found to contain similar n-alkyl substituted compounds (Carlin *et al.*, 1986; Ho *et al.*, 1987). Oxazoles with similar n-alkyl substituents were also found in fried potato, but in roast beef and fried bacon the oxazoles reported had the long n-alkyl substituents in the 4-position. Mechanisms for thiazole and oxazole formation in heated foods have been suggested to involve aldehydes, usually acetaldehyde and Strecker aldehydes (Figure 12.5), and it is reasonable to assume that lipid derived aldehydes can participate in these reactions to yield the long chain 2-alkyl thiazoles and oxazoles, although the formation of the 4-alkyloxazoles requires a modification to this mechanism.

Other heterocyclic compounds with long n-alkyl substituents found in

Table 12.3 Some heterocyclic volatiles identified in meat, which are derived from the interaction of lipid with the Maillard reaction

Compound	Type of meat
2-Butylpyridine	Lamb[1], chicken[2]
2-Pentylpyridine	Beef[3], pork[4], lamb[1], chicken[2]
3-Pentylpyridine	Lamb[1]
2-Pentyl-5-methylpyridine	Lamb[1]
2-Pentyl-5-ethylpyridine	Lamb[1]
2-Hexylpyridine	Lamb[1]
2-Butylpyrrole	Beef[5]
1-Pentylpyrrole	Beef[6]
2-Butylpyrazine	Chicken[2]
2-Butyl-3-methylpyrazine	Chicken[2]
Methylpentylpyrazine	Pork[4]
Dimethylbutylpyrazine	Pork[4]
Dimethylpentylpyrazine	Pork[4]
2-Butyl-4,5-dimethylthiazole	Beef[5], chicken[2], bacon[7]
2-Butyl-4-methyl-5-ethylthiazole	Beef[5], chicken[2]
2-Pentyl-5-methylpyrazine	Beef[5]
2-Pentyl-4,5-dimethylthiazole	Beef[5], chicken[2]
2-Hexyl-4,5-dimethylthiazole	Chicken[2]
2-Heptyl-4,5-dimethylthiazole	Chicken[2]
2-Heptyl-4-methyl-5-ethylthiazole	Chicken[2]
2-Octyl-4,5-dimethylthiazole	Chicken[2]
2-Butylthiophene	Beef[8,9], chicken[2]
2-Pentylthiophene	Beef[8,9], chicken[2]
2-Hexylthiophene	Beef[8]
2-Heptylthiophene	Beef[8]
2-Octylthiophene	Beef[8,9]
Tetradecylthiophene	Beef[9]
2-Butanoylthiophene	Beef[10]
2-Heptanoylthiophene	Beef[10]
2-Octanoylthiophene	Beef[10]
3-Methyl-5-butyl-1,2,4-trithiolane	Chicken[11]
3-Methyl-5-pentyl-1,2,4-trithiolane	chicken[11]
2-Butyloxazole	Bacon[12]
2-Methyl-5-pentyloxazole	Bacon[12]
2,5-Dimethyl-4-butyloxazole	Beef[5], bacon[12]
2,5-Dimethyl-4-pentyloxazole	Bacon[12]
2,5-Dimethyl-4-hexyloxazole	Beef[5]

[1]Buttery *et al.* (1977); [2]Tang *et al.* (1983); [3]Watanabe and Sato (1971); [4]Mottram (1985); [5]Hartman *et al.* (1983); [6]Coppock and MacLeod (1977); [7]Ho *et al.* (1983); [8]Min *et al.* (1979); [9]Wilson *et al.* (1973); [10]Hsu *et al.* (1982); [11]Hwang *et al.* (1986); [12]Ho and Carlin (1989).

meat include butyl and pentyl pyrazines, and it has been suggested that these could be formed from the reaction of pentanal or hexanal with a dihydropyrazine, formed from the condensation of two aminoketone molecules (Figure 12.8).

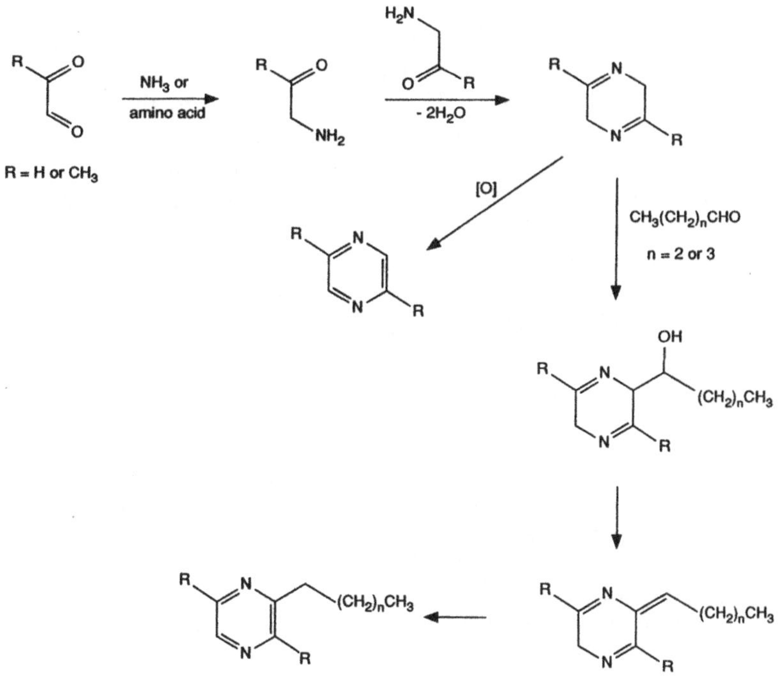

Figure 12.8 Mechanism for the formation of n-butyl and n-pentyl pyrazines. (Ho *et al.*, 1987; Chiu *et al.*, 1990.)

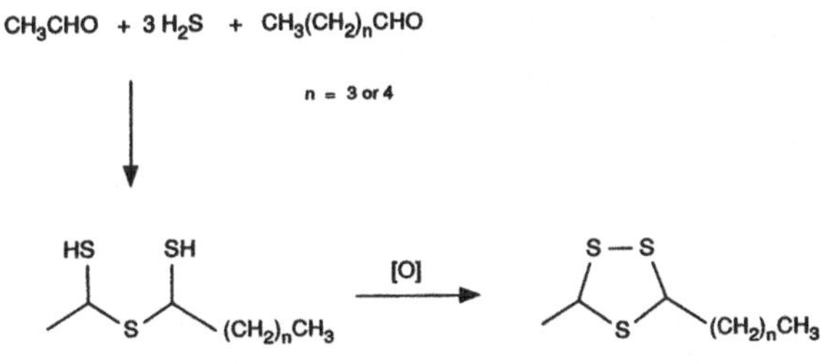

Figure 12.9 Mechanism for the formation of trithiolanes with n-butyl and n-pentyl substituents. (Boelens *et al.*, 1974; Ho *et al.*, 1987.)

Pentanal and hexanal also appear to be involved in the formation of 3-butyl-3-methyl-1,2,4-trithiolane and its 3-pentyl homologue, which have both been reported in fried chicken. Trithiolanes can be formed from aldehydes and hydrogen sulphide (Boelens *et al.*, 1974) and the reaction of hydrogen sulphide, acetaldehyde and pentanal or hexanal (Figure 12.9) has been suggested as the route to these butyl and pentyl trithiolanes (Ho *et al.*, 1987; Hwang *et al.*, 1986).

The aroma characteristics have only been reported for some of these alkyl substituted heterocyclic compounds, but those which have been examined suggest that they may contribute to fatty, fried aromas (Buttery *et al.*, 1977; Ho *et al.*, 1987). The reactions leading to these compounds may influence the aroma of meat in another way by providing competing reactions which modify the extent to which other Maillard reactions can occur and thus affect the balance of aroma compounds produced during the cooking of meat. Thus, lipid has a crucial role in meat flavour, not only producing compounds which give fatty or species characteristics, but also by participating in the complex Maillard-type reactions which are responsible for the characteristic meaty aromas.

References

Boelens, M., van der Linde, L.M., de Valois, P.J., van Dort, H.M. and Takken, H.J. (1974). Organic sulphur compounds from fatty aldehydes, hydrogen sulfide, thiols and ammonia as flavour constituents. *J. Agric. Food Chem.*, **22**, 1071–1076.

Buttery, R.G., Lin, L.C., Teranishi, R. and Mon, T.R. (1977). Roast lamb fat: basic volatile components. *J. Agric. Food Chem.*, **25**, 1227–1229.

Buttery, R.G., Haddon, W.F., Seifert, R.M. and Turnbaugh, J.G. (1984). Thiamin odor and bis(2-methyl-3-furyl)disulphide. *J. Agric. Food Chem.*, **32**, 674–676.

Carlin, J.T., Jin, Q.Z., Huang, T.C. and Ho, C.T. (1986). Identification of alkyloxazoles in the volatile compounds from French-fried potatoes. *J. Agric. Food Chem.*, **34**, 621–623.

Chiu, E.M., Kuo, M.C., Bruechert, L.J. and Ho, C.T. (1990). Substitution of pyrazines by aldehydes in model systems. *J. Agric. Food Chem.*, **38**, 58–61.

Christie, W.W. (1978). The composition, structure and function of lipids in the tissues of ruminant animals. *Prog. Lipid Res.*, **17**, 111–205.

Coppock, B.M. and MacLeod, G. (1977). The effect of ageing on the sensory and chemical properties of boiled beef aroma. *J. Sci Food Agric.*, **28**, 206–214.

Farmer, L.J. and Mottram, D.S. (1990a). Recent studies on the formation of meat aroma compounds. In *Flavour Science and Technology*, ed. Y. Bessiere and A.F. Thomas. Wiley, Chichester, pp. 113–6.

Farmer, L.J. and Mottram, D.S. (1990b). Interaction of lipid in the Maillard reaction between cysteine and ribose: effect of a triglyceride and three phopholipids on the volatile products. *J. Sci. Food Agric.*, **53**, 590–525.

Farmer, L.J. and Patterson, R.L.S. (1991). Compounds contributing to meat flavour. *Food Chem.*, **40**, 201–205.

Farmer, L.J., Mottram, D.S. and Whitfield, F.B. (1989). Volatile compounds produced in Maillard reactions involving cysteine, ribose and phopholipid. *J. Sci. Food Agric.*, **49**, 347–368.

Feather, M.S. and Harris, J.F. (1973). Dehydration reactions of carbohydrates. *Adv. Carbohyd. Chem. Biochem.*, **28**, 161–224.

Flament, I., Kohler, M. and Aschiero, R. (1976). Sur l'arome de viande de boeuff grillee. II. Dihydro-6,7-5H-cyclopenta-[b]pyrazines, identification et mode de formation. *Helv. Chim. Acta.*, **59**, 2308–2313.

Flament, I., Sonnay, P. and Ohloff, G. (1977). Sur l'arome de viande de boeuff grillee. III. Pyrrolo[1,2a]pyrazines, identification et synthese. *Helv. Chim. Acta.*, **60**, 1872–1883.

Fors, S. (1983). Sensory properties of volatile Maillard reaction products and related compounds. In *The Maillard Reaction in Foods and Nutrition*, ed. G.R. Waller and M.S. Feather, American Chemical Society, Washington, DC, pp. 185–286.

Forss, D.A. (1972). Odour and flavour compounds from lipids. *Prog. Chem. Fats other Lipids*, **13**, 181–258.

Frankel, E.N. (1980). Lipid oxidation. *Prog. Lipid Res.*, **19**, 1–22.
Gasser, U. and Grosch, W. (1988). Identification of volatile flavour compounds with high aroma values from cooked beef. *Z. Lebensm. Unters Forsch.*, **186**, 489–494.
Grosch, W. (1982). Lipid degradation products and flavours. In *food Flavours*, ed. I.D. Morton and A.J. MacLeod, Elsevier, Amsterdam, pp. 325–398.
Hartman, G.J., Jin, Q.Z., Collins, G.J., Lee, K.N., Ho, C.T. and Chang, S.S. (1983). Nitrogen-containing heterocyclic compounds identified in the volatile flavour constituents of roast beef. *J. Agric. Food Chem.*, **31**, 1030–1033.
Ho, C.T. and Carlin, J.T. (1989). Formation and aroma characteristics of heterocyclic compounds in foods. In *Flavor Chemistry Trends and Developments*, ed. R. Teranishi, R.G. Buttery and F. Shahidi, American Chemical Society, Washington, DC, pp. 92–104.
Ho, C.T., Lee, K.N. and Jin, Q.Z. (1983). Isolation and identification of flavour compounds in fried bacon. *J. Agric. Food Chem.*, **31**, 336–342.
Ho, C.T., Carlin, J.T. and Huang, T.C. (1987). Flavour development in deep-fat fried foods. In *Flavour Science and Technology*, ed. M. Martens, G.A. Dalen and H. Russwurm, Wiley, Chichester, pp. 35–42.
Hornstein, I. and Crowe, P.F. (1960). Flavour studies on beef and pork. *J. Agric. Food Chem.*, **8**, 494–498.
Hornstein, I. and Crowe, P.F. (1963). Meat flavour: lamb. *J. Agric. Food Chem.*, **11**, 147–149.
Hsu, C.M., Peterson, R.J., Jin, Q.Z., Ho, C.T. and Chang, S.S. (1982). Characterization of new volatile compound in the neutral fraction of roast beef flavour. *J. Food Sci.*, **47**, 2068–2071.
Hurrell, R.F. (1982). Maillard reaction in flavour. In *Food Flavours*, ed. I.D. Morton and A.J. MacLeod, Elsevier, Amsterdam, pp. 399–437.
Hwang, S.S., Carlin, J.T., Bao, Y., Hartman, G.J. and Ho, C.T. (1986). Characterisation of volatile compounds generated from the reactions of aldehydes with ammonium sulphide. *J. Agric. Food Chem.*, **34**, 538–542.
Kramlich, W.E. and Pearson, A.M. (1960). Separation and identification of cooked beef flavour components. *Food Res.*, **25**, 712–719.
Lawrie, R.A. (1985). *Meat Science*, 4th edn. Pergamon Press, Oxford.
Macey, R.L. Jr., Naumann, N.D. and Bailey, M.E. (1964). Water-soluble flavour and odour precursors of meat. *J. Food Sci.*, **29**, 136–148.
MacLeod, G. (1986). The scientific and technological basis of meat flavours. In *Developments in Food Flavours*, ed. G.G. Birch and M.G. Lindley, Elsevier, London, pp. 191–223.
MacLeod, G. and Ames, J.M. (1986). 2-Methyl-3-(methylthio)furan: a meaty character impact aroma compound identified from cooked beef. *Chem. Ind. (London)*, 175–177.
MacLeod, G. and Seyyedain-Ardebili, M. (1981). Natural and simulated meat flavours (with particular reference to beef). *CRC Crit. Rev. Food Sci. Nutr.*, **14**, 309–437.
Mauron, J. (1981). The Maillard reaction in food: a critical review from the nutritional standpoint. In *Maillard Reactions in Food*, ed. C. Eriksson, Pergamon Press, Oxsford, pp. 3–35.
Min, D.B., Ina, K., Peterson, R.J. and Chang, S.S. (1979). Preliminary identification of volatile flavour compounds in the neutral fraction of roast beef. *J. Food Sci.*, **44**, 639–642.
Morton, I.D., Akroyd, P. and May, C.G. (1960). Flavouring substances and their preparation. *Brit. Patent* 836, 694.
Mottram, D.S. (1985). The effect of cooking conditions on the formation of volatile heterocyclic compounds in pork. *J. Sci. Food Agric.*, **36**, 377–382.
Mottram, D.S. (1991). Meat. In *Volatile Compounds in Foods and Beverages*, ed. H. Maarse, Marcel Dekker, New York, pp. 107–177.
Mottram, D.S. and Edwards, R.A. (1983). The role of triglycerides and phospholipids in the aroma of cooked beef. *J. Sci. Food Agric.*, **34**, 517–522.
Mottram, D.S. and Salter, L.J. (1989). Flavour formation in meat-related Maillard systems containing phospholipids. In *Thermal Generation of Aroma*, ed. T.H. Parliment, R.J. McGorrin and C.T. Ho, American Chemical Society, Washington, pp. 442–451.
Mottram, D.S., Edwards, R.A. and MacFie, H.J.H. (1982). A comparison of the flavour volatiles from cooked beef and pork meat systems. *J. Sci. Food Agric.*, **33**, 934–944.
Noleau, I. and Toulemonde, B. (1987). Volatile components of roast chicken fat. *Lebensm. Wiss. Technol.*, **20**, 37–41.

Petit, A.O. and Hruza, D.A. (1974). Comparative study of flavour properties of thiazole derivatives. *J. Agric. Food Chem.*, **22**, 264–269.

Salter, L.J., Mottram, D.S. and Whitfield, F.B. (1988). Volatile compounds produced in Maillard reactions involving glycine, ribose and phospholipid. *J. Sci. Food Agric.*, **46**, 227–242.

Schutte, L. (1974). Precursors of sulphur-containing flavour compounds. *CRC Crit. Rev. Food Technol.*, **4**, 457–505.

Tang, J., Jin, Q.Z., Shen, G.H., Ho, C.T. and Chang, S.S. (1983). Isolation and identification of volatile compounds from fried chicken. *J. Agric. Food Chem.*, **31**, 1287–1292.

van den Ouweland, G.A.M. and Peer, H.G.F. (1975). Components contributing to beef flavour. Volatile compounds produced by the reaction of 4-hydroxy-5-methyl-3(2H)-furanone and its thio analog with hydrogen sulfide. *J. Agric. Food Chem.*, **23**, 501–505.

van den Ouweland, G.A.M., Peer, H.G. and Tjan, S.B. (1978). Occurrence of Amadori and Heyns rearrangement products in processed foods and their role in flavour formation. In *Flavor in Foods and Beverages*, ed. G. Charalambous and G.E. Inglett, Academic Press, New York, pp. 131–143.

van der Linde, L.M., van Dort, J.M., de Valois, P., Boelens, H. and de Rijke, D. (1979). Volatile compounds from thermally degraded thiamin. In *Progress in Flavour Research*, ed. D.G. Land and H.E. Nursten, Applied Science, London, pp. 219–224.

Vernin, G. and Parkanyi, C. (1982). Mechanisms of formation of heterocyclic compounds in Maillard and pyrolysis reactions. In *Chemistry of Heterocyclic Compounds in Flavours and Aromas*, ed. G. Vernin, Ellis Horwood, Chichester, pp. 151–207.

Wasserman, A.E. (1979). Chemical basis for meat flavour: a review. *J. Food Sci.*, **44**, 6–11.

Wasserman, A.E. and Gray, N. (1965). Meat flavour. I. Fractionation of water-soluble flavour precursors of beef. *J. Food Sci.*, **30**, 801–807.

Wasserman, A.E. and Spinelli, A.M. (1972). Effect of some water-soluble components on aroma of heated adipose tissue. *J. Agric. Food Chem.*, **20**, 171–174.

Watanabe, K. and Sato, Y. (1970). Conversion of some saturated fatty acids, aldehydes and alcohols into γ- and δ-lactones. *Agric. Biol. Chem.*, **34**, 464–472.

Watanabe, K. and Sato, Y. (1971). Some alkyl-substituted pyrazines and pyridines in the flavour components of shallow fried beef. *J. Agric. Food Chem.*, **19**, 1017–1019.

Werkhoff, P., Bruning, J., Emberger, R., Guntert, M., Kopsel, M., Kuhn, W. and Surburg, H. (1990). Isolation and characterisation of volatile sulphur-containing meat flavour components in model systems. *J. Agric. Food Chem.*, **38**, 777–791.

Whitfield, F.B., Mottram, D.S., Brock, S., Puckey, D.J. and Salter, L.J. (1988). The effect of phospholipid on the formation of volatile heterocyclic compounds in heated aqueous solutions of amino acids and ribose. *J. Sci. Food Agric.*, **42**, 261–272.

Wilson, R.A., Mussinan, C.J., Katz, I. and Sanderson, A. (1973). Isolation and identification of some sulphur chemicals present in pressure-cooked beef. *J. Agric. Food Chem.*, **21**, 873–876.

Wong, E., Nixon, L.N. and Johnson, C.B. (1975). Volatile medium chain fatty acids and mutton fat. *J. Agric. Food Chem.*, **23**, 495–498.

13 Instrumental methods of meat flavour analysis

A.J. MACLEOD

13.1 Introduction

In any flavour analysis, once a valid sample has been obtained the problem can be stated quite simply: (1) to separate the mixture into its individual components, and (2) to identify and quantify the separated components. Ideally, the aroma contributions of the constituents should also be assessed, but this is not entirely within the scope of instrumental analysis, and so will not be considered here.

For such a dual problem, having to cope with minute quantities of solutes in highly complex mixtures (especially the case in meat flavour analysis), tandem analysis is virtually essential—that is, linking the analytical instrumentation directly on-line beyond the separatory device, rather than applying the two techniques independently and isolating all the components of the mixture in a pure state, one by one, before their analytical investigation.

Of all possible methods of tandem analysis, gas chromatography combined with mass spectrometry (GC–MS) has undoubtedly been *the* technique of choice for all flavour and aroma analysis for many years—indeed, since the very birth of the technique when the two instruments were first successfully integrated. There are many reasons for this pre-eminence, including the fact that GC provides the best performance of all separatory procedures, and in dealing with solutes in the vapour phase it is ideally suited to the study of flavour volatiles. Mass spectrometry is one of the most powerful techniques for identification of unknown compounds, and although nuclear magnetic resonance spectroscopy is probably superior for most applications, this is broadly irrelevant here as it cannot operate in a tandem mode, on-line beyond a separatory device. Furthermore, its sensitivity, which is obviously very important in trace analysis, is generally inferior to that of mass spectrometry.

For three decades, GC–MS has reigned supreme in flavour analysis, and continues to do so. The vast majority of meat flavour papers published in the last few years have still described the use of GC–MS, and only GC–MS, to study the particular problem and to identify components. The technique is thus so standard and so routine in meat flavour analysis that there is no need to describe it nor to discuss it any further here.

However, there are other ways of tackling the problem which can, in certain circumstances, provide valuable additional and/or complementary information to GC–MS. They can be classified under four main headings:

1. Refinements to routine GC in GC–MS.
2. Refinements to routine MS in GC–MS.
3. Alternatives to GC as a method of separation prior to identification.
4. Alternatives to MS as a method of identification following separation.

Before dealing with these, perhaps it is appropriate to emphasize the detail of what is here considered to be the current, standard routine GC–MS used in flavour analysis—that is, fused silica, capillary column GC, providing high resolution, combined with fast scanning, high sensitivity MS operating in electron impact ionization mode (EI).

13.2 Refinements to routine GC in GC–MS

The main refinements to routine GC in GC–MS which have been used in flavour analysis are:

1. Multidimensional gas chromatography (MDGC).
2. Chiral phase gas chromatography.
3. Preparative gas chromatography.

13.2.1 Multidimensional gas chromatography (MDGC)

With samples as complex as those obtained in typical flavour analysis, even using the best high-resolution GC columns, components sometimes co-elute, producing mixed mass spectra on GC–MS which are difficult to interpret. MDGC, in which typically two different GC columns are used, a pre-column and an analytical column, can overcome this problem, and the technique can be applied in a number of ways, e.g. foreflushing, backflushing, heartcutting. It is easy to construct one's own apparatus for MDGC, but commercial instrumentation is available. Although this useful procedure has been quite widely used in flavour analysis (Nitz and Julich, 1984; Schomburg *et al.*, 1984; Nitz, 1985; Flament *et al.*, 1988; Nitz *et al.*, 1988; Van Wassenhove *et al.*, 1988; Hener *et al.*, 1990; Homatidou *et al.*, 1990), it has not been extensively applied to studies on meat.

13.2.2 Chiral phase gas chromatography

Another refinement in GC technique which is increasingly being used in general flavour analysis, but has not yet been taken up to any significant extent by meat researchers, is the use of a chiral stationary phase. As is well known, different enantiomers of the same compound can have different aroma properties, so properly characterizing the configuration of optically active flavour constituents can be important. Using an appropriate chiral stationary phase in GC the components of a racemic mixture will form diastereoisomeric species which, due to their different physical properties, can then be resolved. The same objective can be achieved by derivatizing with chiral reagents prior to conventional GC on an achiral phase (Tressl *et al.*, 1985a; Engel, 1988; Krammer *et al.*, 1988), but this procedure can be more difficult and, as a variation in sample preparation, does not fall into the category of refinements in GC. Modified β-cyclodextrins and various organo-metallic complexes are common chiral GC phases, and they have been employed in a variety of flavour analyses (Bricout, 1987; Schurig, 1988; Guntert *et al.*, 1990; Hener *et al.*, 1990; Mosandl *et al.*, 1990; Werkhoff *et al.*, 1990a).

13.2.3 Preparative gas chromatography

Although techniques of tandem analysis are almost universally employed in flavour analysis, preparative GC followed by off-line analysis is possible in certain simple situations. This does then provide the immense advantage of it being possible to analyse the collected solute at leisure by a variety of techniques, including the powerful NMR. Preparative GC is, however, extremely difficult, and requires great skill and technical expertise, which partly explains its limited use. Nevertheless it has been applied in studies of meat flavour compounds, but in the simpler model systems (Tressl *et al.*, 1985b, 1986; Werkhoff *et al.*, 1989).

13.3 Refinements to routine MS in GC–MS

The main refinements to routine electron impact (EI) MS in GC–MS which have been used in flavour analysis are:

1. High resolution mass spectrometry.
2. Selected ion monitoring mass spectrometry.
3. Chemical ionization mass spectrometry.
4. Negative ion chemical ionization mass spectrometry.

13.3.1 High resolution mass spectrometry

The ability of the modern high resolution mass spectrometer to yield precise elemental compositions in the spectra of compounds separated by capillary GC has not yet been widely exploited in flavour analysis. However, with the continuing improvements in the performance of commercial magnetic sector mass spectrometers, especially with regard to sensitivity at high resolution, there is little doubt that this valuable facility will become more readily available and hence more widely used in the future.

13.3.2 Selected ion monitoring mass spectrometry

In selected ion monitoring, sometimes known as 'mass fragmentography', only selected ions representative of a specific compound, or group of compounds, being searched for are recorded during GC–MS. The technique is extremely useful in enabling a very high sensitivity assay for the *known* component or types of components in question, but it does not contribute to the identification of unknown compounds, since full spectra are not recorded. A number of examples of its use in flavour analysis can be quoted (Eberhardt *et al.*, 1981; ten Noever de Brauw and van Ingen, 1981; Hirvi and Honkanen, 1983, 1984; Liddle and Bossard, 1984; Whitfield and Shaw, 1984; Garcia-Regueiro and Diaz, 1989).

A related approach is 'mass chromatography', which is useful for deconvoluting seemingly single GC peaks which nevertheless produce mixed mass spectra (Thomas *et al.*, 1984). The difference here, however, is that complete mass spectra have been recorded throughout the GC–MS run, rather than selected ions as in mass fragmentography. Data systems can then be instructed to select appropriate specific ions from the full recorded spectra through the GC peak, with the objective of artificially resolving and recognizing the two (or more) components of the peak.

Mass chromatography can also be considered as retrospective selected ion monitoring. Thus the selected ion or ions from the full spectra can be output as single ion chromatograms throughout the whole GC run, rather than just one peak, hence pinpointing the compound or group of compounds of interest (e.g. selecting m/z 93 and 136 will isolate most monoterpene hydrocarbons from the full total ion current trace). This approach has the advantage over genuine selected ion monitoring that full spectra are available for detailed interpretation, but on the other hand it is, of course, far less sensitive.

13.3.3 Chemical ionization mass spectrometry

One of the main frustrations in conventional EI MS is when no molecular ion peak is obtained in the mass spectrum. This is due to the instability of the molecular ion under the excess energy imparted by electron impact (an energy of 70 eV is usually employed in EI MS). However, this is not so much a problem in meat flavour analysis as in other areas, since many of the aroma components are aromatic (in the chemical sense), including many of the heterocyclic compounds, and such compounds generally yield molecular ions of reasonable abundance due to aromatic stabilization. Nevertheless, not all meat flavour compounds fall into this category, and use of a 'softer' ionization method than electron impact, which imparts less energy to the molecular ion and hence limits its fragmentation, can be very useful. Chemical ionization (CI) is the most common alternative, softer approach, in GC–MS. In CI MS a reagent gas, such as methane, isobutane or ammonia, is introduced into the mass spectrometer source to be ionized by broadly conventional electron impact. A range of positive ions, such as $C_2H_5^+$ from methane, is produced. Sample molecules are then ionized by ion–molecule reactions with the reagent gas species, with the result that so-called pseudo-molecular ions are produced, such as $(M+H)^+$ by proton transfer. Typically an energy of only 5 eV is imparted to sample molecules, so usually very little fragmentation is observed under these circumstances.

The value of CI MS in flavour analysis is to complement and supplement the data provided by EI MS. It is not often used on its own, although there are exceptions (Lange and Schultze, 1988a,b), for the very reason that few fragmentation data for interpretation are usually obtained.

13.3.4 Negative ion chemical ionization mass spectrometry

As well as the above positive ion CI MS, negative ion CI MS is also possible, in which negatively charged reagent gas ions, such as OH^-, undergo similar ion–molecule interactions with sample molecules, but with the result that negatively charged pseudo-molecular ions are obtained, such as $(M-H)^-$, for example, produced by proton abstraction. In many respects, negative ion CI MS can be superior to positive ion CI MS, both in terms of sensitivity and 'softness'. A good example of the latter is that isobornyl acetate (Mol. wt 196) still fragments under positive ion CI MS to yield only one main peak in the spectrum at m/z 137 due to $(M-CH_3COO)^+$, whereas under negative ion CI MS an intense peak at m/z 195, due to $(M-H)^-$, is obtained (Bruins, 1985). Negative ion CI MS has not been widely used in flavour analysis (Bruins, 1979; Hendriks and Bruins, 1980, 1983; George, 1984), but it has great potential and is an underused technique.

There are some other methods of 'soft' ionization in mass spectrometry, but they are either not applicable to GC–MS (e.g. fast atom bombardment, or FAB) or they have been very little used in flavour analysis (e.g. photoionization MS (Adamczyk *et al.*, 1987).

13.4 Alternatives to GC as a method of separation prior to identification

There are four main alternatives to GC as the method of separation prior to identification:

1. High-performance liquid chromatography (HPLC).
2. Supercritical fluid chromatography (SFC).
3. Capillary zone electrophoresis (CZE).
4. Mass spectrometry in MS–MS.

13.4.1 High-performance liquid chromatography

An obvious question is why even consider HPLC, when GC offers superior performance, is a commercially more highly developed technique, and by definition is more suited to dealing with volatile components. Answers to those specific points include the fact that modern HPLC is, in fact, a more *efficient* chromatographic process than even the best GC, routinely providing far greater numbers of theoretical plates per unit length and hence superior HETP values. GC, however, provides far better *performance* overall, by virtue of the much longer open tubular columns which can routinely be used, e.g. 50 m, whereas typical HPLC columns are only 25 cm in length. Although GC is undoubtedly the more developed technique, it did have a start of about two decades, and HPLC is rapidly catching up, with recent developments in instrumentation and column packings. Finally, it must not be overlooked that HPLC can also cope with analysis of volatiles in the same way as GC, in addition to being able to deal with non-volatiles. It is also better suited to the analysis of thermally labile flavour components, although clearly this is not such a significant advantage in meat flavour analysis as in some other areas.

The main problem which has held back HPLC as a viable alternative to GC in combination with mass spectrometry has been the great difficulty in satisfactorily and efficiently interfacing the two techniques. There are now, however, a number of ways in which this can be accomplished, including the following:

1. moving belt interface
2. direct liquid introduction
3. thermospray ionization
4. momentum interface
5. particle beam interface
6. electrospray ionization
7. continuous flow–fast atom bombardment (CF–FAB).

It is not possible to give further details of these interfaces here, but each has its advantages and disadvantages. Some are perhaps rather old-fashioned now, e.g. the moving belt, and some are still rather more at the research and development stage, e.g. electrospray ionization. HPLC–MS has not yet been widely exploited in flavour analysis, although it has been used by some authors (Games *et al.*, 1984; van der Greef *et al.*, 1984;

Hartman *et al.*, 1989; Werkhoff *et al.*, 1990b). As improved commercial instrumentation becomes more available its use should increase.

13.4.2 *Supercritical fluid chromatography*

When a gas, such as CO_2, is heated above its critical temperature and pressurized above a certain level, it becomes a fluid with liquid-like density and solvating power, which broadly has properties between those of a gas and a liquid. Use of such a supercritical fluid as the mobile phase in chromatography provides a technique somewhat intermediate between GC and HPLC. GC provides better chromatographic performance, but supercritical fluid chromatography (SFC), like HPLC, is also capable of dealing with solutes not amenable to GC (e.g. those of low volatility, high polarity or thermal instability) as well as volatile components. SFC is superior to HPLC in a number of respects. For example, as just explained, the integration of HPLC with MS is somewhat problematical, whereas the combination of SFC with MS is easier. In particular, mobile phase elimination after chromatography before MS is clearly less of a problem with a supercritical fluid such as CO_2 than with a conventional liquid solvent. Commercial instrumentation for SFC exists and has been successfully interfaced with mass spectrometry. Both packed and capillary column SFC are possible; supercritical CO_2 is the most common mobile phase.

In many respects, SFC combines the best features of GC and HPLC, namely the excellent resolving power of the former and the mild operating conditions of the latter, but to date it has not been extensively used in flavour analysis. Some examples can be quoted (Hellgeth *et al.*, 1986; Skelton *et al.*, 1986; Flament *et al.*, 1987; Werkhoff *et al.*, 1990), but this certainly is a technique that should be used to a far greater extent than it presently is.

Interestingly, as well as being combined with MS as the identification technique, SFC has also been linked to Fourier transform infrared spectroscopy (Morin *et al.*, 1986, 1987), and used in flavour analysis (Morin *et al.*, 1987).

13.4.3 *Capillary zone electrophoresis*

In capillary zone electrophoresis (CZE), a thin-walled fused silica capillary, about 1 m in length, is filled with a buffer solution and one end held in a buffer reservoir at ground potential with the other end in the buffer at a high voltage potential, typically 30 kV. Under the influence of the applied electric field, ionic species have a tendency to migrate electrophoretically to the appropriate electrode. The speed at which these species migrate is dependent on the strength of the applied electric field and the mass to charge ratio of analyte molecules. Thus, a small singly charged species migrates at a faster rate than a large multiply charged molecule, so that the latter will have a longer retention time.

In addition, however, there is also an overwhelming electro-osmotic flow that sweeps all solutes through the capillary from positive to negative electrode, but without itself promoting any separation. This effect is caused by the interaction of the buffer and the negatively charged capillary wall, that arises from the ionization of surface silanol groups. As a result, all components actually flow in the same direction, and although not true chromatography, CZE can then be used chromatographically. Thus, positively charged species migrate towards the cathode at a speed corresponding to the sum of the electro-osmotic flow and electrokinetic migration. Neutral components migrate under the influence of electro-osmotic flow alone, and negative species flow towards the cathode at a rate determined by the difference between the electro-osmotic flow and the electrokinetic migration experienced by the molecule. CZE thus provides a complex and highly efficient separatory process, and close to 1 million theoretical plates per metre have been obtained.

Furthermore, CZE has also been interfaced successfully with mass spectrometry, usually either via electrospray ionization (Olivares *et al.*, 1987; Smith *et al.*, 1988a,b) or via CF/FAB (Caprioli *et al.*, 1989; Moseley *et al.*, 1989). CZE/MS constitutes a powerful analytical technique which, although as yet very much in its infancy, is causing great enthusiasm and excitement amongst biological scientists. It is ideally suited to the analysis of a very wide range of labile biological molecules. It is included here mainly on the basis of its potential, since it has not as yet been used to any great extent in any flavour analysis. Impressive results have, however, very recently been obtained using CZE/MS to analyse products from Maillard model systems (Tomlinson, 1991).

13.4.4 Mass spectrometry in MS–MS

MS–MS is now taken to refer to mass spectrometry–mass spectrometry, although the acronym was originally devised for mass separation–mass spectrometry (McLafferty and Bockoff, 1978) and this is, in fact, far more appropriate. Nevertheless, in MS–MS there are, in effect, two mass spectrometers linked together, but with the first achieving separation of a mixture on the basis of individual selection of constituents, and the second providing the conventional mass spectrometry process. The mixture is introduced into MS-1 and the molecular ions (or pseudo-molecular ions from CI or FAB MS) which are produced are separated in the normal manner, according to their different masses. At the exit of MS-1 the separated molecular ion is subjected to collisionally activated dissociation (CAD) by collision with a neutral gas in a high pressure region, as a result of which the molecular ion fragments into characteristic daughter ions.

These are then mass analysed conventionally in MS-2 to provide a pure spectrum. It should be emphasized that there are other forms of MS–MS, but the preceding concept is the one that is the more relevant here.

Although this type of approach has been known for a long time, for example in the study of metastable ions and in 'mass analysed ion kinetic energy spectrometry' (MIKES), it is only during the last decade that MS–MS has been realized as a simple and effective procedure for analysing mixtures as such. It is not, of course, necessary to buy two mass spectrometers, and a wide range of different types of multiple sector instruments is now commercially available for MS–MS. Triple sector instruments provide an extra analyser (electrostatic, magnetic or quadrupole) beyond the CAD cell after MS-1, and four-sector instruments with a variety of geometries are also possible. Perhaps the most common configuration at present is a hybrid triple sector, consisting of a quadrupole analyser (as MS-2) beyond an otherwise conventional double focussing instrument (as MS-1), although triple quads are also possible for less demanding work.

Obviously MS–MS fails if isomers are present or if at low resolution other different components in the mixture give the same unit mass parent ion in MS-1. The latter can, of course, be overcome at high resolution. A problem which is less easy to overcome is if a compound does not yield a molecular or pseudo-molecular ion in sufficient abundance for significant CAD, which itself is not always an efficient process.

The main advantages of MS–MS over GC–MS are its simplicity and the fact that, by its very nature, the necessity for extensive sample preparation and handling is reduced. Nevertheless, it is highly unlikely that MS–MS will ever replace GC–MS in flavour analysis, although it is certain that as commercial instrumentation becomes more widely available, its use will increase. At present, MS–MS is employed much more widely in other analytical fields, and of the relatively limited applications in flavour analysis (Davis and Cooks, 1982; Labows and Shushan, 1983; Walther *et al.*, 1983; Fraisse *et al.*, 1984; Busch and Kroha, 1985), the majority has been in determining specific compounds or groups of compounds, which is one of the strengths of the technique.

13.5 Alternatives to MS as a method of identification following separation

There is really only one instrumental method of analysis, other than MS, which can satisfactorily be combined, on-line in tandem beyond a separatory procedure, and that is Fourier transform infrared spectroscopy.

13.5.1 Fourier transform infrared spectroscopy (FTIR)

Although the only viable possibility in this category, FTIR has been very widely and very commonly used in flavour analysis, more so than any of the other variations considered here. However, that is not to say that it is the most useful. Nor is it a rival to MS in GC–MS; rather it provides valuable, complementary information.

A major problem with GC–FTIR is that it possesses significantly lower sensitivity (and smaller dynamic range) than GC–MS, but on the other hand, when IR spectra can be obtained they can provide important structural information which is either lacking or less obvious in the mass spectrum. The most valuable attribute of IR spectroscopy, of course, is in providing information regarding functional groups and their environment, but it can also sometimes be extremely useful, in flavour studies, in enabling discrimination between isomers, especially geometric isomers. In addition, since the IR spectrum of a compound is a virtually unique 'fingerprint', the technique provides a powerful single method of identification, in combination with an appropriate library of reference spectra. In this latter context, however, another major limitation of GC–FTIR at present is the lack of adequate, extensive data collections of vapour phase IR spectra, but doubtless this situation will improve rapidly as the technique is more used.

Many excellent descriptions of the apparatus for GC–FTIR exist in the literature (Schreier and Idstein, 1985a), which relates specifically to its application in flavour analysis, but in summary the equipment functions as follows. The effluent from the GC is passed through a so-called lightpipe, which is a heated capillary gas cell internally coated with a layer of gold and capped at both ends with KBr windows. IR radiation, typically in the range 4000–750 cm^{-1}, passes through the cell, interacting with components as they elute from the GC, to be detected at the other end by a suitable device (e.g. a cooled mercury cadmium telluride (MCT) semiconductor detector). Interferograms result which are processed by computer to yield typical IR spectra. Full spectra can readily be obtained from 1 s 'scans' showing a resolution of 4 cm^{-1}. Many different commercial GC–FTIR instruments are available.

A newer type of GC–FTIR uses a matrix isolation interface, and offers improved sensitivity of about 100-fold. Again, commercial equipment is available. In this system, the effluent stream from the capillary GC is mixed with a matrix gas (usually 1% argon) and sprayed onto a cryogenic surface (often a slowly rotating disc) at about 12°K, where it immediately freezes on the surface. Helium carrier gas is removed by vacuum, while the argon, which is then in excess, forms a solid matrix completely surrounding and trapping solute molecules. FTIR spectra can then be taken at leisure. If the cold surface is transparent to IR (e.g. a CsI or CsBr window) then the IR beam passes through the sample and surface to yield an absorption spectrum; the frozen matrix gas is transparent to IR and therefore does not interfere. If the surface is a mirror (e.g. gold coated), then the IR beam passes through the sample and is then reflected back off the mirror surface to give a reflectance spectrum. The use of cryogenic matrix isolation GC–FTIR in flavour analysis has been well described (Williams *et al.*, 1987).

A considerable number of papers has been published describing the use of GC–FTIR in flavour analysis (Herres *et al.*, 1983; Schreier and Idstein, 1984, 1985b; Idstein *et al.*, 1984; Idstein and Schreier, 1985a,b; Purcell and Magidman, 1984; Nykanen *et al.*, 1985; Williams *et al.*, 1987; Fischboeck *et al.*, 1987, 1988; Le Quere *et al.*, 1987; Fehl and Marcott, 1989; Kempfert, 1990; Pfannhauser *et al.*, 1990) and many others could be quoted in addition. Reviews have appeared on the use of the technique in flavour analysis (Herres, 1984; Werkhoff *et al.*, 1990c) and some publications have specifically dealt with the important problem of compiling and searching GC–FTIR library data bases (Fields and White, 1987). As previously mentioned, FTIR has also been successfully combined with SFC and used in flavour analysis (Morrin *et al.*, 1986, 1987a,b).

An interesting advance is to link together GC with both MS and FTIR, and hence to obtain both sets of spectral data at the same time rather than in two separate runs. In many respects this would seem relatively easy, since using light-pipe GC–FTIR the system is non-destructive, and separated solutes can then be passed from the FTIR to the MS. However, most successful integration has utilized the more sensitive matrix isolation GC–FTIR system, which is, of course, destructive in this context, so GC effluents have to be split 50:50 before being fed to both the FTIR and the MS (Croasmun and McGorrin, 1989). It is uncertain whether this 'in parallel' approach is more effective or sensible than the obvious 'in series' one, but useful results from an analysis of chicken volatiles have been reported, illustrating the value of having data from both analytical techniques (Croasmun and McGorrin, 1989).

There are no other useful analytical instruments which can be used on-line, in tandem, beyond a GC for flavour analysis, but it is worth commenting that there are various sophisticated GC detectors which can provide more information than that from a routine FID or any other conventional detector. For example, an atomic emission detector provides simultaneous detection of up to four specified elements (dependent on the characteristic wavelength of emission) in a complex sample matrix. It has been used in a study of ham flavour to enable heterocyclic compounds to be pinpointed to facilitate GC–MS (Baloga *et al.*, 1990).

13.6 Conclusion

In conclusion, none of the refinements or alternatives discussed here is likely either now or in the future to prove superior for meat flavour analysis to top class GC–MS. However, many of the procedures described can provide extremely valuable additional and/or complementary data to those obtained by GC–MS. Indeed, for certain specific problems, alternative approaches may sometimes be superior. In addition, with a problem as difficult and complex as studying and analysing meat flavour components, it is absolutely essential to use all possible techniques and procedures which are available and which might yield constructive information.

References

Adamczyk, B., Genuit, W. and Boon, J.J. (1987). A simple approach to the dynamic headspace analysis of volatile flavours using a gas chromatography–photoionization mass spectrometer. *Biomed. Environ. Mass Spectrom.*, **161**, 373–375.

Baloga, D.W., Reineccius, G.A. and Miller, J.W. (1990). Characterization of ham flavor using an atomic emission detector. *J. Agric. Food Chem.*, **38**, 2021–2026.

Bricout, J. (1987). Characterization of natural constituents of fruit by enantioselective gas chromatography. In *Flavour Science and Technology*, ed. M. Martens, G.A. Dalen and H. Russwurm Jr. Wiley, Chichester, pp. 187–194.

Bruins, A.P. (1979). Negative ion chemical ionization mass spectrometry in the determination of components in essential oils. *Anal. Chem.*, **51**, 967–972.

Bruins, A.P. (1985). GC/MS employed with chemical ionisation sources for studies of positive and negative ions. In *Proc. 10th Int. Mass Spectrometry Conference*, Swansea.

Busch, K.L. and Kroha, K.J. (1985). Tandem mass spectrometry applied to the characterization of flavor compounds. In *Characterization and Measurement of Flavor Compounds*, ed. D.D. Bills and C.J. Mussinan. American Chemical Society, Washington, DC, pp. 121–137.

Caprioli, R.M., Moore, W.T., Martin, M., DaGue, B.B., Wilson, K. and Moring, S. (1989). Coupling capillary zone electrophoresis and continuous-flow fast atom bombardment mass spectrometry for the analysis of peptide mixtures. *J. Chromatogr.*, **480**, 247–257.

Croasmun, W.R. and McGorrin, R.J. (1989). Gas chromatography–matrix isolation infrared spectroscopy–mass spectrometry for analysis of thermally generated aroma compounds. In *Thermal Generation of Aromas*, ed. T.H. Parliment, R.J. McGorrin and C.-T. Ho. American Chemical Society, Washington, DC, pp. 61–72.

Davis, D.V. and Cooks, R.G. (1982). Direct characterisation of nutmeg constituents by mass spectrometry–mass spectrometry. *J. Agric. Food Chem.*, **30**, 495–504.

Eberhardt, R., Woidich, H. and Pfannhauser, W. (1981). Analysis of natural and artificial coconut flavouring in beverages. In *Flavour '81*, ed. P. Schreier. de Gruyter, Berlin, pp. 377–383.

Engel, K.-H. (1988). Investigation of chiral compounds in biological systems by chromatographic micromethods. In *Bioflavour '87*, ed. P. Schreier. de Gruyter, Berlin, pp. 75–88.

Fehl, A.J. and Marcott, C. (1989). Capillary gas chromatography/Fourier transform infrared spectroscopy using an injector/trap and liquid–liquid extraction. *Anal. Chem.*, **61**, 1596–1598.

Fields, R.E. and White, R.L. (1987). Real-time library search of vapor-phase spectra for gas chromatography/Fourier transform infrared spectrometry eluents. *Anal. Chem.*, **59**, 2709–2716.

Fischboeck, G., Pfannhauser, W. and Kellner, R. (1987). Characterization of citrus and kiwi flavors and their degradation processes by combination of GC/FTIR and GC/MS. *Mikrochim. Acta*, **1**, 27–30.

Fishboeck, G., Pfannhauser, W. and Kellner, R. (1988). GC/FTIR as a powerful tool for the characterization of flavor components in kiwi. *Mikrochim. Acta*, **3**, 249–257.

Flament, I. and Chevallier, C. (1988). Analysis of volatile constituents of coffee aroma. *Chem. Ind.*, 592–596.

Flament, I., Chevallier, C. and Keller, U. (1987). Extraction and chromatography of food constituents with supercritical CO_2. In *Flavour Science and Technology*, ed. M. Martens, G.A. Dalen and H. Russwurm Jr. Wiley, Chichester, pp. 151–163.

Fraisse, D., Maquin, F., Stahl, D., Suon, K. and Tabet, J.C. (1984). Analyse d'extraits de vanille. *Analusis*, **12**, 63–71.

Games, D.E., Alcock, N.J., van der Greef, J., Nyssen, L.M., Maarse, H. and ten Noever de Brauw, M.C. (1984). Analysis of pepper and capsicum oleoresins by high performance liquid chromatography/mass spectrometry and field desorption mass spectrometry. *J. Chromatgr.*, **294**, 269–279.

Garcia-Regueiro, J.A. and Diaz, I. (1989). Evaluation of the contribution of skatole, indole, androstenone and androstenols to boar-taint in back fat of pigs by HPLC and capillary gas chromatography. *Meat Sci.*, **25**, 307–316.

George, G. (1984). The negative ion technique in mass spectrometry. Application to problems of aroma. *Labo-Pharma—Probl. Tech.*, **343**, 479–481.

Guntert, M., Emberger, R., Hopp, R., Kopsel, M., Silberzahn, W. and Werkhoff, P. (1990). Chiral analysis in flavor and essential oil chemistry. Part A. Filbertone—the character impact compound of hazel-nuts. In *Flavour Science and Technology*, ed. Y. Bessiere and A.F. Thomas. Wiley, Chichester, pp. 29–32.

Hartman, T.G., Ho, C.-T., Rosen, J.D. and Rosen, R.T. (1989). Modern techniques in mass spectrometry for the analysis of nonvolatile or thermally labile flavor compounds. In *Thermal Generation of Aromas*, ed. T.H. Parliment, R.J. McGorrin and C.-T. Ho. American Chemical Society, Washington, DC, pp. 73–92.

Hellgeth, J.W., Jordan, J.W., Taylor, L.T. and Khorassani, M.A. (1986). Supercritical fluid chromatography of free fatty acids with on-line FTIR detection. *J. Chromatogr. Sci.*, **24**, 183–188.

Hendriks, H. and Bruins, A.P. (1980). Study of three types of essential oil of *Valeriana officinalis* L. s.l. by combined gas chromatography–negative ion chemical ionization mass spectrometry. *J. Chromatogr.*, **190**, 321–330.

Hendriks, H. and Bruins, A.P. (1983). A tentative identification of components in the essential oil of *Cannabis sativa* L. by a combination of gas chromatography negative ion chemical ionization mass spectrometry and retention indices. *Biomed. Mass Spectrom.*, **10**, 377–381.

Hener, U., Hollnagel, A., Kreis, P., Maas, B., Schmarr, H.-G., Schubert, V., Rettinger, K., Weber, B. and Mosandl, A. (1990). Direct enantiomer separation of chiral volatiles from complex matrices by multidimensional gas chromatography. In *Flavour Science and Technology*, ed. Y. Bessiere and A.F. Thomas. Wiley, Chichester, pp. 25–28.

Herres, W. (1984). Capillary GC–FTIR analysis of volatiles: HRGC–FTIR. In *Analysis of Volatiles*, ed. P. Schreier. de Gruyter, Berlin, pp. 183–217.

Herres, W., Idstein, H. and Schreier, P. (1983). Flavor analysis by HRGC/FT–IR; cherimoya (*Annona cherimolia*, Mill.) fruit volatiles. *J. High Resolut. Chromatogr.*, **6**, 590–594.

Hirvi, T. and Honkanen, E. (1983). Application of the mass fragmentographic SIM technique to the analysis of volatile compounds of berries, especially of genera *Vaccinium* and *Fragaria*. In *Instrumental Analysis of Food, Vol. 1*, ed. G. Charalambous and G. Inglett. Academic Press, New York, pp. 85–97.

Hirvi, T. and Honkanen, E. (1984). Selected ion monitoring technique and sensory analysis in the evaluation of the aroma of berries. In *Proc. Alko. Symp. on Flavour Research of Alcoholic Beverages*, ed. L. Nykanen and P. Lehtonen. Foundation for Biotechnical and Industrial Fermentation Research, pp. 275–278.

Homatidou, V., Karvouni, S. and Dourtoglou, V. (1990). Determination of characteristic aroma components of 'Cantaloupe' *Cucumis melo* using multidimensional gas chromatography (MDGC). In *Flavors and Off-Flavors*, ed. G. Charalambous. Elsevier, Amsterdam, pp. 1011–1023.

Idstein, H. and Schreier, P. (1985a). Volatile constituents from guava (*Psidium guajava* L.) fruit. *J. Agric. Food Chem.*, **33**, 138–143.

Idstein, H. and Schreier, P. (1985b). High-resolution gas chromatography–Fourier transform IR spectroscopy in flavor analysis. In *Characterization and Measurement of Flavor Compounds*, ed. D.D. Bills and C.J. Mussinan. American Chemical Society, Washington, pp. 109–120.

Idstein, H., Herres, W. and Schreier, P. (1984). High resolution gas chromatography–mass spectrometry and Fourier transform–infrared analysis of cherimoya (*Annona cherimolia*, Mill.) volatiles. *J. Agric. Food Chem.*, **32**, 383–389.

Kempfert, K.D. (1990). Analysis of aroma compounds by combined GC–FTIR. *Lebensm.-Biotechnol.*, **7**, 132–133.

Krammer, G., Frohlich, O. and Schreier, P. (1988). Chirality evaluation of 1,4-deca- and 1,4-dodecanolide in strawberry. In *Bioflavour '87*, ed. P. Schreier. de Gruyter, Berlin, pp. 89–95.

Labows, J.N. and Shushan, B. (1983). Direct analysis of food aromas. *Amer. Lab.*, **15**, 56–61.

Lange, G. and Schultze, W. (1988a). Studies on terpenoid and non-terpenoid esters using chemical ionization mass spectrometry in GC/MS coupling. In *Bioflavour '87*, ed. P. Schreier. de Gruyter, Berlin, pp. 105–114.

Lange, G. and Schultze, W. (1988b). Differentiation of isopulegol isomers by chemical ionization mass spectrometry. In *Bioflavour '87*, ed. P. Schreier. de Gruyter, Berlin, pp. 115–122.

Le Quere, J.L., Samon, E., Latrasse, A. and Etievant, P. (1987). Gas chromatography–Fourier transform infrared spectrometry. Applications in flavor analysis. *Sci. Aliments*, **7**, 93–109.

Liddle, P.A.P. and Bossard, A. (1984). Volatile naturally occurring restricted compounds derived from flavouring and their determination in food and beverages. In *Progress in Flavour Research*, ed. J. Adda. Elsevier, Amsterdam, pp. 467–476.

McLafferty, F.W. and Bockoff, F.M. (1978). Separation/identification system for complex mixtures using mass separation and mass spectral characterisation. *Anal. Chem.*, **50**, 69–76.

Morin, P., Caude, M., Richard, H. and Rosset, R. (1986). Carbon dioxide supercritical fluid chromatography–Fourier transform infrared spectrometry. *Chromatographia*, **21**, 523–530.

Morin, P., Caude, M., Richard, H. and Rosset, R. (1987). Supercritical fluid chromatography–Fourier transform infrared spectrometry coupling applied to polycyclic aromatic and sesquiterpene hydrocarbon analysis. *Analusis*, **15**, 117–127.

Morin, P., Caude, M., Richard, H. and Rosset, R. (1987) On-line carbon dioxide SFC–FTIR in aroma research. Perspectives and limits. In *Flavour Science and Technology*, ed. M. Martens, G.A. Dalen and H. Russwurm Jr. Wiley, Chichester, pp. 165–174.

Mosandl, A., Hener, U., Schmarr, H.G. and Rautenschlein, M. (1990). Chirospecific flavor analysis by means of enantioselective gas chromatography, coupled on-line with isotope ratio mass spectrometry. *J. High Resolut. Chromatogr.*, **13**, 528–531.

Moseley, M.A., Deterding, L.J., Tomer, K.B. and Jorgenson, J.W. (1989). Coupling of capillary zone electrophoresis and capillary liquid chromatography with coaxial continuous-flow fast atom bombardment tandem sector mass spectrometry. *J. Chromatogr.*, **480**, 197–209.

Nitz, S. (1985). Multidimensional gas-chromatography in aroma research. In *Topics in Flavour Research*, ed. R.G. Berger, S. Nitz and P. Schreier. Eichhorn, Marzling–Hangenham, pp. 43–57.

Nitz, S. and Julich, E. (1984). Concentration and GC–MS analysis of trace volatiles by sorption–desorption techniques. In *Analysis of Volatiles*, ed. P. Schreier. de Gruyter, Berlin, pp. 151–170.

Nitz, S., Kollmannsberger, H. and Drawert, F. (1988). Analysis of flavours by means of combined cryogenic headspace enrichment and multimensional GC. In *Bioflavour '87*, ed. P. Schreier, de Gruyter, Berlin, pp. 123–135.

Nykanen, L., Savolahti, P. and Nykanen, I. (1985). First experiences with HR–GC–FT–IR analysis of flavour compounds in distilled alcoholic beverages. In *Topics in Flavour Research*, ed. R.G. Berger, S. Nitz and P. Schreier. Eichhorn, Marzling–Hangenham, pp. 109–123.

Olivares, J.A., Nguyen, N.T., Yonker, C.R. and Smith, R.D. (1987). On-line mass spectrometric detection for capillary zone electrophoresis. *Anal. Chem.*, **59**, 1230–1232.

Pfannhauser, W., Kellner, R. and Fischboeck, G. (1990). GC/FTIR and GC/MS analysis of kiwi flavors. In *Flavors and Off-Flavors*, ed. G. Charalambous. Elsevier, Amsterdam, pp. 357–373.

Purcell, J.M. and Magidman, P. (1984). Analysis of the aroma of intact fruit of *Coffea arabica* by GC–FT–IR. *Appl. Spectrosc.*, **38**, 181–184.

Schomburg, G., Husmann, H., Podmaniczky, L. and Weeke, F. (1984). Coupled gas chromatographic methods for separation, identification and quantitative analysis of complex mixtures: MDGC, GC–MS, GC–IR, LC–GC. In *Analysis of Volatiles*, ed. P. Schreier. de Gruyter, Berlin, pp. 121–150.

Schreier, P. and Idstein, H. (1984). The use of HRGC–FTIR in tropical fruit flavour analysis. In *Analysis of Volatiles*, ed. P. Schreier. de Gruyter, Berlin, pp. 293–306.

Schreier, P. and Idstein, H. (1985a). Advances in the instrumental analysis of food flavours. *Z. Lebensm.-Unters. Forsch.*, **180**, 1–14.

Schreier, P. and Idstein, H. (1985b). High-resolution gas chromatography–Fourier transform infrared spectroscopy in flavor analysis. *Z. Lebensm.-Unters. Forsch.*, **181**, 183–188.

Schurig, V. (1988). Enantiomer separation by complexation gas chromatography—applications in chiral analysis of pheromones and flavours. In *Bioflavour '87*, ed. P. Schreier. de Gruyter, Berlin, pp. 35–54.

Skelton, R.J., Johnson, C.C. and Taylor, L.T. (1986). Sampling considerations in supercritical fluid chromatography. *Chromatographia*, **21**, 3–8.

Smith, R.D., Olivares, J.A., Nguyen, N.T. and Udseth, H.R. (1988a). Capillary zone electrophoresis–mass spectrometry using an electrospray ionization interface. *Anal. Chem.*, **60**, 436–441.

Smith, R.D., Baringa, C.J. and Udseth, H.R. (1988b). Improved electrospray ionization interface for capillary zone electrophoresis–mass spectrometry. *Anal. Chem.*, **60**, 1948–1952.

ten Noever de Brauw, M.C. and van Ingen, C. (1981). Mass spectrometry. In *Isolation, Separation and Identification of Volatile Compounds in Aroma Research*, ed. H. Maarse and R. Belz. Akademie–Verlag, Berlin, pp. 155–225.

Thomas, A.F., Wilhalm, B. and Flament, I. (1984). Some aspects of GC–MS in the analysis of volatile flavours. In *Chromatography and Mass Spectrometry in Nutrition Science and Food Safety*, ed. A. Frigerio and H. Milon. Elsevier, Amsterdam, pp. 47–65.

Tomlinson, A.J. (1991). Analysis of the coloured products of the Maillard reaction. PhD thesis, London University, London.

Tressl, R., Engel, K.-H., Albrecht, W. and Bille-Abdullah, H. (1985a). Analysis of chiral aroma components in trace amounts. In *Characterization and Measurement of Flavor Compounds*, ed. D.D. Bills and C.J. Mussinan. American Chemical Society, Washington, pp. 43–60.

Tressl, R., Grunewald, K.G., Kersten, E. and Rewicki, D. (1985b). Formation of pyrroles and tetrahydroindolizin-6-ones as hydroxyproline-specific Maillard products from glucose and rhamnose. *J. Agric. Food Chem.*, **33**, 1137–1142.

Tressl, R., Grunewald, K.G., Kersten, E. and Rewicki, D. (1986). Formation of pyrroles and tetrahydroindolizin-6-ones as hydroxyproline-specific Maillard products from erythrose and arabinose. *J. Agric. Food Chem.*, **34**, 347–350.

van der Greef, J., Nyssen, L.M., Maarse, H., ten Noever de Brauw, M.C., Games, D.E. and Alcock, N.J. (1984). The applicability of field desorption mass spectrometry and liquid chromatography/mass spectrometry for the analysis of the pungent principles of capsicum and black pepper. In *Progress in Flavour Research*, ed. J. Adda. Elsevier, Amsterdam, pp. 603–612.

Van Wassenhove, F., Dirinck, P. and Schamp, N. (1988). Analysis of the key components of celery by two dimensional capillary gas chromatography. In *Bioflavour '87*, ed. P. Schreier. de Gruyter, Berlin, pp. 137–144.

Walther, H., Schlunegger, U.P. and Friedli, F. (1983). Quantitative determination of compounds in mixtures by B^2E=constant linked scans. *Org. Mass Spectrom.*, **18**, 572–574.

Werkhoff, P., Emberger, R., Guntert, M. and Kopsel, M. (1989). Isolation and characterization of volatile sulfur-containing meat flavor components in model systems. In *Thermal Generation of Aromas*, ed. T.H. Parliment, R.J. McGorrin and C.-T. Ho. American Chemical Society, Washington, DC, pp. 460–478.

Werkhoff, P., Bretschneider, W., Guntert, M., Hopp, R. and Surburg, H. (1990a). Chiral analysis in flavor and essential oil chemistry. Part B. Direct enantiomer resolution of (*E*)-α-ionone and (*E*)-α-damascone by inclusion gas chromatography. In *Flavour Science and Technology*, ed. Y. Bessiere and A.F. Thomas. Wiley, Chichester, pp. 33–36.

Werkhoff, P., Bretschneider, W., Herrman, H.J. and Schreiber, K. (1990b). Progress in analysis of odorous substances. Part 10. Techniques for structure analysis of aromas and flavors. *LaborPraxis*, **14**, 151–160.

Werkhoff, P., Bretschneider, W., Herrmann, H.J. and Schreiber, K. (1990c). Progress in analysis of odorous substances. Part 11. Techniques for structure determination of odorous and flavor compounds. *LaborPraxis*, **14**, 256–264.

Whitfield, F.B. and Shaw, K.J. (1984). Analysis of food off-flavours. In *Progress in Flavour Research*, ed. J. Adda. Elsevier, Amsterdam, pp. 221–238.

Williams, A.A., Tucknott, O.G., Lewis, M.J., May, H.V. and Wachter, L. (1987). Examples of cryogenic matrix isolation GC/IR in the analysis of flavour extracts. In *Flavour Science and Technology*, ed. M. Martens, G.A. Dalen and H. Russwurm Jr. Wiley, Chichester, pp. 259–270.

14 Assessment of lipid oxidation and off-flavour development in meat and meat products

F. SHAHIDI

14.1 Introduction

Lipid oxidation is a major cause of rancidity and flavour deterioration of muscle foods. Many methods for its evaluation have been employed and these measure: (1) primary changes such as those in the structure of original fatty acids and formation of free radical intermediates, oxygen uptake and weight gain; (2) formation of lipid hydroperoxides, the primary products of oxidation; and (3) formation of secondary products of lipid oxidation (secondary changes) such as aldehydes production. However, the ultimate criterion for the suitability of any of these markers as an index of lipid oxidation is its adequate correlation with sensory data. This chapter discusses some of the commonly used methods of assessment of lipid oxidation and off-flavour development in meat and meat products.

A major cause of quality deterioration of foods, in general, and of muscle foods, in particular, is due to changes in their lipid components by both enzymatic and non-enzymatic lipid oxidation processes. Oxidation of lipids results not only in the loss of essential fatty acids but it generally brings about undesirable changes in flavour, colour, texture and functional properties, as well as causing the destruction of fat-soluble vitamins and formation of cholesterol oxides. The susceptibility and rate of oxidation of fatty acids in the lipids depend on the degree of their unsaturation (Belitz and Grosch, 1987). Thus, autoxidation of major fatty acids of meat follows the order given below.

$$C18:0 < C18:1 < C18:2 < C18:3$$

Seafoods, containing a large proportion of C20:5 and C20:6 fatty acids (eicosapentaenoic acid, EPA, and docosahexaenoic acid, DHA, respectively) in their lipids, are even more prone to autoxidation than the C18 acids listed above. Tichivangana and Morrissey (1985) have shown that the autoxidation of muscle foods occurs in the following order:

$$\text{fish} > \text{poultry (chicken and turkey)} > \text{pork} > \text{beef} \geqslant \text{lamb}$$

Autoxidation of lipids proceeds via a free-radical chain mechanism and is catalysed by many factors such as the presence of heat, light, ionizing radiation, metal ions and metalloporphyrins (Wong, 1989). It involves initiation, propagation and termination steps, as given below.

Reaction		Step
$RH \longrightarrow R^{\bullet} + H^{\bullet}$		Initiation
$R^{\bullet} + O_2 \longrightarrow ROO^{\bullet}$	}	Propagation
$ROO^{\bullet} + RH \longrightarrow ROOH + R^{\bullet}$	}	Propagation
$R^{\bullet} + R^{\bullet} \longrightarrow R\text{–}R$	}	Termination
$R^{\bullet} + ROO^{\bullet} \longrightarrow ROOR$	}	Termination
$ROO^{\bullet} + ROO^{\bullet} \longrightarrow ROOR + O_2$	}	Termination

Hydroperoxides (ROOH) are the primary products of lipid autoxidation. Each fatty acid may produce several hydroperoxides upon oxidation; however, hydroperoxides do not possess any flavour. Their breakdown into secondary oxidation products produces a wide array of low molecular weight organic compounds. Some of these degradation products are potently flavour-active and impart off-flavour to cooked, stored muscle foods (Figure 14.1) (Chang and Peterson, 1977). One way to retard lipid oxidation is by using natural or synthetic antioxidants.

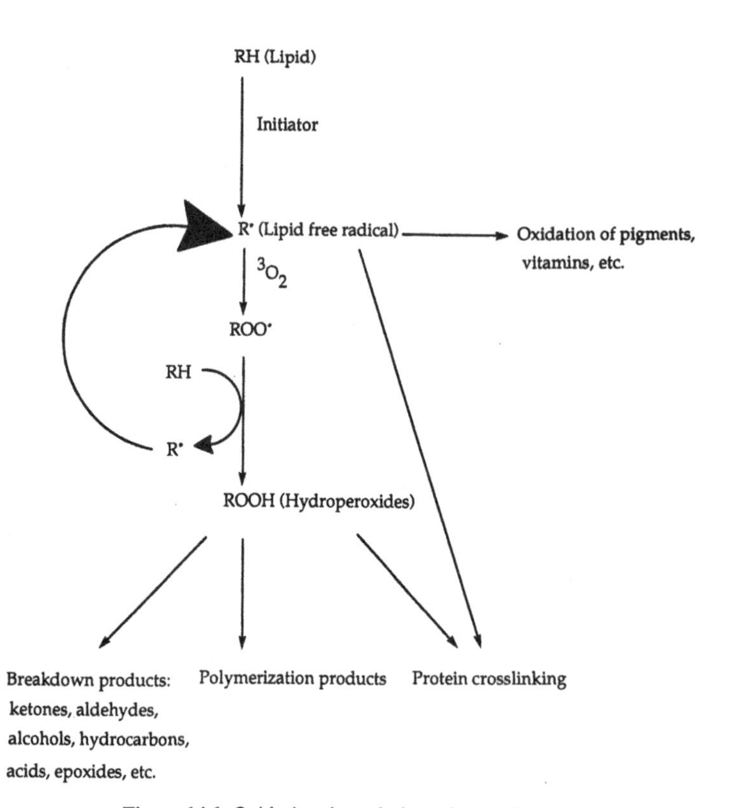

Figure 14.1 Oxidative degradation of meat lipids.

Primary antioxidants act as scavengers or terminators of free radicals by an electron- or a hydrogen-donating mechanism (Labuza, 1971). Phenolic antioxidants such as butylated hydroxyanisole (BHA), butylated hydroxytoluene (BHT), propyl gallate (PG), and *t*-butyl hydroquinone (TBHQ) are common food additives used in processed products. Secondary antioxidants are compounds that act as deactivators of catalysts of autoxidation such as chelating agents or those which show a synergistic effect when used in combination with common antioxidants (Labuza, 1971; Roozen, 1987). Thus, for extending the shelf-life of muscle foods, elimination or inactivation of pro-oxidants and/or oxygen from the system and termination of free radicals are common methods employed. Polyphosphates, citric acid and ethylenediaminetetraacetic acid (EDTA) are examples of typical secondary antioxidants used in the food industry.

In meats, nitrite curing is an effective means of prevention of flavour deterioration (Gray *et al.*, 1981; Gray and Pearson, 1984). The exact role of nitrite in inhibiting lipid autoxidation is not fully understood. However, possible mechanisms for both its antioxidant and chelating properties, as well as its possible role in stabilizing meat membrane lipids have been suggested (MacDonald *et al.*, 1980; Zubillaga *et al.*, 1984; Igene *et al.*, 1985).

The techniques employed for assessing lipid oxidation in muscle foods include analysis of the following: changes in the fatty acids, presence and type of free radical intermediates, conjugated dienes and trienes, polar components, polymers and *trans*-isomers. Determination of oxygen uptake, total or individual carbonyl compounds, peroxide value (PV), 2-thiobarbituric acid (TBA) value and anisidine value (AnV) are also commonplace. Furthermore, the Kries test, the oxirane test, fluorescence tests, and polarographic and chromatographic determinations are employed.

In order to assess the quality of muscle foods, one must be able to quantify the extent of oxidation. Different methods for measurement of the oxidative state of foods have been employed as listed above. These methods may measure one of the following parameters:

1. Measurement of changes in the concentration of one or more substrates. Such methods include analysis of lipid fatty acids, formation of conjugated dienes/trienes and/or *trans*-isomers, and oxygen uptake. Formation of lipid free radicals, or their derivatives, may also be measured.
2. Measurement of the primary products of lipid oxidation. In this respect, estimation of PV is practised.
3. Measurement of the secondary products of lipid oxidation. Such methods commonly employed include analysis of the content of malonaldehyde (MA), 2-thiobarbituric acid reactive substances (TBARS), and total carbonyls or selected volatile constituents.
4. Indirect methods. These methods may include measurements of texture, functional properties and colour.

5. Sensory analysis. This is a method of great importance and generally all objective methods of quantification of lipid oxidation must be correlated with it.

Apart from sensory evaluation for estimation of the oxidative state of muscle foods, other methods are categorized as either chemical or physical. Sensory evaluation techniques are simple, but are time-consuming, costly and often poorly reproducible. To avoid human bias and error, several measurements are performed and statistical analysis is used to process the results. The present chapter reviews methods of assessment of the oxidative changes of meat and meat products.

14.2 Fatty acid analysis

An indirect method to measure the susceptibility of meat lipids to oxidation is to monitor changes that occur in their fatty acid composition (Keller and Kinsella, 1973). A 25% decrease in the concentration of C20:4 of phosphatidylethanolamine in hamburgers upon cooking was noticed and this corresponded well with an increase in the TBA values and total content of carbonyl compounds. However, measurement of fatty acids is not a very sensitive method as they must possess three or more double bonds to be very susceptible to oxidation (Moerck and Ball, 1974). In meats, these fatty acids reside primarily in the phospholipid fraction. Thus, separation of phospholipids from other lipid components prior to their determination is required.

Dimick and MacNeil (1970) and Moerck and Ball (1974) reported that the level of polyunsaturated fatty acids (PUFA) in chicken phospholipids decreases rapidly during refrigerated or frozen storage. Similar results have been reported by Igene *et al.* (1980) in beef and poultry meat model systems and by Gokalp *et al.* (1981) for beef patties. However, in intact muscles, lipids were less susceptible to changes in their content of PUFA. Furthermore, variations in the content of α-tocopherol as well as the level of haemoproteins may influence the degree of changes that might occur in muscle foods (Kunsman *et al.*, 1978; Yamauchi *et al.*, 1980). Nonetheless, use of chromatographic methods such as HPLC procedures for separation and quantification of individual phospholipids may present a fast and accurate method for assessing lipid autoxidation of muscle foods (Bolton *et al.*, 1985; Yeo and Horrocks, 1985).

14.3 Oxygen uptake

The uptake of oxygen has been used as a method of assessing the susceptibility of muscle tissues to lipid oxidation (Fisher and Deng, 1977; Lee *et al.*, 1975; Silberstein and Lillard, 1978). Although oxidation of proteins may contribute to oxygen uptake results by muscle tissue lipids, existing differences in the rate of their oxidation makes effective utilization of this method viable (Rhee, 1978b). Our own work on seal muscle tissues lends support to this latter finding (Shahidi, unpublished results).

14.4 Conjugated dienes

Sklan *et al.* (1983) determined the content of conjugated dienes, trienes and tetraenes, referred to as total conjugated products of oxidation, in total lipid extracts of turkey meat during a 60-day storage at 4°C and over an 18-month storage at −18°C. The content of conjugated products resulting from oxidation was found to increase at both temperatures. Similar studies were made by Ahmad and Augustin (1985) for fried fish during a 40-day storage at −60°C. These authors indicated that the level of both dienes and trienes increased with increasing storage time. Therefore, this method may offer a practical procedure for assessing lipid oxidation of meat and meat products.

14.5 Peroxide value

Hydroperoxides are primary products of lipid autoxidation. A common method for monitoring oxidative changes in muscle foods is via determination of peroxide value (PV) of their lipids. Peroxides in meat and meat products may be measured by a variety of techniques as described by Gray (1978). Peroxide value is commonly reported in meq of iodine per kg of fat and is generally determined by the use of an iodometric method (Fiedler, 1974; AOCS, 1989). Although interferences from the presence of oxygen in the assay sample and uptake of iodine by lipid double bonds are possible, this method nonetheless has been used extensively for assessing the oxidative state of fatty foods.

The time to reach a certain PV may be used as an index of oxidative stability of meat lipids. The effect of antioxidants and food processing on lipids is often monitored in this way. Thus, a longer period to reach a certain PV generally indicates a better antioxidant activity for the additive under examination.

Bailey *et al.* (1973) have used PV for evaluation of pork fat quality during frozen storage, and the oxidative state of beef muscle tissues during a 10-week storage at −10°C was reported by Owen *et al.* (1975). Jeremiah (1980) reported the keeping quality of frozen pork samples in different packaging materials. In general, a significant relationship between the PV and sensory flavour scores was noted. However, breakdown of hydroperoxides to secondary oxidation products may result in a decrease in the PV during the storage period (Awad *et al.*, 1968; Noble, 1976). Due to the aforementioned considerations, PV has not been used extensively in evaluation of the oxidative state of meat and meat products (Pearson *et al.*, 1977; Melton, 1983).

14.6 2-Thiobarbituric acid test

The 2-thiobarbituric acid (TBA) test is the most widely used method for assessing oxidative state of muscle foods (Tarladgis *et al.*, 1960; Gray, 1978; Melton, 1983). Malonaldehyde (MA) is a decomposition product of lipid peroxides formed in meats. It has been extensively studied due to its reactivity with biological molecules such as amino groups of amino acids, proteins, nucleic acids and with sulphydryl groups (Chio and Tappel, 1969a, 1969b; Draper *et al.*, 1986). Malonaldehyde itself is generally bound to biological materials and therefore, prior to determination, it must be released from muscle tissues by acid treatment.

Kohn and Liversedge (1944) were the first to use the 2-thiobarbituric acid (TBA) test for evaluation of the oxidative state of meats. The specificity of this method for estimating malonaldehyde, as its TBA derivative, was then reported by Sinnhuber and Yu (1958), as formulated in Figure 14.2. The pink-coloured chromophore formed between the TBA reagent and malonaldehyde in a 2:1 ratio (Figure 14.2) (Nair and Turner, 1984) has an absorption maximum at 532 nm. However, researchers have shown that alkenals and 2,4-alkadienals may also react with the TBA reagent to form a coloured complex with absorption at 532 nm (Marcuse and Johansson, 1973; Kosugi *et al.*, 1987, 1988). Therefore, the TBA test is generally used to quantify the content of TBA reactive substances (TBARS).

The TBA test may be performed directly on a food product (Wills, 1965), and this may be followed by extraction of the coloured pigment into butanol or a butanol–pyridine mixture (Placer *et al.*, 1966; Uchiyama and Mihara, 1978; Ohkawa *et al.*, 1979). The test may also be carried out on an aliquot of an acid extract of food (usually in 7.5 to 28% trichloroacetic acid (TCA) solution) (Siu and Draper, 1978) or on a portion of a steam distillate of the sample under investigation (Tarladgis *et al.*, 1960; 1964). The distillation procedure developed by Tarladgis *et al.* (1960) is the method frequently used and the TCA extraction procedure has also been used by many researchers (Shahidi and Hong, 1991).

Quantification of MA and its precursor(s), using TBA colorimetry, has been followed by heating the assay sample with the TBA reagent in an acidic solution. However, the various procedures used differ from one another. Therefore, sample preparation, types of acidifying agents and

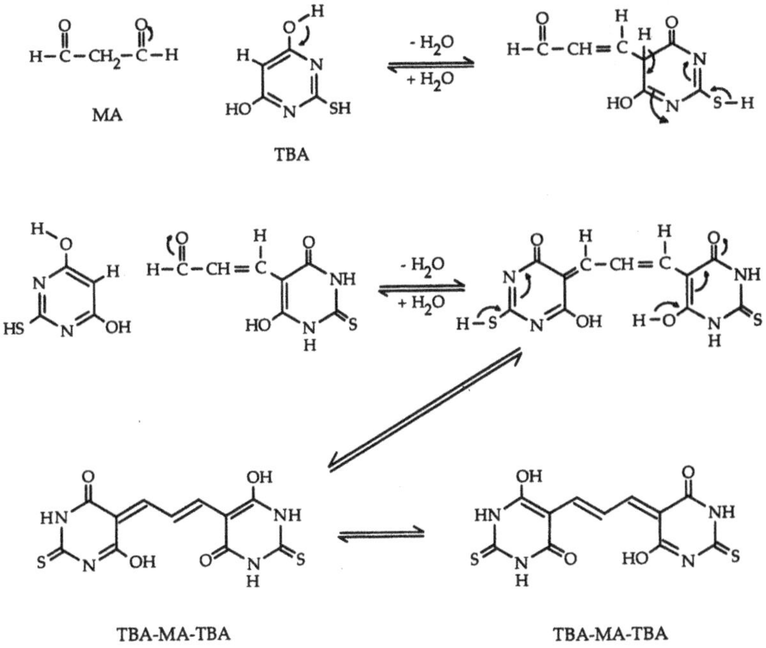

MA

TBA

$- H_2O$

$+ H_2O$

TBA-MA-TBA

TBA-MA-TBA

Figure 14.2 Mechanism of formation of 2-thiobarbituric acid (TBA)–malonaldehyde (MA) adduct.

their concentration in the reaction mixture, pH of the reaction mixture, composition of the TBA reagent, length of the TBA reaction time and possible use of antioxidants and chelators in the systems, are factors which may influence results reported by researchers from various laboratories. For example, Moerck and Ball (1974) suggested that Tenox II be added to the distillation mixture prior to heating in order to retard further oxidation and consequently artefact formation during this step, while Ke *et al.* (1977) reported the use of propyl gallate (PG) and ethylenediaminetetraacetic acid (EDTA) during distillation for this purpose. However, researchers have pointed out that some phenolic antioxidants such as butylated hydroxyanisole (BHA) used in order to retard further oxidation of samples may in fact enhance the decomposition of lipid peroxides during distillation (Rhee, 1978a).

Recently Shahidi and Hong (1991) showed that addition of commonly used antioxidants and chelators has a marginal effect in prevention of further lipid oxidation of meat lipids during the distillation step. It was further demonstrated that while the distillation procedure generally affords results that are numerically higher than those obtained by the acid extraction procedure, each method has its own advantages and drawbacks. Nonetheless, the relative, rather than the absolute TBA values, should be compared against one another in such determinations. The extraction procedure has a further disadvantage in that the presence of coloured ingredients, such as the cooked cured-meat pigment (CCMP), may cause overestimation of the numerical TBA values. Furthermore, a similar effect may be observed due to the occurrence of turbidity in the extraction solution owing to dissolution of certain proteins or the presence of fat emulsions in the systems.

Many modifications to improve the TBA test, proposed by Tarladgis *et al.* (1960), have been reported by researchers. The use of whole-tissue homogenates without deproteinization (Wills, 1965) for the TBA test is commonplace; however, some deproteinization prior to analysis may afford more realistic results. Placer *et al.* (1966) modified the TBA test for whole-tissue homogenates and used an alkaline pyridine–butanol mixture to extract the TBA–MA complex. The TBA–MA adduct was extracted into butanol and the absorbance of the solution was measured at both 520 and 535 nm. The difference in absorbance values at these two wavelengths was then taken as the TBA value. Quantification of malonaldehyde alone by a high-performance liquid chromatographic (HPLC) method has been reported (Kakuda *et al.*, 1981; Csallany *et al.*, 1984; Bull and Marnett, 1985).

14.6.1 Advantages and limitations of the TBA test

Lipid oxidation in muscle foods is conveniently assessed by the classical TBA methodology as mentioned earlier. This method is simple and generally correlates well with sensory data. However, there are some limitations associated with it. TBA values of cooked muscle foods tend to increase over the storage period, reach a maximum, and then start to decline. Interaction of MA with available amino groups of meat constituents forming 1-amino-3-imino propene structures has been suggested to occur (Chio and Tappel, 1969b). Consequently, a given TBA value may correspond with two points during the storage period (i.e. one value during the early stages of storage corresponding to an increase in MA concentration and another equal value during its decrease over a longer storage period. Nonetheless, TBA values do correspond well with a single point during the early stages of storage as studied by many researchers (Spanier and Taylor, 1991).

In nitrite-cured products, use of sulphanilamide (SA) is recommended (Zipser and Watts, 1962). SA reacts with residual nitrite present in the sample, leading to the formation of a diazonium salt, thus preventing the nitrosation of malonaldehyde which would be otherwise unreactive towards the TBA reagent (Figure 14.3). However, in the absence of residual nitrite, SA itself may react with MA to form an amino-iminopropene derivative. Formation of this latter complex has been positively

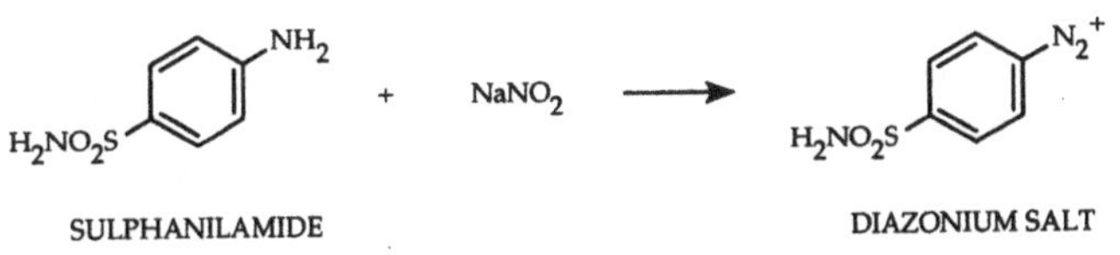

H_2NO_2S–C_6H_4–NH_2 + $NaNO_2$ ⟶ H_2NO_2S–C_6H_4–N_2^+

SULPHANILAMIDE DIAZONIUM SALT

Figure 14.3 Interaction of sulphanilamide with nitrite.

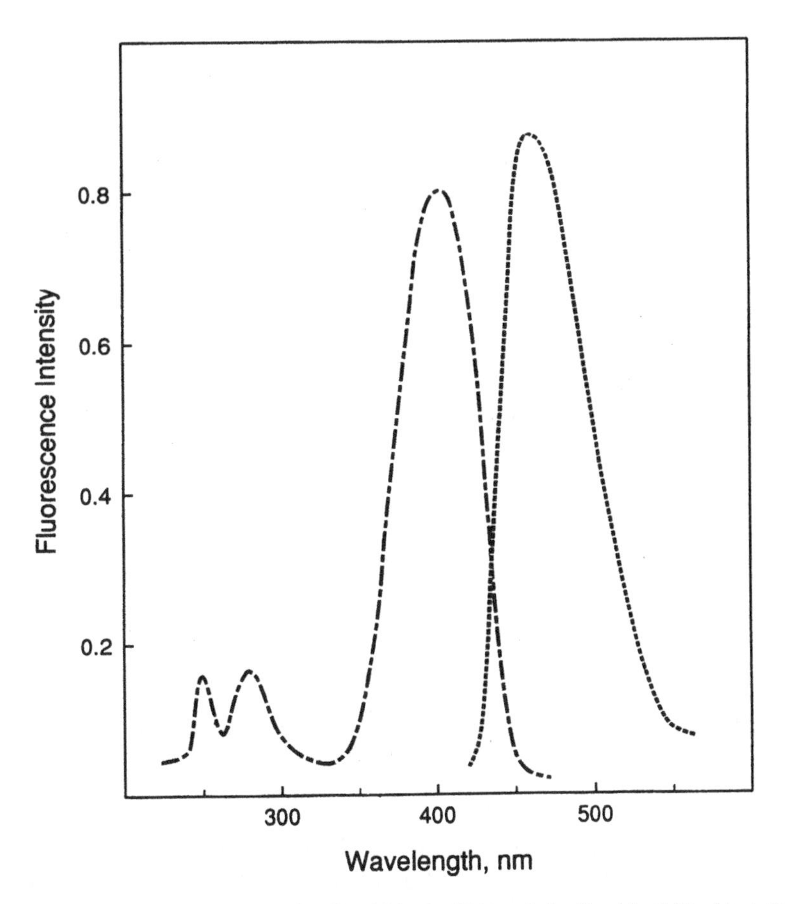

Figure 14.4 Fluorescence spectra of malonaldehyde (MA)–sulphanilamide (SA) adduct. Excitation, – – – – – -, and fluorescence, - - - - - - -.

detected by fluorescence spectroscopy (Figure 14.4) (Shahidi *et al.*, 1991). Furthermore, in the presence of TBA, a new complex of SA–MA–TBA (SMT, 1:1:1, mol basis) was detected and elucidated (Figure 14.5) (Pegg *et al.*, 1992).

Use of an extraction, rather than a distillation procedure, as well as its

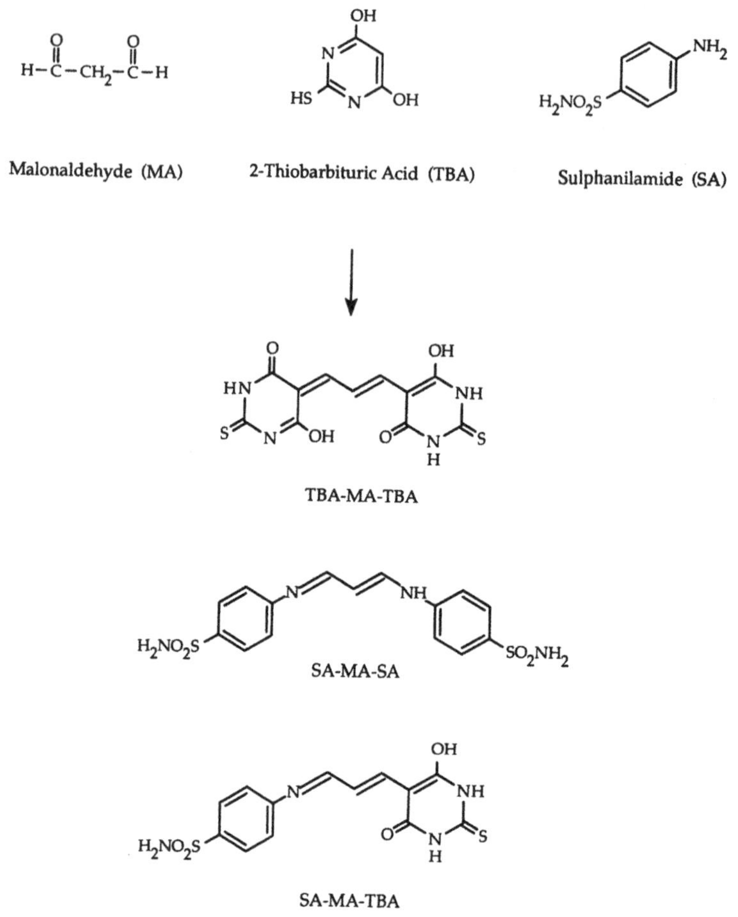

Malonaldehyde (MA) 2-Thiobarbituric Acid (TBA) Sulphanilamide (SA)

TBA-MA-TBA

SA-MA-SA

SA-MA-TBA

Figure 14.5 Interaction of malonaldehyde (MA) with sulphanilamide (SA) and 2-thiobarbituric acid (TBA).

advantages and disadvantages in the TBA test such as extraction of coloured components and dissolution of some proteins or dispersion of fat droplets which may interfere with the determination has been discussed earlier. Bound malonaldehyde also may not be released in the absence of heat treatment or acidic conditions. Therefore, the TBA methodology, although simple and adequate for general estimation of oxidative state of meats, is not without its drawbacks when accurate quantitation is required. Nonetheless, the TBA test, due to its simplicity and acceptability, remains the most widespread procedure for estimation of the oxidative state of meat and meat products.

14.7 The Kries test

This is a rapid test which may be performed on lipid extracts from meat (see Robards *et al.*, 1988). It is one of the first methods of evaluation of the oxidative state of lipids and involves the formation of a red-coloured adduct when phloroglucinol reacts with oxidized lipids under acidic conditions. Although the Kries test is simple, it does not always parallel the development of rancidity in foods. Nonetheless, it has been reported that careful experimentation may provide both qualitative and quantitative information (Rossell, 1991). Therefore, lipids may be reacted with phloroglucinol in diethyl ether followed by extraction with a hydrochloric acid solution. The red colour so produced may be measured with a Lovibond colorimeter. However, interference from certain colorants and food additives may be noted (Rossell, 1991).

14.8 Anisidine value

The peroxides in food lipids are transitory species that decompose into various carbonyl and other products. Reaction of aldehydes of oxidized lipids with the *p*-anisidine reagent in isooctane is followed by measuring the concentration of reaction products spectrophotometrically at 350 nm (Figure 14.6). Various factors influencing the accuracy and reproducibility

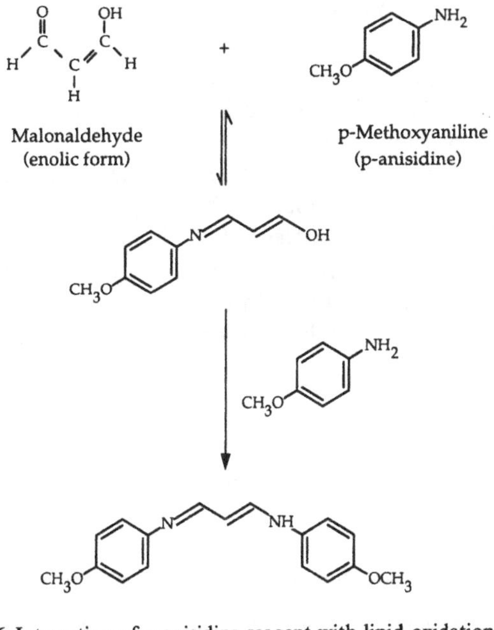

Malonaldehyde (enolic form) + p-Methoxyaniline (p-anisidine)

Figure 14.6 Interaction of *p*-anisidine reagent with lipid oxidation products.

of the method have been reviewed by Rossell (1986) and Robards *et al.* (1988). This method has not usually been employed for meat products but has been used for evaluation of the quality of fish meal as feed for aquacultured species (Cho, 1990).

14.9 Totox value

The anisidine value (AnV) is often used together with the PV to calculate the so called total oxidation value or totox value. Totox value is defined as 2PV + AnV. It is generally considered that extracted lipids with a totox value of less than 10 are of good quality. This parameter has been used extensively in the evaluation of fish meal quality. Nonetheless, its theoretical significance is questionable as parameters with different dimensions are added together.

14.10 Carbonyl compounds

Carbonyl compounds in meat and meat products may be determined by converting them to their 2,4-dinotrophenylhydrazones. Carbonyl compounds together with meat triacylglycerols are generally extracted first from muscle tissue into hexane and then derivatized on a column of activated celite impregnated with 2,4-dinitrophenylhydrazine, prior to measurement at 340 nm (Schwartz *et al.*, 1963). Alternatively, total carbonyls may be converted to their 2,4-dinotrophenylhydrazones in an aqueous medium and then extracted with hexane prior to their measurement at 340 nm (Lawrence, 1965; Keller and Kinsella, 1973). The general reaction involved is given in Figure 14.7.

Evaluation of the oxidative state of muscle foods from beef, lamb, poultry and fish by this procedure has been reported extensively (Schwartz *et al.*, 1963; Dimick and MacNeil, 1970; Dimick *et al.*, 1972; Sink and Smith, 1972; Keller and Kinsella, 1973; Caporaso and Sink, 1978; Kunsman *et al.*, 1978; Mai and Kinsella, 1979; and Misock *et al.*, 1979). In the column separation of hydrazones from lipid components, hexane

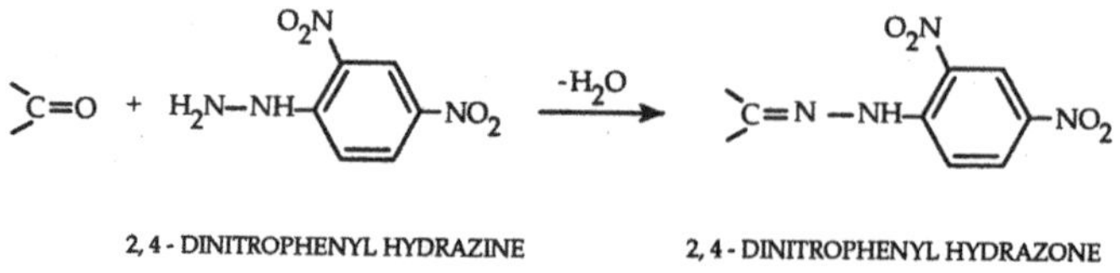

$$>C{=}O + H_2N{-}NH{-}C_6H_3(NO_2)_2 \xrightarrow{-H_2O} >C{=}N{-}NH{-}C_6H_3(NO_2)_2$$

2,4 - DINITROPHENYL HYDRAZINE 2,4 - DINITROPHENYL HYDRAZONE

Figure 14.7 Interaction of carbonyl compounds with 2,4-dinitrophenylhydrazine.

was used to elute the lipid fraction whereas chloroform was employed to elute the hydrazone derivatives. Ketoglycerides unintentionally derivatized may interfere with the analysis and their removal on an alumina column with benzene–hexane as eluent is necessary. Monocarbonyl compounds from fresh muscle foods may be isolated by similar techniques (Thomas *et al.*, 1971). Monocarbonyls may also be separated from unreacted lipids and dicarbonyls and their content determined spectrophotometrically at 365 nm. The monocarbonyl fraction may be separated further into alkanals, 2-alkanones, 2-alkenes and 2,4-alkadienals by the chromatographic procedure of Schwartz *et al.* (1963) or its modification (Caporaso and Sink, 1978). It has been reported that the total monocarbonyls served as a better indicator of lipid oxidation in mechanically deboned meats than the total carbonyl content (Kunsman *et al.*, 1978). However, individual constituents of the monocarbonyls did not serve as good indicators of oxidative deterioration of muscle foods (Dimick and MacNeil, 1970; Dimick *et al.*, 1972).

Regeneration of the original carbonyl compounds from their hydrazone derivatives is another interesting aspect (Dartey and Kinsella, 1971). However, regeneration efficiency of carbonyl compounds decreased as their chain length increased. Hydrolysis of hydrazones with 4N HCl followed by the extraction of the liberated carbonyl compounds in pentane is recommended (Buttery, 1973).

14.11 Hexanal and other carbonyl compounds

Hexanal is a seemingly ubiquitous component of food, both fresh and stored. This stems from the fact that practically all foods have some percentage of linoleate, the fatty acid which undergoes oxidation to produce hexanal (Frankel *et al.*, 1981). Linoleic acid may form a large number of volatile compounds during its autoxidation (Ullrich and Grosch, 1987). Hexanal, pentanal and 2,4-decadienal have been proposed as indicators for the development of off-flavours in vegetable oils (Dupuy *et al.*, 1976; Warner *et al.*, 1978).

Initially the autoxidation of linoleic acid produces 9-hydroperoxy-10,12-octadecadienoic (9-HPOD) and 13-hydroperoxy-9,11-octodecadienoic (13-HPOD) acids. Hexanal may be formed from both of these hydroperoxides (Schieberle and Grosch, 1981). It has also been shown that hexanal may be formed via degradation of secondary oxidation products of lipids such as 2-octenal and 2,4-decadienal. Hexanal itself may be oxidized to hexanoic acid. Nonetheless, the content of hexanal in muscle foods has been shown to be a good indicator for evaluation of the oxidative state of meat lipids (Shahidi *et al.*, 1987).

The protocols used for hexanal quantification may involve the isolation of carbonyls, together with all other volatile compounds by distillation and/or solvent extraction, followed by their analysis by gas chromatography or gas chromatography–mass spectrometry without any derivitization (Shahidi *et al.*, 1987). A direct headspace analysis may also be used. Alternatively, the carbonyl compounds may be reacted with 2,4-dinitrophenylhydrazine. The hydrazones so obtained may then be separated by a reversed-phase high-performance liquid chromatographic method and the hexanal content quantified as its hydrozone derivative (Morrissey and Apte, 1988).

Hexanal has been successfully used for evaluation of the oxidative state of red meat from different species (Cross and Ziegler, 1965; Ruenger *et al.*, 1978; Shahidi *et al.*, 1987; Morrissey and Apte, 1988; Torres *et al.*, 1989; Lamikanra and Dupuy, 1990; Drum and Spanier, 1991). Shahidi *et al.* (1987) reported that a linear relationship existed between the content

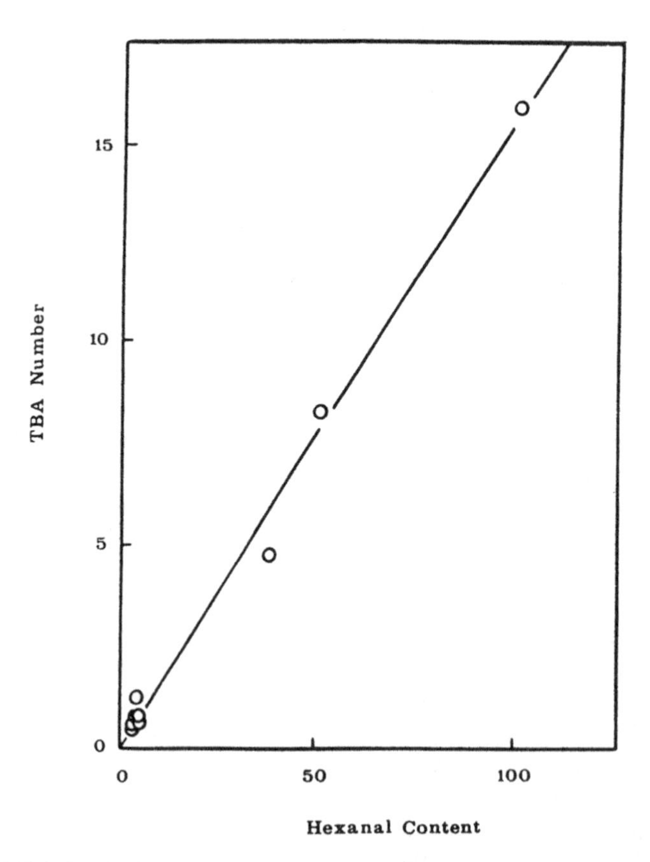

Figure 14.8 Relationship between TBA numbers and relative hexanal content of cooked pork (Shahidi *et al.*, 1987).

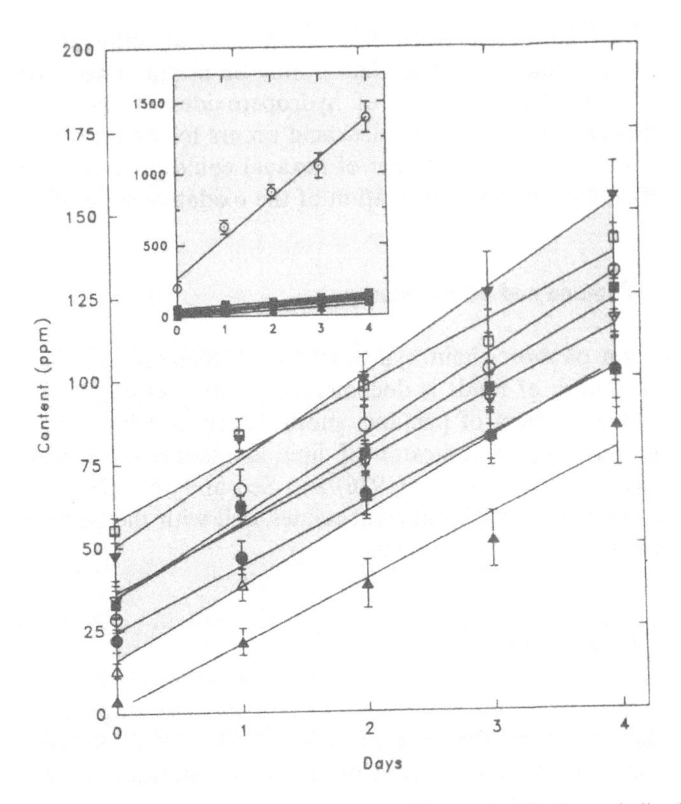

Figure 14.9 Relationship between hexanal (main plot), and hexanal and other volatiles (insert plot), and storage period of beef patties (Drumm and Spanier, 1991).

of hexanal and both the TBA values and sensory acceptability scores of cooked ground pork (Figure 14.8). Similar results were reported for fresh beef, pork and fish (Morrissey and Apte, 1988), for chevon (Lamikanra and Dupuy, 1990), for salted and dried beef (Torres *et al.*, 1989) and for cooked beef (Drumm and Spanier, 1991). The latter authors have also shown that hexanal concentration increases much faster than any other aldehyde during the storage of cooked beef (Figure 14.9). A similar conclusion was reached by Shahidi (1991) in a recent study. Therefore, hexanal appears to be a sensitive and reliable indicator for evaluation of the oxidative state and flavour quality of meat and meat products. Furthermore, it should be noted that hexanal has a distinctive odour described by many researchers as 'green' or 'grassy' (MacLeod and Ames, 1986; Ullrich and Grosch, 1987; Gasser and Grosch, 1988). It has a very low odour threshold concentration and is detectable at levels as low as 100 ppm (Bengtsson *et al.*, 1967; Schieberle and Grosch, 1987).

In muscle foods rich in omega-3 fatty acids such as fish and poultry meat from birds fed extensively on fish meal, ethanal, propanal and propenal as indicators of oxidation may be useful. These compounds are typical breakdown products of hydroperoxides of omega-3 fatty acids. Nonetheless, the fact that linoleic acid occurs to some degree in all muscle foods means that measurement of hexanal could provide a rapid, sensitive and reliable method for evaluation of the oxidative state of muscle foods.

14.12 Pentane and other alkanes

Formation of short-chain hydrocarbons, particularly ethane and pentane, from oxidation of lipids is documented (Horvat *et al.*, 1964; Evans *et al.*, 1967). Measurement of pentane, another dominant breakdown product of linoleic acid, as an indicator of lipid oxidation in freeze-dried muscle foods was reported by Seo (1976) and Seo and Joel (1980). Bigalli (1977) showed that the pentane content relates well with the development of off-flavours in several food systems.

14.13 Conclusions

Although a large number of procedures for assessing lipid oxidation in meat and meat product are available, both the TBA test and hexanal assay seem to provide the best means for monitoring oxidative deterioration of muscle foods during storage. Absolute TBA values obtained by the test do not always provide accurate results; nonetheless in uncured meats the TBA test always offers a reliable trend when comparing several systems under investigation. In cured meat products, however, the presence of residual nitrite in samples may require addition of sulphanilamide to the assay sample in order to prevent the nitrosation of malonaldehyde, which results in its underestimation. Use of hexanal as an indicator of lipid oxidation during early stages of oxidative changes in meats has proved beneficial.

Acknowledgments

Financial support from the Natural Sciences and Engineering Research Council (NSERC) of Canada is gratefully acknowledged.

References

Ahmad, S. and Augustin, M.A. (1985). Effect of triarylbutylhydroquinone on lipid oxidation in fish crackers. *J. Sci. Food Agric.*, **36**, 393–401.

AOCS (1989). Official Methods and Recommended Practices of the American Oil Chemists' Society. Fourth Ed., Champaign, Ill.

Awad, A., Powrie, W.D. and Fennema, O. (1968). Chemical deterioration of frozen bovine muscle at -4°C. *J. Food Sci.*, **33**, 227–235.

Bailey, C., Cutting, C.L., Enser, M.B. and Rhodes, D.N. (1973). The influence of slaughter weight on the stability of pork sides in frozen storage. *J. Sci. Food Agric.*, **24**, 1299–1304.

Belitz, H.-D. and Grosch, W. (1987). Lipids. In *Food Chemistry*. Springer–Verlag, Berlin. pp. 128–200.

Bengtsson, B.L., Bosund, I. and Rasmussen, I. (1967). Hexanal and ethanol formation in peas in relation to off-flavor development. *Food Technol.* **21**(3), 160A–164A.

Bigalli, G. (1977). Determination of pentane formed during autoxidation of oils contained in solid samples. *J. Am. Oil Chem. Soc.*, **54**, 229–232.

Bolton, J.C., Melton, S.L., Riemann, M.J. and Backus, W.R. (1985). An isocratic HPLC method for analysis of bovine phospholipids. Presented at the 77th Annual Meeting of the *American Society of Animal Sciences*, University of Georgia, August 13–16.

Bull, A.W. and Marnett, L.J. (1985). Determination of malondialdehyde by ion-pairing high-performance liquid chromatography. *Anal. Biochem.*, **149**, 284–290.

Buttery, R.G. (1973). Some unusual volatile carbonyl components of potato chips. *J. Agric. Food Chem.*, **21**, 31–33.

Caporaso, F. and Sink, J.D. (1978). Lipid-soluble carbonyl components of ovine adipose tissue. *J. Food Sci.*, **43**, 1379–1381.

Chang, S.S. and Peterson, R.J. (1977). Symposium: The basis of quality in muscle foods. Recent developments in the flavour of meat. *J. Food Sci.*, **42**, 298–305.

Chio, K.S. and Tappell, A.L. (1969a). Inactivation of ribonuclease and other enzymes by peroxidizing lipids and by malonaldehyde. *Biochemistry*, **8**, 2827–2832.

Chio, K.S. and Tappel, A.L. (1969b). Synthesis and characterization of the fluorescent products derived from malonaldehyde and amino acids. *Biochemistry*, **8**, 2821–2827.

Cho, C.Y. (1990). Fish nutrition, feeds, and feeding: with special emphasis on salmonid aquaculture. *Food Reviews International*, **6**, 333–357.

Cross, C.K. and Ziegler, P. (1965). A comparison of the volatile fractions from cured and uncured meat. *J. Food Sci.*, **30**, 610–614.

Csallany, A.S., Guan, M.D., Manwaring, J.D. and Addis, P.B. (1984). Free malonaldehyde determination in tissues by high-performance liquid chromatography. *Anal. Biochem.*, **142**, 277–283.

Darty, C.K. and Kinsella, J.E. (1971). Rate of formation of methyl ketones during blue cheese ripening. *J. Agric. Food Chem.*, **19**, 771–774.

Dimick, P.S. and MacNeil, J.H. (1970). Poultry product quality. 2. Storage time–temperature effects on carbonyl composition of cooked turkey and chicken skin fractions. *J. Food Sci.*, **35**, 186–190.

Dimick, P.S., MacNeil, J.H. and Grunden, L.P. (1972). Poultry product quality. Carbonyl composition and organoleptic evaluation of mechanically deboned poultry meat. *J. Food Sci.*, **37**, 544–546.

Draper, H.H., McGirr, L.G. and Hadley, M. (1986). The metabolism of malondialdehyde. *Lipids*, **21**, 305–307.

Drumm, T.D. and Spanier, A.M. (1991). Changes in the content of lipid autoxidation and sulfur-containing compounds in cooked beef during storage. *J. Agric. Food Chem.*, **39**, 336–343.

Dupuy, H.P., Rayner, E.T. and Wadsworth, J.I. (1976). Correlations of flavour scores with volatiles of vegetable oils. *J. Am. Oil Chem. Soc.*, **53**, 628–631.

Evans, C.D, List, G.R., Dolev, A., McConnell, D.G. and Hoffman, R.L. (1967). Pentane from thermal decomposition of lipoxidase-derived products. *Lipids*, **2**, 432–434.

Fielder, U. (1974). A coulometric method for the determination of low peroxide values of fats and oils. *J. Am. Oil Chem. Soc.*, **51**, 101–103.

Fischer, J. and Deng, J.C. (1977). Catalysis of lipid oxidation: a study of mullet (*Mugil cephalus*) dark flesh and emulsion model system. *J. Food Sci.*, **42**, 610–614.

Frankel, E.N., Neff, W.E. and Selke, E. (1981). Analysis of autoxidized fats by gas chromatography–mass spectrometry: VII. volatile thermal decomposition products of pure hydroperoxides from autoxidized and photosensitized oxidized methyl oleate, linoleate and linolenate. *Lipids* **16**, 279–285.

Gasser, U. and Grosch, W. (1988). Flavor deterioration of soya bean oil: identification of intense odour compounds during flavour reversion. *Fett. Wiss. Technol.*, **90**, 332–336.

Gokalp, H.Y., Ockerman, H.W., Plimpton, R.F. and Peng, A.C. (1981). Qualitative alteration of phospholipid in beef patties cooked after vacuum and nonpackaged, frozen storage. *J. Food Sci.*, **46**, 19–22.

Gray, J.I. (1978). Measurement of lipid oxidation: a review. *J. Am. Oil Chem. Soc.*, **55**, 539–546.

Gray, J.I. and Pearson, A.M. (1984). Cured meat flavor. *Adv. Food Res.*, **29**, 1–86.

Gray, J.I., MacDonald, B., Pearson, A.M. and Morton, I.D. (1981). Role of nitrite in cured meat flavor: a review. *J. Food Protect.* **44**, 302–312.

Horvat, R.J., Lane, W.G., Ng., H. and Shepherd, A.D. (1964). Saturated hydrocarbons from autoxidizing methyl linoleate. *Nature*, **203**, 523–524.

Igene, J.O., Pearson, A.M., Dugan, L.R., Jr. and Price, J.F. (1980). Role of triglycerides and phospholipids on development of rancidity in model meat systems during frozen storage. *Food Chem.*, **5**, 263–276.

Igene, J.O., Yamauchi, K., Pearson, A.M., Gray, J.I. and Aust, S.D. (1985). Mechanisms by which nitrite inhibits the development of warmed-over flavor (WOF) in cured meat. *Food Chem.*, **18**, 1–18.

Jeremiah, L.E. (1980). Effect of frozen storage and protective wrap upon the cooking losses, palatability, and rancidity of fresh and cured pork cuts. *J. Food Sci.* **45**, 187–196.

Kakuda, Y., Stanley, D.W. and Van de Voort, F.R. (1981). Determination of TBA number of high performance liquid chromatography. *J. Am. Oil Chem. Soc.*, **58**, 773–775.

Ke, P.J., Ackman, R.G., Linke, B.A. and Nash, D.M. (1977). Differential lipid oxidation in various parts of frozen mackerel. *J. Food Technol.*, **12**, 37–47.

Keller, J.D. and Kinsella, J.E. (1973). Phospholipid changes and lipid oxidation during cooking and frozen storage of raw ground beef. *J. Food Sci.*, **38**, 1200–1204.

Kohn, H.I. and Liversedge, M. (1944). On a new aerobic metabolite whose production by brain is inhibited by apomorphine, emetine, epinephrine and menadione. *J. Pharmacol.*, **82**, 292–295.

Kosugi, H., Kato, T. and Kikugawa, K. (1987). Formation of yellow, orange, and red pigments in the reaction of alk-2-enals with 2-thiobarbituric acid. *Anal. Biochem.*, **165**, 456–464.

Kosugi, H., Kato, T. and Kikugawa, K. (1988). Formation of red pigment by a two-step 2-thiobarbituric acid reaction of alka-2,4-dienals. Potential products of lipid oxidation. *Lipids*, **23**, 1024–1031.

Kunsman, J.E., Field, R.A. and Kazantzis, D. (1978). Lipid oxidation in mechanically deboned red meat. *J. Food Sci.*, **43**, 1375–1378.

Labuza, T.P. (1971). Kinetics of lipid oxidation in foods. *CRC Crit. Rev. Food Technol.*, **2**, 355–405.

Lamikanra, V.T. and Dupuy, H.P. (1990). Analysis of volatiles related to warmed over flavor of cooked chevon. *J. Food Sci.*, **55**, 861–862.

Lawrence, R.C. (1965). Use of 2,4-dinitrophenylhydrazine for the estimation of micro amounts of carbonyls. *Nature*, **205**, 1313–1314.

Lee, Y.B., Hargus, G.L., Kirkpatrick, J.A., Berner, D.L. and Forsythe, R.H. (1975). Mechanism of lipid oxidation in mechanically deboned chicken meat. *J. Food Sci.*, **40**, 964–967.

MacDonald, B., Gray, J.I. and Gibbins, L.N. (1980). Role of nitrite in cured meat flavor: Antioxidant role of nitrite. *J. Food Sci.*, **45**, 893–897.

MacLeod, G. and Ames, J. (1986). The effect of heat on beef aroma: comparisons of chemical composition and sensory properties. *Flav. Frag. J.* **1**, 91–104.

Mai, J. and Kinsella, J.E. (1979). Changes in lipid composition of cooked mince carp (*Cyprinus carpio*) during frozen storage. *J. Food Sci.*, **44**, 1619–1624.

Marcuse, R. and Johansson, L. (1973). Studies on the TBA test for rancidity grading: II. TBA reactivity of different aldehyde classes. *J. Am. Oil Chem. Soc.*, **50**, 387–391.
Melton, S.L. (1983). Methodology for following lipid oxidation in muscle foods. *Food Technol.*, **37**(7), 105–111, 116.
Misock, J.P., Kunsman, J.E. and Field, R.A. (1979). Lipid oxidation in bologna containing mechanically deboned beef. *J. Food Sci.*, **44**, 151–153.
Moerck, K.E. and Ball, H.R., Jr. (1974). Lipid autoxidation in mechanically deboned chicken meat. *J. Food Sci.*, **39**, 876–879.
Morrissey, P.A. and Apte, S. (1988). Influence of species, haem and non-haem iron fraction and nitrite on hexanal production in cooked muscle systems. *Sci. Alimentes*, **8**, 3–14.
Nair, V. and Turner, G.A. (1984). The thiobarbituric acid test for lipid peroxidation: structure of the adduct with malondialdehyde. *Lipids*, **19**, 804–805.
Noble, A.C. (1976). Effect of carbon dioxide and sodium chloride on oxidative stability of frozen mechanically deboned poultry meat. *Can. Inst. Food Sci. Technol. J.*, **9**, 105–107.
Ohkawa, H., Ohishi, N. and Yagi, K. (1979). Assay for lipid peroxides in animal tissues by thiobarbituric acid reaction. *Anal. Biochem.*, **95**, 351–358.
Owen, J.E., Lawrie, R.A. and Hardy, B. (1975). Effect of dietary variation, with respect to energy and crude protein levels, on the oxidative rancidity exhibited by frozen porcine muscles. *J. Sci. Food Agric.*, **26**, 31–41.
Pearson, A.M., Love, J.D. and Shorland, F.B. (1977). 'Warmed-over' flavor in meat, poultry, and fish. *Adv. Food Res.*, **23**, 1–74.
Pegg, R.B., Shahidi, F. and Jablonski, C.R. (1992). Interactions of sulfanilamide and 2-thiobarbituric acid with malonaldehyde: structure of adducts and implications in determination of oxidative state of nitrite-cured meats. *J. Agric. Food Chem.*, **40**, 1826–1832.
Placer, Z.A., Cushman, L.L. and Johnson, B.C. (1966). Estimation of product of lipid peroxidation (malonyl dialdehyde) in biochemical systems. *Anal. Biochem.* **16**, 359–364.
Rhee, K.S. (1978a). Minimization of further lipid peroxidation in the distillation 2-thiobarbituric acid test of fish and meat. *J. Food Sci.*, **43**, 1776–1778, 1781.
Rhee, K.S. (1978b). Factors affecting oxygen uptake in model systems used for investigating lipid peroxidation in meat. *J. Food Sci.*, **43**, 6–9.
Robards, K., Kerr, A.F. and Patsalides, E. (1988). Rancidity and its measurement in edible oils and snack foods. *Analyst*, **113**, 213–224.
Roozen, J.P. (1987). Effects of types I, II and III antioxidants on phospholipid oxidation in a meat model for warmed over flavour. *Food Chem.*, **24**, 167–185.
Rossell, J.B. (1986). Measurement of rancidity. In *Rancidity in Foods*, 2nd edition, ed. J.C. Allen and R.J. Hamilton, Elsevier Science publishers, London, pp. 23–52.
Rossell, J.B. (1991). How to measure oxidative rancidity in fats and fatty foods. *Lipid Technol.* **3**, 122–126.
Ruenger, E.L., Reineccius, G.A. and Thompson, D.R. (1978). Flavor compounds related to the warmed-over flavor of turkey. *J. food Sci.*, **43**, 1198–1200.
Schieberle, P. and Grosch, W. (1981). Model experiments about the formation of volatile carbonyl compounds. *J. Am. Oil Chem. Soc.*, **58**, 602–607.
Schieberle, P. and Grosch, W. (1987). Evaluation of the flavour of wheat and rye bread crusts by aroma extra dilution analysis. *Z. Lebensm. Unters. Forsch.* **185**, 111–113.
Schwartz, D.P., Haller, H.S. and Keeney, M. (1963). Direct quantitative isolation of monocarbonyl compounds from fats and oils. *Anal. Chem.*, **35**, 2191–2194.
Seo, C.W. (1976). Hydrocarbon production from freeze-dried meats. *J. Food Sci.*, **41**, 594–597.
Seo, C.W. and Joel, D.L. (1980). Pentane production as an index of rancidity in freeze-dried pork. *J. Food Sci.*, **45**, 26–28, 92.
Shahidi, F. (1991). Prevention of lipid oxidation in muscle foods by nitrite and nitrite-free compositions. In *Lipid Oxidation in Food.* ed. A.J. St. Angelo. ACS Symposium Series, American Chemical Society, Washington, DC, pp. 161–182.
Shahidi, F. and Hong, C. (1991). Evaluation of malonaldehyde as a marker of oxidative rancidity in meat products. *J. Food Biochem.*, **15**, 97–105.
Shahidi, F., Yun, J., Rubin, L.J. and Wood, D.F. (1987). The hexanal content as an indicator of oxidative stability and flavour acceptability in cooked ground pork. *Can. Inst. Food Sci. Technol. J.*, **20**, 104–106.

Shahidi, F., Pegg, R.B. and Harris, R. (1991). Effects of nitrite and sulfanilamide on the 2-thiobarbituric acid (TBA) values in aqueous model and cured meat systems. *J. Muscle Foods*, **2**, 1–9.

Silberstein, D.A. and Lillard, D.A. (1978). Factors affecting the autoxidation of lipids in mechanically deboned fish. *J. Food Sci.*, **43**, 764–766.

Sink, J.D. and Smith, P.W. (1972). Changes in the lipid soluble carbonyls of beef muscle during aging. *J. Food Sci.*, **37**, 181–182.

Sinnhuber, R.O. and Yu, T.C. (1958). 2-Thiobarbituric acid method for the measurement of rancidity in fishery products. II. The quantitative determination of malonaldehyde. *Food Technol.*, **12**(1), 9–12.

Siu, G.M. and Draper, H.H. (1978). A survey of the malonaldehyde content of retail meats and fish. *J. Food Sci.*, **453**, 1147–1149.

Sklan, D., Tenne, Z. and Budowski, P. (1983). The effect of dietary fat and tocopherol on lipolysis and oxidation of turkey meat stored at different temperatures. *Poultry Sci.*, **62**, 2017–2021.

Spanier, A.M. and Traylor, R.D. (1991). A rapid, direct chemical assay for the quantitative determination of thiobarbituric acid reactive substances in raw, cooked, and cooked/stored muscle foods. *J. Muscle Foods*, **2**, 165–176.

Tarladgis, B.G., Watts, B.M., Younathan, M.T. and Dugan, L.R., Jr. (1960). A distillation method for the quantitative determination of malonaldehyde in rancid foods. *J. Am. Oil Chem. Soc.*, **37**, 44–48.

Tarladgis, B.G., Pearson, A.M. and Dugan, L.R., Jr. (1964). Chemistry of the 2-thiobarbituric acid test for determination of oxidative rancidity in foods. II. Formation of the TBA–malonaldehyde complex without acid-heat treatment. *J. Sci. Food Agric.*, **15**, 602–607.

Thomas, C.P., Dimick, P.S. and MacNeil, J.H. (1971). Poultry product quality. 4. Levels of carbonyl compounds in fresh, uncooked chicken and turkey skin. *J. Food Sci.*, **36**, 527–531.

Tichivangana, J.Z. and Morrissey, P.A. (1985). Metmyoglobin and inorganic metals as pro-oxidants in raw and cooked muscle systems. *Meat Sci.*, **15**, 107–116.

Torres, E., Pearson, A.M., Gray, J.I., Ku, P.K. and Shimokomaki, M. (1989). Lipid oxidation in charqui (salted and dried beef). *Food Chem.* **32**, 257–268.

Uchiyama, M. and Mihara, M. (1978). Determination of malonaldehyde precursor in tissues by thiobarbituric acid test. *Anal. Biochem.*, **86**, 271–278.

Ullrich, F. and Grosch, W. (1987). Identification of the most intense volatile flavour compounds formed during autoxidation of linoleic acid. *Z. Lebensm. Unters. Forsch* **184**, 277–282.

Warner, K., Evans, C.D., List, G.R., Dupuy, H.P., Wadsworth, J.I. and Goheen, G.E. (1978). Flavor score correlation with pentanal and hexanal contents of vegetable oil. *J. Am. Oil Chem. Soc.*, **55**, 252–256.

Wills, E.D. (1965). Mechanisms of lipid peroxide formation in tissues. Role of metals and haematin proteins in the catalysis of the oxidation of unsaturated fatty acids. *Biochim. Biophys. Acta*, **98**, 238–251.

Wong, D.W.S. (1989). Lipids. In *Mechanism and Theory in Food Chemisry*. Van Nostrand Reinhold, New York, pp. 1–47.

Yamauchi, K., Nagai, Y. and Ohashi, T. (1980). Quantitative relationship between alpha-tocopherol and polyunsaturated fatty acids and its connection to development of oxidative rancidity in porcine skeletal muscle. *Agric. Biol. Chem.*, **44**, 1061–1067.

Yeo, Y.K. and Horrocks, L.A. (1985). Decrease in plasmalogin in beef muscle by electrical stimulation. Presented at the 45th Ann. Meet. of the *Inst. of Food Technologists*, Atlanta, GA, June 9–12.

Zipser, M.W. and Watts, B.M. (1962). A modified 2-thiobarbituric acid (TBA) method for the determination of malonaldehyde in cured meats. *Food Technol.*, **16**, 102–104.

Zubillaga, M.P., Maerker, G. and Foglia, I.A. (1984). Antioxidant activity of sodium nitrite in meat. *J. Am. Oil Chem. Soc.*, **61**, 772–776.

15 Sensory and statistical analyses in meat flavour research

A.J. ST. ANGELO, B.T. VINYARD and
K.L. BETT

15.1 Introduction

In 1985, the Southern Regional Research Centre (SRRC) established a new research unit to establish sensory and objective definitions of food flavours in meat, catfish, eggs and peanut products. Included in this research was a focus on the chemical and biological origins of flavour compounds as well as causes of flavour deterioration and undesirable flavour development. Of course, the ultimate goal was to serve both the consumer and the food industry through quality assessment.

One part of the research mission deals with the aetiology and control of undesirable flavours of meat, predominantly beef, or more specifically the so-called 'warmed-over flavour' (WOF) phenomenon. Tims and Watts (1958) described WOF as the rapid development of oxidized flavour in cooked, uncured meat that has been stored refrigerated over a period of a few days. Actually, the meaty flavour can deteriorate after only a few hours of refrigeration. Since first described by Tims and Watts in 1958, the original definition has been expanded to include raw meat that is ground and exposed to air (Green, 1969; Sato and Hegarty, 1971). The process in which meat develops WOF has recently been called 'meat flavour deterioration', or MFD (Spanier *et al.*, 1990; St. Angelo *et al.*, 1992). To investigate the WOF (or rather MFD, which is the term that will be used in this report) phenomenon, we set out to establish a sensory panel of trained experts that could judge flavour intensities by descriptive analysis, and to correlate those findings with chemical and instrumental analyses. These correlations were made by appropriate use of statistical methods, such as multivariant analysis by principal components analysis. The purpose of this report is to present an overview of this multi-disciplinary approach used to solve this problem.

15.2 Sensory evaluation

The sensory laboratory for evaluating meat samples requires different considerations from those for evaluating non-heated or non-cooked samples, as will be explained in this section. There are several different requirements that a laboratory must have to minimize bias among panelists and among samples.

15.2.1 Odour control

A meat sensory laboratory should consist of a minimum of two rooms: one for preparation of samples and the other for evaluation of samples. The preparation of meat samples generates odours that will interfere with flavour evaluation; therefore, the two processes need to be separated. More rooms for other functions may be necessary, such as office space, storage room, training room, entrance and exit areas, or computer space. The air conditioning system in the evaluation room should have positive pressure compared to surrounding rooms to keep odours flowing out of the room. In addition, activated carbon filters installed in the air conditioning system can prevent outside odours from entering into the evaluation room. Care should be taken when setting up the evaluation room that no highly odorous materials are used. For example, vinyl floors should be used instead of carpet.

15.2.2 Lighting

The room should have an adequate level of illumination such as that provided by fluorescent lights. To hide colour differences in the samples a special lighting system is required. This system can include light fixtures with red, green or blue bulbs or theatre-type light filters in frames over recessed lighting (Meilgaard *et al.*, 1987). Care should be taken to provide adequate lighting when coloured lights are used, because persons with poor eyesight will have difficulty. Pangborn (1967) suggested presenting samples singly to prevent comparison of colour. This works well in descriptive flavour analysis, but not for forced-choice methods.

15.2.3 General comfort

The temperature of the evaluation room should be comfortable for the panelists. If it is too warm or too cool the panelists will be distracted by their discomfort. The ideal temperature should be 22°C and 45 to 50% humidity (Meilgaard *et al.*, 1987). In New Orleans, Louisiana, the relative humidity is normally between 80 and 100%. We find that 21.5°C (71°F) with 67% R.H. is comfortable for our panelists and feasible with our air conditioning system in the humid climate of southeastern Louisiana.

The evaluation room should be large enough to comfortably accommodate the number of panelists present. Booths can be constructed to reduce auditory and visual distractions. They should be 68.5 to 81 cm wide with dividers that extend about 46 cm beyond the counter top. Sinks should not be installed in booths because of inherent odour problems. Counter tops should be 46 to 56 cm deep and 74 to 79 cm from the floor. Comfortably padded operator chairs (that swivel and roll) are easiest for panelists to manoeuvre and turn around to listen to instructions.

If a computer ballot system is used, the design (placement of monitor, keyboard, etc.) should be comfortable for the panelists to use. If light pens or touch screens are used, the screen should be low enough that panelists do not have to hold their arm in the air for long periods of time, but yet at a level that makes it easy to see. A software package that is user friendly allows the panelists to concentrate on the sample rather than the computer system.

15.2.4 Preparation area

The preparation area should include appropriate cooking equipment. A survey by Cross (1977) indicated that most researchers broil or roast steaks. This may be true today also. Convection ovens are most appropriate for roasting. Broiling can be done in a conventional oven or on an electric grill. Braising (cooking slowly in a moist atmosphere) is another common method of cooking meat.

Intact meat samples are cooked to an internal temperature to determine doneness. Thermocouples with wire diameters less than 0.02 cm connected to a monitoring device are recommended. The thermocouple should not be encased in metal sheaths, because the metal conducts heat into the sample. This causes the sample in contact with the metal sheath to reach the end-point temperature before the rest of the sample. The constant time of cook method can be used for ground or flaked samples when a temperature monitoring system is not available (AMSA, 1987).

The preparation area should be constructed of easy-to-clean materials. Equipment and counter tops should be made of stainless steel or a comparable material that will not transfer volatiles to the sample. Plastic materials are unsuitable for this reason. Porous materials are not suitable because they absorb odours, which subsequently can be transferred to the sample, and they harbour bacteria that can contaminate the sample (Meilgaard *et al.*, 1987).

15.2.5 Sample preparation and serving

The objectives of the experiment determine the best preparation method. If one wants to compare the effect of two or more treatments on flavour descriptors, then a ground meat model (85 g ground patties) results in a more uniform sample. If one wants to compare the descriptive analysis data with consumer data, then a model that is patterned after the consumer samples is necessary. The method of cooking, whether roasted, broiled, braised or grilled, depends on the objectives of the experiment and the serving model. Ground patties can be grilled or broiled. Whole roasts are roasted or braised. Meat should be cooked to a specified end-point temperature.

Reheating of meat samples can be accomplished by several methods. Johnsen and Civille (1986) used three methods: baking, broiling and boiling. To bake, patties were rewarmed in a 150°C preheated oven until they reached an internal temperature of 65°C. To broil, patties were rewarmed in a preheated broiler for 5 min on each side. to steam-cook, patties were sealed in polyethylene pouches and boiled for 10 min in 1.14 l of water.

Microwaving can also be used to rewarm samples. Cremer microwaved beef samples on high power until the internal temperature reached 74°C (Cremer, 1982). Length of time of exposure was shown to be dependent upon the weight of sample microwaved. Times needed to be determined before the start of the experiment. Cremer found that beef patties reheated in a convection oven had higher sensory quality ratings for flavour, appearance and general acceptability than those reheated in a microwave. From these data, one can conclude that the objectives of any experiment should determine the reheating method, otherwise heating in a convection oven is preferred.

At SRRC, we trim the visible fat, grind the meat and form 85 g patties. Making ground patties increases the uniformity of the samples. We use a Farbarware electric grill (Model 455ND, Subsidiary of W. Kidde Co., Bronx, New York) to cook the patties 7 min/side to a well done state. Frequent monitoring of internal temperature assures that the end-point remains constant. Patties are cut into wedges and placed in glass Petri dishes (60 mm diameter), which are placed in Hobart warming drawers (Troy, Ohio) set at 51.7°C to keep them warm (for the few minutes) until they are served to the panelists. Cooking or reheating of samples is staggered so they can be served as soon as possible after cooking or reheating to minimize time in the warming drawers.

After receiving a sample, a panelist will cut the wedge of ground meat into two pieces, lengthwise, to ensure having two equal portions. This allows for a second chance to evaluate. Meat samples are identified with a three digit random number. The order of serving is randomly selected before cooking, so that no clues are obtained from the serve order. Panelists receive between four and six experimental samples at a session. Each session begins with a warm-up (non-experimental) freshly grilled sample. Consensus (mean scores) are taken on that sample and the panelists use those sample scores to compare the other samples to, instead of comparing between samples. The interval of time between presentation of samples is 5 min. This allows time to evaluate the sample and to allow the senses to recover. Unsalted crackers (i.e. biscuits) and reverse osmosis treated, deionized water are available at all times to cleanse the palate.

15.3 Sensory analysis of meat

15.3.1 Descriptive flavour panel

Descriptive flavour panelists should be selected on the basis of having 'normal' abilities to taste and smell and on their availability. Testing for normal abilities to taste and smell should include tests for rating or ranking intensity, tests that determine a person's ability to discriminate between aromas and between tastes, and tests to determine the ability to describe what is being perceived (Meilgaard *et al.*, 1987).

15.3.2 Descriptor development

Once the panelists have been selected and orientated in the basic principles of descriptive flavour analysis, they can start descriptor development. A range of samples that includes the flavours and off-flavours similar to those of interest are presented, and the panelists describe the flavours that they perceive. A trained panel leader then guides the discussion to eliminate redundant descriptors. The panelists need to come to a consensus on the descriptors that eventually will be placed on the ballot.

Johnsen and Civille (1986) worked with seven experts in the meat industry to develop descriptors for warmed-over flavour in meats. Love (1988) reported on the development of beef descriptors at SRRC. The panel at SRRC also developed descriptors for lamb (St. Angelo *et al.*, 1991).

The descriptor beefy/brothy, defined by Love (1988), was divided into beefy and brothy by the beef descriptive analysis panel to aid in studies on desirable beef flavours at SRRC. It was hypothesized that these were two unique descriptors, but it was to be determined if the descriptive panel could evaluate them. At the first session the panel was presented with drippings from roasted chicken, roasted pork, roasted veal and roasted top round beef, and broth from boiled chicken, boiled pork, boiled veal and boiled beef (top round). They also received a sample of grilled ground beef in which to discern these two flavours. At this session, the panel decided that two terms were feasible and developed definitions. 'Beefy' was defined as the aromatic commonly associated with matured cooked beef muscle products. The reference sample is prepared by boiling cubes of top round roast in water until well done. The liquid is drained off and served to the panelists. It contains a distinct beefy flavour. The broth from chicken tasted distinctly like chicken and the broth from pork tasted dis-

Table 15.1 Sums of squares and coefficients of variability from the analysis of variance

Source	df	STY	BEF	BTH	PTY	SER	BRC	CKL	CBD	SOU	SWT	BTR
Cut	3	0.16*	6.55**	0.71	0.51	1.13**	5.78**	1.46**	0.20	1.03**	0.66**	0.51*
Storage	1	0.11**	0.25	0.26	0.20	0.15**	0.23	0.01	0.12	0.03	0.03	0.07
Cut, storage*	3	0.01	0.26	0.23	0.05	0.04	0.10	0.03	0.03	0.10	0.03	0.02
Error	16	0.25	4.66	1.67	0.95	0.40	1.94	1.29	1.81	0.32	0.40	0.67
Total	23	0.53	11.71	2.27	1.71	1.73	8.05	2.79	2.19	1.68	1.13	1.28
C. V.		9.3	16.8	15.8	39.8	17.3	15.0	14.8	44.1	19.1	14.1	29.9
Grand mean		1.3	3.2	2.0	0.6	0.9	2.3	1.9	0.8	0.9	1.1	0.7

Abbreviations: Beef cut (Cut), salty (STY), beefy (BEF), brothy (BTH), painty (PTY), serum/raw (SER), browned/carramel (BRC), cooked liver (CKL), cardboardy (CBD), sour (SOU), sweet (SWT), bitter (BTR).
**Means are significantly different at $p < 0.01$.
*Means are significantly different at $p < 0.05$.

Table 15.2 Table of means of descriptive flavour analysis by beef cuts

Descriptor	Veal	T-bone	Top round	Chuck
Salty	1.5^{a}	$1.3^{a,b}$	$1.3^{a,b}$	1.2^{b}
Beefy	2.4^{c}	$3.6^{a,b}$	3.7^{a}	$3.1^{b,c}$
Brothy	2.1^{a}	2.1^{a}	2.0^{a}	1.9^{a}
Painty	$0.6^{a,b}$	0.4^{b}	0.5^{b}	0.8^{a}
Serum/raw	1.1^{a}	0.6^{c}	0.9^{b}	1.1^{a}
Browned/caramel	1.5^{b}	2.7^{a}	2.7^{a}	2.3^{a}
Cooked liver	1.5^{b}	2.0^{a}	2.1^{a}	2.0^{a}
Cardboardy	0.7^{a}	0.7^{a}	0.7^{a}	0.9^{a}
Sour	1.3^{a}	0.7^{b}	0.9^{b}	0.9^{b}
Sweet	1.0^{b}	1.4^{a}	1.1^{b}	1.0^{b}
Bitter	0.9^{a}	0.5^{b}	0.6^{b}	$0.7^{a,b}$

[a,b,c] Means with the same superscript are not significantly ($p < 0.05$) different based on L.S.D. mean comparison test.

tinctly like pork, etc. 'Brothy' was defined as the aromatic associated with the drippings from roasted meat which is characteristic of all meats (i.e. poultry, beef and pork). The reference is the drippings from top round roasted in a 177°C oven until done. It has a brothy note that is characteristic in the drippings from beef, poultry, veal and pork. In the second session, panelists evaluated beefy and brothy along with the other descriptors in ground patties made of top round, chuck, veal and T-bone. This further tested the feasibility of these revised descriptors.

An experiment was designed to test panelists' ability to evaluate flavour with the revised descriptors. Three cuts of beef and one cut of veal were individually ground and formed into 85 g patties. One half of patties in each group were frozen at −11°C for 3 days and the other halves were refrigerated at 4°C for 3 days. At the end of 3 days, the patties were grilled and served to the panel as described above. Each preparation of beef or veal under a particular storage regimen is called an experimental unit. The actual beef sample tasted by an individual panelist during a session is a subsample of the experimental unit prepared for that session. The analysis of variance (Table 15.1) was accomplished on means across panelists for each experimental unit of samples within a session. Analysing the means instead of individual panelists' scores removes most of the panelist effect. This training exercise indicated that the brothy flavour intensity did not significantly differ between cuts. Conversely, beefy flavour was highest in intensity in top round, lowest in veal (Table 15.2). The other descriptors that vary among beef cuts are cooked liver, serum/raw, brown/caramel, salty, sour, sweet and bitter.

15.4 Chemical and instrumental parameters

15.4.1 Thiobarbituric acid reactive substances

The thiobarbituric acid (TBA) test is one of the most popular and widely accepted methods for measuring lipid oxidation in food products. The extent of rancidity is usually expressed in terms of TBA number (μg malonaldehyde/g sample). One mol of malonaldehyde (MDA), a secondary decomposition product from peroxidized unsaturated fatty acids, can complex with two mols of TBA reagent to form a chromophore measurable by spectrophotometry. However, there have been problems related to the TBA test. Its sensitivity has been questioned (Tsoukala and Grosch, 1977); its reliability has been questioned (Witty *et al.*, 1970); interfering compounds have been identified, which led to a modification of the method to overcome their effect (Sinnhuber and Yu, 1977). More recently, limitation of the TBA test for oxidized lipids containing alka-2,4-dienals was claimed since the test was found to be nonspecific to malonaldehyde (Kosugi *et al.*, 1988). Nevertheless, in spite of the problems related to the TBA/MDA reaction, the method is still used throughout the food industry. The distillation method of Tarladgis (1960) seems to be the most popular. In addition to measuring rancidity by the Tarladgis procedure, our samples were also analysed by a second objective method, i.e. gas chromatography.

15.4.2 Direct gas chromatography

As lipids are oxidized, they form hydroperoxides, which can decompose to produce volatile flavour components such as aldehydes, ketones, alcohols, etc. Over the years, enrichment techniques were developed to obtain a sufficient concentration of these volatile compounds for analysis by gas chromatography (GC). One of the most novel approaches to enrichment of flavour compounds for analysis on GC was developed by Dupuy *et al.* (1971). They described their method as a simple and very sensitive technique to analyse volatiles in vegetable oils by direct gas chromatography (DGC). Briefly, the method involved adding a large sample (500 mg) of oil directly onto glass wool, which is in a glass liner. Next, the glass liner is inserted into the injection port, which is heated. A carrier gas then sweeps the volatiles onto a column for analysis. The method requires no extractions, distillations or formation of derivatives. Later, Legendre and coworkers modified the procedure by inventing an external closed inlet device (ECID) that can be attached to almost any gas chromatograph (Legendre *et al.*, 1979). Whereas the original method was first used with packed columns, the method was later modified to utilize capillary columns (Dupuy *et al.*, 1985). This method is neither an example of the

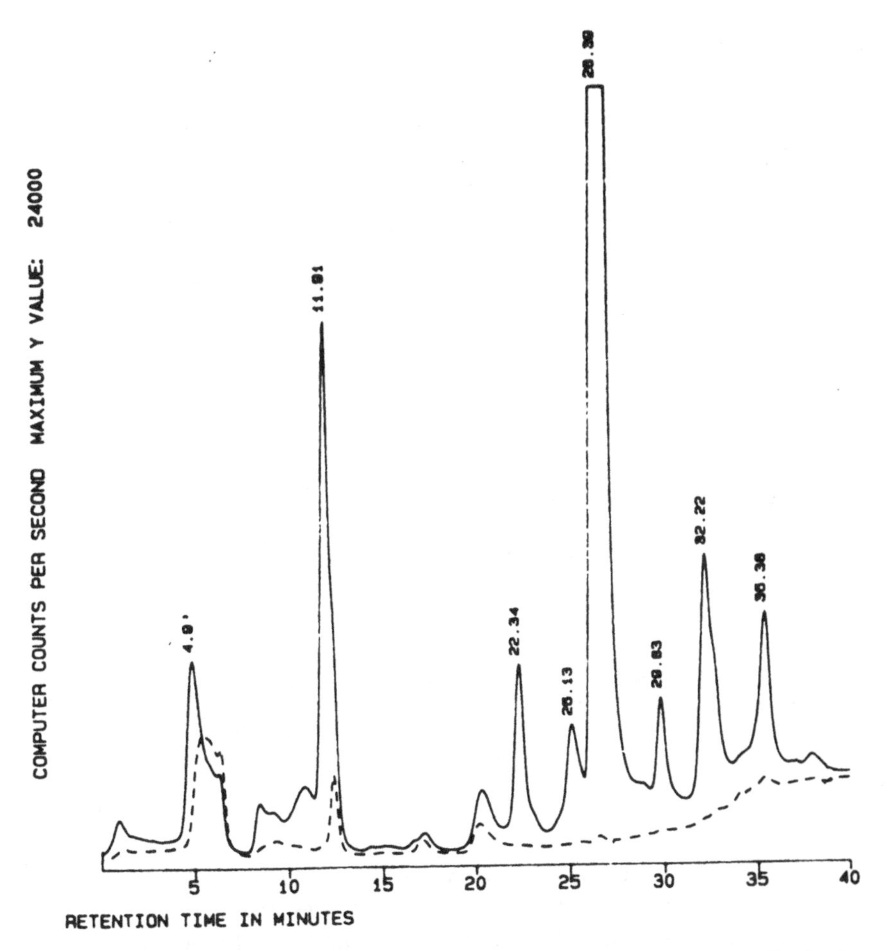

Figure 15.1 Volatile profiles of cooked beef obtained with packed column GC. The broken line graph represents freshly cooked beef; the solid line graph represents cooked beef after storage at 4°C for 8 h. Compounds identified were propanol (11.9 min retention time), butanal (17), pentanal (22.3), 2,3-hydroxybutanone (25.1), hexanal (26.4), heptanal (29.8), 2,3-octanedione (32.2), nonanal (35.4).

classical static head space technique nor the dynamic head space method as defined by Wampler *et al.* (1985). Hence, the name 'direct' is assigned to the 'Dupuy' GC method.

Although first employed to assess vegetable oil quality, this methodology was utilized during the past two decades to assess the quality of many foods, including meats. For example, bacon (Dupuy *et al.*, 1978); beef, ham and pepperoni (Bailey *et al.*, 1980); beef, chicken and turkey (Dupuy *et al.*, 1987); and more recently lamb (St. Angelo *et al.*, 1991). A typical profile of freshly cooked and WOF beef ground patties using packed

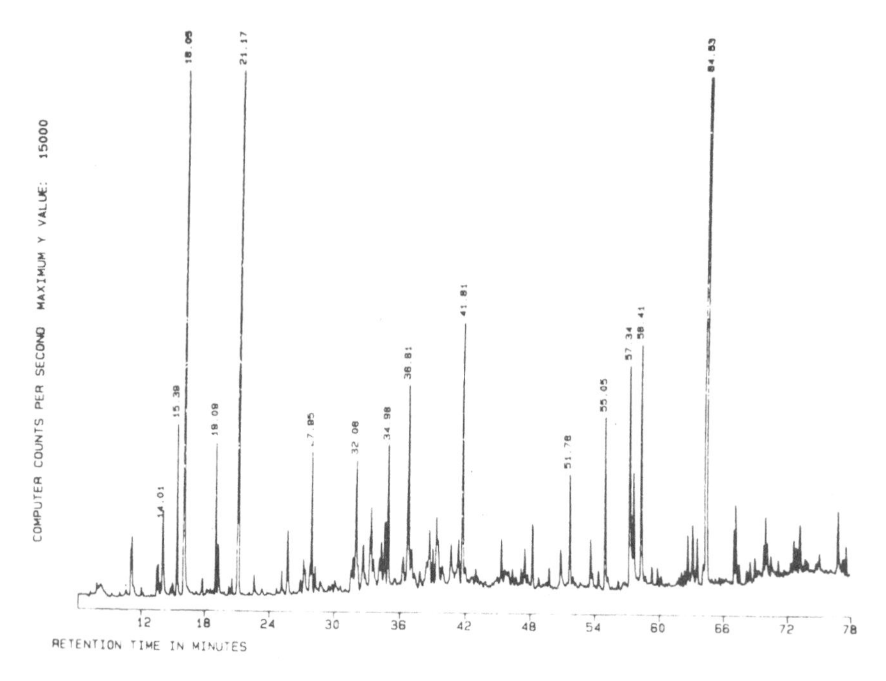

Figure 15.2 Volatile profile of freshly cooked beef obtained with capillary column GC. Compounds identified were pentanal (15.4 min retention time), 2,3-hydroxybutanone (19.1), hexanal (21.2), heptanal (27.9), 2,3-octanedione (32.1), 2-pentylfuran (34.9), nonanal (41.8), *trans*-2, *cis*-4-decadienal (57.3), *trans*-2, *trans*-4-decadienal (58.4).

columns is shown in Figure 15.1. A capillary column profile is shown in Figure 15.2 for fresh cooked ground beef patties and in Figure 15.3 for WOF patties stored for 24 h at 4°C. The primary lipid oxidation markers are the aldehydes, such as hexanal, pentanal, nonanal and the decadienals. As oxidation progresses, the intensity of these compounds increases proportionally.

15.5 Correlations among sensory, chemical and instrumental analyses

When comparisons were made between volatile compounds obtained from DGC and TBA numbers, from both fresh and cooked/stored ground beef patties, the results correlated very well (St. Angelo *et al.*, 1987; Bailey *et al.*, 1987). In fact, the major volatile compounds (e.g. butanal, pentanal, hexanal, heptanal, 2,3-octanedione, 2,4-heptadienal, nonanal, and *trans*, *cis*- and *trans*, *trans*-2,4-decadienal) that were found in fresh and WOF meat samples by DGC were also found in the distillate prepared for the

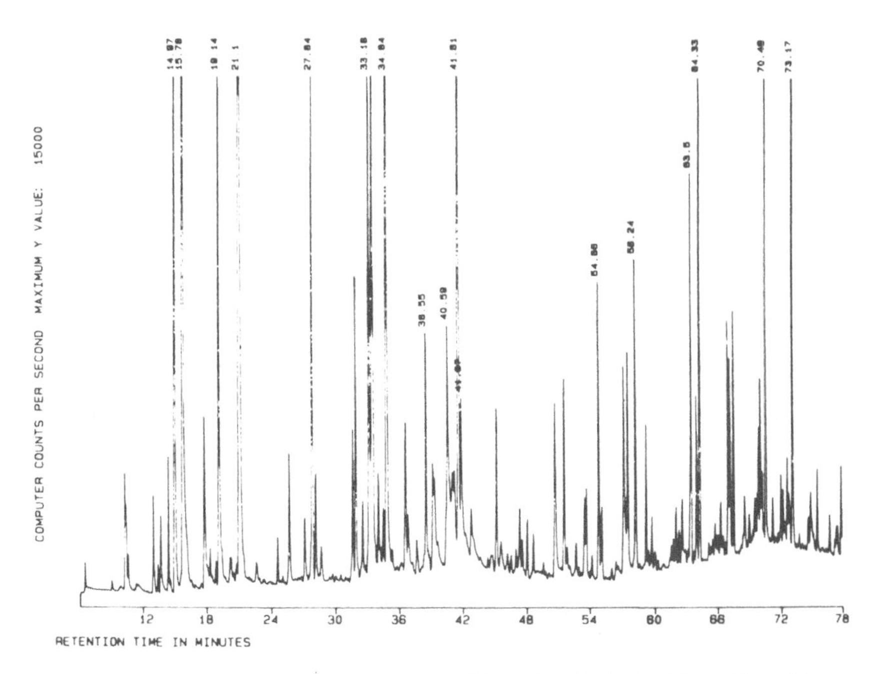

Figure 15.3 Volatile profile of cooked/stored WOF patties obtained with capillary GC; compounds identified had comparable retention times to those listed in Figure 15.2.

TBA reaction. Similar correlations were also made between DGC data or TBA numbers and data obtained from sensory panels for fresh and WOF ground beef patties (St. Angelo *et al.*, 1987). More recently, Spanier *et al.* (1991) used multivariate principal components analysis to show statistical correlations among the sensory (cooked beef/brothy, browned caramel, serumy, sweet, painty, cardboardy, sour, bitter and salty), chemical (TBA) and instrumental attributes (namely, hexanal, 2,3-octanedione, nonanal and pentanal).

The collection of sensory, chemical and instrumental data on meat samples, as discussed above, provides a profile of each meat sample that can facilitate identification of relationships among these characteristics. Such relationships can subsequently provide insight regarding the mechanisms at work in meat samples that do or do not contribute to meat flavour deterioration (MFD). Two experiments, a replicated lamb study and an unreplicated beef additive study, will be discussed to illustrate the appropriate statistical techniques for examining these relationships.

15.5.1 Experimental designs

An experimental design should be developed for each experiment. If collection of data is accomplished without regard to a design, then relationships of interest may become confused with day-to-day variation. Designing an experiment requires: (1) the construction of a schedule by which the experimental treatments are applied to a design structure (i.e. panel sessions), as discussed below; and (2) a decision regarding how much replication can be conducted, as discussed later.

Application of treatments to a design structure. The manner by which treatments are appropriately applied to panel sessions depends mainly upon the experimental objective(s) and, to a lesser degree, upon the number of treatments being considered in the experiment. It is also essential that at least one identically treated sample be administered during every panel session to serve as a blind control for use in reducing session-to-session variability.

Consideration of experimental objective and treatments. Ideally, it is recommended that the number of experimental treatments chosen for consideration in an experiment be limited to the number of samples that can be presented in one sensory panel session. This restriction prevents the effect of treatments from becoming confused with the day-to-day variation in meat samples and panel performance. Each sensory panel session is limited to a maximum of six experimental treatments and a control treatment to calibrate the panelists.

Practically, an experiment limited to the examination of six experimental treatments often lacks scope for providing sufficiently useful information. Hence, the experimental factors are assigned a degree of importance so that levels of the experimental factor of primary interest are administered to panelists in the same panel session.

For example, St. Angelo *et al.* (1991) compared the effects of five distinct tenderization treatments applied to lamb carcasses immediately after slaughter on fresh frozen and 2-day stored samples. These treatments were as follows: Sample 1, Control (standard slaughter); Sample 2, ES (electrical stimulation); Sample 3, ES + Ca (ES plus calcium chloride); Sample 4, ES + Ca + M (as Sample 3, plus maltol); and Sample 5, ES + Ca + SA (as Sample 3, plus sodium ascorbate). Freshly ground leg of lamb was used as a standard.

Assessing the tenderness of meat was the primary objective of this experiment and the consistency of response to the tenderization treatments with storage was secondary. Hence, a design was constructed to randomly administer samples from lamb treated with all five tenderization treatments in the same panel session. Each panel session was then completely composed of either fresh frozen samples or 2-day stored samples totally confusing the variability of secondary interest (due to storage), with session-to-session variability. Reduction of session effects to allow for more accurate examination of variability due to storage only is discussed in the next section. One replicate was comprised of the two sessions described above.

In the current beef additive study, we initially investigated the effect of six chemical additives (each administered to beef samples at three concentrations, e.g. 0 ppm, 50 ppm and 125 ppm) on the sensory, chemical and instrumental responses of fresh frozen, and 2-day stored beef samples. The objective of primary importance was, for each of the six additives, to characterize the MFD trends occurring in stored samples relative to fresh frozen samples and to determine if these trends exhibited a significant reduction in rate of MFD development with the application of varying concentrations of chemical additive to the beef samples. Secondary to the study was a comparison among the six additives.

A design was constructed so that each session was composed of both fresh frozen and 2-day stored samples, each treated with all three concentrations of a single additive. The schedule of sample administration for one such panel session (i.e. Session 1) constitutes a single replicate of all six storage times (0 and 2-day) and concentration (0%, higher %, highest %) combinations.

Analysis of data collected using this design indicated no significant reduction in MFD with increased additive concentrations for any of the six additives. Hence, the experiment was repeated using fresh frozen and 3-day stored beef samples. Session 2 included the use of a sample presentation schedule identical to that of the first session, except for the replacement of 2-day samples with 4-day samples. One replicate was comprised of Sessions 1 and 2.

A design constructed to initially consider all three levels of stored samples (fresh frozen, 2-day and 4-day) would have been very similar to the design comprised of Sessions 1 and 2. Based on the experimental objective, a preferred design would administer samples stored for all three storage times but only two of the three concentrations of an additive during the same session. The differences between this preferred design and the design resulting from the combination of the two experiments actually conducted can, under certain assumptions, be assumed negligible, as discussed in the next section.

Including a blind control. The session effect, caused by session-to-session variability can, under certain assumptions, be effectively eliminated from the sensory data prior to its analysis, by ensuring that the experiment be designed to include a blind control in every panel session.

Before discussing the blind controls for our two experiments we will describe how the session-to-session variability (i.e. session effect) can be removed from the sensory data. The average rating assigned to a sample for a particular flavour characteristic by the panelists is considered to be one replicate measurement of the treatment assigned to that sample. Hence, each panel session provides one replicate of each treatment represented in that session. If an identical blind control treatment is applied to one sample in each session then data collected from all panel sessions can be adjusted to one another respective to the ratings received by this common sample. The session effect can be practically eliminated under the assumptions that none of the blind control samples are outliers in any manner and that any variability observed from session-to-session can be identified by differences among the blind control samples. The elimination of session effect is accomplished by comparing the average panelist rating of this blind control to the overall average rating of the blind control from all panel sessions. If a session's blind control was rated above (or below) the overall average rating then the average rating observed for each treatment in the session is adjusted downwards (or upwards) by this difference.

St. Angelo *et al.* (1991) required an adjustment for the session effect to allow comparison of fresh frozen samples with samples stored for 2 days. Owing to the limited amount of experimental sample no blind control was administered. Hence, the 'standard' samples used to calibrate the panelists at the beginning of each session were used, less effectively, to eliminate session effects.

In the current beef additive study interest only lay in direct comparisons between the storage time trends occurring at three concentrations of each particular additive. Hence, for analytical purposes, a separate experiment was conducted for each additive, and elimination of session effect was conducted separately for each additive. The blind control for this experiment was actually a collection of the experimental treatments. One replicate, as described above, indicates that each session contains six experimental samples and that three of the six samples are fresh frozen (0-day stored) samples, each representing one of the three additive concentrations. Each of these three fresh frozen samples receives an average rating with respect to a particular flavour characteristic. An average of these three sample averages then provides a 'comprehensive' blind control for eliminating session effect.

Session effect would have been eliminated in data collected according to the 'preferred' design, discussed above, by using the mean of values from fresh frozen, 2-day and 4-day stored samples with 0% additive. Hence, under the assumption that this adjustment technique effectively eliminates session effects, the design that was actually used in conducting the experiment provides information comparable to that provided by the preferred design.

Summary—applications of treatment to a design structure. Techniques for applying treatments to panel sessions have been discussed. These techniques: (1) emphasize the application of treatments directly related to the primary objective of the experiment to the same panel session; and (2) mandate the inclusion of an identical blind control sample in each panel session. These concepts are equally applicable to replicated and unreplicated experiments.

Replication—how much is possible? The amount of replication that can be conducted in any experiment is primarily dependent upon the experimental material, the time, and physical, monetary, and human resources available. Experimental material is usually a valuable and limited commodity. However, if sufficient experimental material is available to conduct five replicate panel sessions for each treatment (considered optimal, based on the sensory history at SRRC), then it must be determined whether the sensory panel can be convened for the required number of sessions within time-frame limitations. If these requirements are feasible, then there must 'ideally' remain a sufficient amount of experimental material, time, and physical and human resources to conduct all chemical and instrumental analyses on a sample of the same experimental material, representing each treatment, administered to the panel during each session.

For instance, St. Angelo *et al.* (1991) conducted four complete replicates of the five tenderization treatments evaluated in fresh frozen and 2-day stored lamb samples. The experiment required eight panel sessions, each session consisting of samples representing one of the two storage times and all five experimental treatments.

Practically, owing to the time-intensive nature of many chemical analyses, the researcher must sometimes settle for a single or duplicate chemical analysis for each treatment. If this is the case, the resulting data cannot be combined with sensory data for use in a multivariate analysis such as that discussed below. Rather, a separate univariate analysis must be conducted for each chemical response variable to identify any differences among experimental treatments.

Ideally, each experimental treatment should be replicated as many times as there are sensory, chemical and physical characteristics of interest. Practically, if relationships among these characteristics are to be identified using multivariate statistical analyses, then: (1) each treatment must have been replicated; and (2) the number of treatments multiplied by the amount of treatment replication must be at least three times the total number of sensory, chemical and instrumental responses of interest (Tabachnick and Fidell, 1983). Hence, the number of sensory, chemical, and instrumental measures that can be considered in a multivariate statistical analysis is limited by the amount of replication possible.

For example, in the lamb study (St. Angelo *et al.*, 1991) four replicates were conducted on each of the ten tenderization and storage time treatments to obtain a data set containing a total of 40 observations. Hence, a multivariate analysis can be conducted using an absolute maximum of 13 response variables collected in the experiment.

The scope or exploratory nature of an experiment may sometimes limit each treatment to a single replicate. Although informative univariate statistical analyses can always be performed on the data collected from unreplicated experiments, no multivariate statistical analyses can be conducted. The current beef additive study provided duplicates for fresh frozen (0-day stored) samples only at each concentration of additive. This partial replication provided sufficient information to conduct a univariate analysis of variance (ANOVA) (Steel and Torrie, 1980) for each sensory, chemical and instrumental characteristic measured. Milliken and Johnson (1989) present techniques for analysing completely unreplicated experiments.

Summary—experimental design. Techniques have been discussed for applying experimental and blind control treatments to panel sessions based on the primary experimental objective(s). The dependency of various types of statistical analyses on the amount of replication conducted and the dependency of replication on resource and practical constraints were examined.

15.5.2 Statistical analysis

The data collected according to an experiment's design can easily be placed in a format to facilitate the univariate and multivariate statistical analyses of the sensory, chemical and instrumental responses of interest.

Data preparation. Prior to its analysis, sensory data can be subjected to noise reduction techniques of Crippen *et al.* (1991) to identify panelists who are either super-sensitive or cannot discern the presence of a particular flavour characteristic and to normalize such data to clarify the information contained in the data. The averages of panelist responses for each treatment in a session are then used as data for univariate and multivariate statistical analysis, as shown in Table 15.3.

If the chemical and/or instrumental measurements were not recorded for every replicate sample administered to the sensory panel, such as the TBA value for the control treatment in Session 7 of Table 15.3 and the values for all three chemical variables for the Es + Ca + SA treatment in Session 8, then one of two options can be utilized. The first option simply omits these two lines of data (i.e. treatment replicates) from use in the multivariate analysis and conducts the analysis based on 38 rather than 40

Table 15.3 Data format for univariate and multivariate analyses of sensory and chemical data from lamb study

Session	STOR	TRT	HEX	TVOL	TBA	CBD	MTH	MTY	PTY	SWT
1	2	Control	347	841.0	12.96	2.6	1.5	3.7	2.5	1.3
1	2	ES	66	161.5	8.63	1.9	1.4	3.9	2.0	1.2
1	2	ES+Ca	218	456.0	12.30	2.5	1.5	4.2	2.6	1.0
1	2	ES+Ca+M	25	216.0	7.84	0.9	1.7	4.1	0.8	1.1
1	2	ES+Ca+SA	44	87.0	1.91	1.8	1.1	4.6	1.6	0.8
2	0	Control	2	48	2.37	1.4	0.9	4.3	1.1	1.0
2	0	ES	46	201	2.39	0.3	1.1	4.7	0.1	1.5
2	0	ES+Ca	30	134	3.61	0.5	1.5	4.9	0.3	1.4
2	0	ES+Ca+M	32	94	1.34	0.3	1.4	4.7	0.1	1.3
2	0	ES+Ca+SA	1	30	0.57	1.0	1.3	4.8	0.7	1.1
7	0	Control	1	57	–	2.2	0.4	4.5	2.8	0.6
7	0	ES	10	59	2.43	0.3	1.3	4.9	0.1	1.3
7	0	ES+Ca	48	117	5.35	0.3	1.6	4.6	0.4	1.3
7	0	ES+Ca+M	13	100	1.18	0.7	1.7	4.3	0.4	1.5
7	0	ES+Ca+SA	4	65	0.66	1.2	1.2	4.3	0.9	1.3
8	2	Control	260	451	8.06	1.6	2.0	3.2	2.3	2.0
8	2	ES	187	300	7.05	2.3	1.4	3.5	2.1	1.1
8	2	ES+Ca	415	700	22.51	2.1	1.3	3.7	2.3	1.1
8	2	ES+Ca+M	107	235	7.78	0.9	1.6	4.3	0.6	1.2
8	2	ES+Ca+SA	–	–	–	0.9	1.1	3.6	0.7	2.1

Abbreviations: STOR, days storage; TRT, treatments; HEX, hexanal; TVOL, total volatiles; TBA, thiobarbituric acid; CBD, cardboardy; MTH, muttony/herby; MTY, meaty; PTY, painty, SWT, sweet; ES, electrical stimulation; Ca, calcium chloride; M, maltol; SA, sodium ascorbate.

observations. The second option requires a separate analysis for the sensory and the chemical response variables. Because of interest in relationships among chemical and sensory variables, the first option is the most desirable. The collected data are now ready for individual and simultaneous analysis of sensory, chemical and instrumental responses.

Univariate analyses of variance. A univariate analysis of variance (ANOVA) (Steel and Torrie, 1980) is conducted using data collected for a single response variable. ANOVA can identify experimental factors (such as storage time, additive and additive concentration) and combinations of these factors that affect statistically distinct behaviour with respect to a particular sensory, chemical or instrumental response. Once all statistically significant treatment effects have been identified for a particular response variable, appropriate mean comparisons (i.e. contrasts) or orthogonal polynomial contrasts (Steel and Torrie, 1980) can be conducted. These contrasts identify, specifically, which experimental treatments or combinations of treatments are statistically distinct with respect to a response variable.

St. Angelo *et al.* (1991) conducted contrasts, some of which are shown in Table 15.4, to compare results from specific pairs and specific linear combinations of the five tenderizing treatments, respective to treatment effects found significant in the ANOVA for each individual sensory, chemical and instrumental response. This approach, however, for comparing responses measured on distinct treatments could not be applied to the analysis of data from the current beef additive study.

In the current beef additive study, interest focussed on a response's trend across 0-, 2- and 4-day storage times at various concentrations of a particular additive. Hence, the objective was to compare the effect on a response resulting from a numeric change in the amount of a single treatment applied to the experimental material. Alternatively, the objective of the lamb study was to compare the effect on a response of the application of several distinct treatments.

The appropriate technique for identifying differences among the numeric levels of treatments such as storage times and additive concentrations is to fit a response surface (Milliken and Johnson, 1989) (i.e. regression model) to the data observed at all combinations of storage times and concentrations considered. The response surface model and plot for the data collected on the sensory descriptor 'painty' using the maltol additive is shown in Figure 15.4. This figure illustrates that an increased concentration of maltol significantly reduces the rate at which the off-flavour descriptor 'painty' increases with sample storage time.

At minimum, partial replication (such as the 0-day duplicates observed for all three concentrations of an additive) is strongly recommended for any experiment. Without replication, there is no source by which to estimate error variability in the data for the subsequent testing for significant differences among experimental treatments. How much replication is actually conducted can be determined as discussed above.

Multivariate principal factor analysis. Multivariate statistical analyses utilize the relationships among the response variables included in the analysis to better describe differences occurring among the experimental treatments. Specifically, a principle factor analysis utilizes the linear correlation structure existing among the observed responses to create a new set of 'factors'. Each factor identifies a unique source of common variability observed among the response variables respective to the experimental treatments.

Before conducting a principal factor analysis, it is essential that any of the sensory, chemical and/or instrumental analysis variables be omitted from this analysis that, by means of a mathematical relationship among one or more of the other response variables, theoretically reproduce identical information. This problem can be avoided by choosing a set of sensory, chemical and instrumental analysis variables that are of key

Table 15.4 Means, standard errors and p values of treatment contrasts for responses with significant treatment by storage time interaction

RESPONSE	DAY	Means/standard error			p value		
		Treatment			Contrast		
		ES (1)	ES + $CaCl_2$ (2)	(2) + SA[a] (3)	(1) versus (2)	(2) versus (3)	(1) versus (3)
Hexanal	0	18.25/6.97	47.25/6.97	5.25/6.97	0.0101*	0.0007**	0.2068
(area counts $\times 10^3$)	2	126.50/38.13	261.75/38.13	24.67/44.03	0.0251*	0.0011**	0.1023
Total volatiles	0	84.50/21.67	133.25/21.67	70.25/21.67	0.1325	0.0576	0.6486
(area counts $\times 10^3$)	2	230.63/66.27	481.25/66.27	123.67/76.52	0.0181*	0.0033**	0.3986
TBARS	0	2.02/0.44	4.50/0.44	1.18/0.44	0.0013**	0.0001**	0.1940
(mg MDA/kg sample)	2	10.10/1.68	19.15/1.68	1.56/1.95	0.0020**	0.0001**	0.0051**
Meaty[b]	0	4.88/0.11	4.73/0.12	4.56/0.11	0.3502	0.2852	0.0682
	2	3.74/0.15	3.72/0.15	3.97/014	0.9417	0.2277	0.2935
Cardboardy[b]	0	0.51/0.15	0.57/0.15	1.42/0.14	0.7631	0.0001**	0.0018**
	2	1.96/0.19	2.33/0.19	1.54/0.18	0.1590	0.0028**	0.0484**
Painty[b]	0	0.31/0.15	0.35/0.16	1.13/0.14	0.8495	0.0002**	0.0135*
	2	2.05/0.20	2.59/0.20	1.52/0.20	0.0557	0.0002**	0.0485*

[a]Sodium ascorbate
[b]Intensity values: 0, lowest; 15, highest.
*Mean differences significant at the 0.05 level.
**Mean differences significant at the 0.01 level.

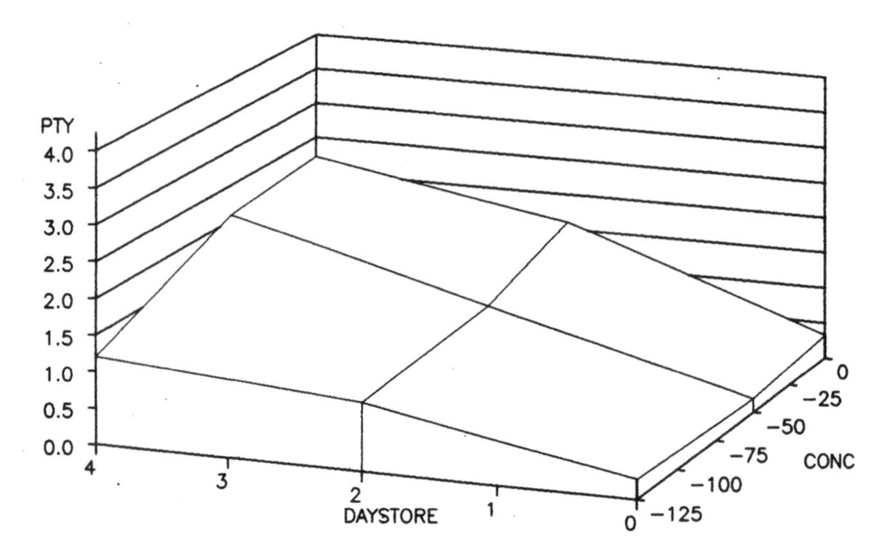

Figure 15.4 Univariate factor analysis; effect of maltol on the intensity of 'painty'. Model: Painty = 0.29 + (0.53 − 0.0022 × concentration) × daystore

importance to the experiment and are known to provide unique information.

The data from St. Angelo *et al.* (1991) can be used to illustrate a principal factor analysis. Five key sensory variables are chosen from a set of 15 measured variables and all three chemical responses are included for a total of eight variables to be utilized in the analysis. The 38 data observations, for which all eight response variables had values, provided sufficient replication, as discussed above, to conduct the factor analysis.

The first step in choosing an appropriate factor solution (i.e. result from a principal factor analysis) is determining the number of factors that can be extracted from the variability among treatments exhibited in the data set. This choice depends upon the variance of each extracted factor. The variance (i.e. eigenvalue) of each principal factor indicates its ability to describe variation occurring in the data. If the eigenvalue of a principal factor is less than 1 then an individual response variable is capable of explaining more of the data variability than is this particular principal factor (Tabachnick and Fidell, 1983). Hence, only those principal factors with eigenvalues of at least 1 need be considered in examining the relationship among the experimental treatments.

A principal factor analysis is only one of several techniques for extracting factors from a set of data. The factor solution (i.e. factors extracted) resulting from a principal factor analysis is often readily interpretable

(Tabachnick and Fidell, 1983). In general, a desirable factor solution will exhibit three characteristics. First, each factor must 'load' (i.e. be highly correlated with) a minimum of two response variables. Secondly, each response variable used in the analysis should have a much higher loading (i.e. stronger) correlation with one factor than with any of the other factors. Finally, the factor solution should be rotated to produce mutually uncorrelated factors.

The factor solution resulting from conducting a principal factor analysis of the data collected in the lamb study consisted of two factors. Of the total treatment variance (6.436), the variances of Factor 1 and Factor 2 were 4.587 and 1.504, respectively. All other extracted factors in this factor solution were able to explain 94.6% of the total variability among the tenderization and storage treatment combinations.

The rotated factor pattern shown in Table 15.5 illustrates which of the eight response variables exhibits the strongest correlation with each factor. Rotation of a factor pattern is simply a technique for removing any correlation among the factors in a factor solution. The response variables hexanal, total volatiles, TBA, cardboardy, and painty all exhibit a high positive 'loading' onto (i.e. correlation with) Factor 1 while the sensory descriptor, meaty, exhibits a high negative loading onto Factor 1. This indicates that hexanal, total volatiles, TBA, cardboardy, and painty are all directly related but are inversely related to meaty. The common variability represented by Factor 1 can be labelled as 'desirability' of the meat samples. The two response variables, muttony-herby and sweet, exhibit positive loadings onto Factor 2 of similar magnitude. However, there is no apparent interpretation of the common variability represented by Factor 2.

The factor pattern (Table 15.5) can be used to graphically examine the relationship among the experimental treatments. Each principal factor produces a score for each observation in the data set used for the analysis. The principal factor scores resulting from the observations associated with

Table 15.5 Rotated factor pattern from a principal factor analysis for the lamb tenderization study

Response variable	Factor 1	Factor 2
Hexanal	0.868*	0.302
Total volatiles	0.847*	0.312
TBA	0.823*	0.184
Cardboardy	0.906*	−0.317
Muttony/herby	0.310	0.702*
Meaty	−0.797*	−0.221
Painty	0.937*	−0.193
Sweet	−0.100	0.776*

*Response variable loads onto this factor.

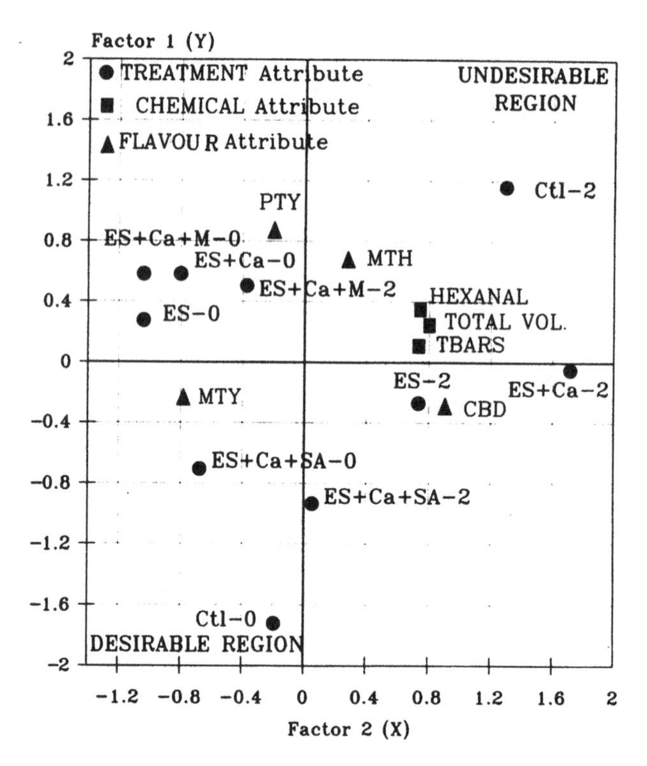

Figure 15.5 Multivariate factor analysis; infusion of lamb muscle experiment.

a particular treatment and principal factor can be averaged. These factor score averages for each treatment can be plotted (Figure 15.5), together with the factor loadings for each response variable, to facilitate graphical interpretation of relationships existing among the response variables and the experimental treatments. Notice in Figure 15.5 that the undesirable sensory descriptor and the chemical volatile loadings appear in the upper portion of the plot while the loadings for the more desirable sensory descriptors appear in the lower portion of the plot. Hence, scores for tenderization and storage treatments appearing in the upper or lower portion of the plot associate each treatment, respectively, with either undesirable or desirable sensory and chemical characteristics.

The researcher should note that a factor solution is specific to the data from which it was created. Hence, the validity of any generalization of relationships between response variables and common behaviour (i.e. principal factors) is a direct function of the amount of replication conducted. Most factor analyses are considered 'exploratory' rather than 'confirmatory' analyses (Tabachnick and Fidell, 1983).

15.6 Summary

Aspects of appropriate experimental designs and statistical analyses for both replicated and unreplicated experiments have been discussed. The design of an experiment applies the treatments under consideration to experimental material in a manner to allow statistically correct analyses of the collected data. The types of treatments, whether distinct treatments or a single treatment observed at various levels, together with the amount of replication allowed by the resources and practical limitations of the experiment determine an appropriate statistical analysis for the collected data.

References

AMSA (1987). Guidelines for cookery and sensory evaluation of meat. American Meat Science Association in cooperation with the National Livestock and Meat Board, Chicago, Illinois.

Bailey, M.E., Dupuy, H.P. and Legendre, M.G. (1980). Undesirable meat flavor and its control. In *The Analysis and Control of Less Desirable Flavors in Foods and Beverages*, ed. G. Charalambous, Academic Press, Orlando, Florida, pp. 31–52.

Bailey, M.E., Shin-Lee, S.Y., Dupuy, H.P., St. Angelo, A.J. and Vercellotti, J.R. (1987). Inhibition of warmed-over flavor by Maillard reaction products. In *Warmed-Over Flavor of Meat*, ed. A.J. St. Angelo and M.E. Bailey, Academic Press, Orlando, Florida, pp. 237–266.

Cremer, M.L. (1982). Sensory quality and energy use for scrambled eggs and beef patties in institutional microwave and convection ovens. *J. Food Sci.*, **47**, 871–874.

Crippen, K.L., Shaffer, G.P., Vercellotti, J.R., Sanders, T.H. and Blankship, P.D. (1991). Reducing the noise contained in descriptive sensory data. (Personal communication.)

Cross, H.R. (1977). A survey of meat cookery and sensory evaluation methods among AMSA meat scientists. Paper presented at Reciprocal Meat Conf., Auburn, Alabama.

Dupuy, H.P., Fore, S.P. and Goldblatt, L.A. (1971). Elution and analysis of volatiles in vegetable oils by gas chromatography. *J. Amer. Oil Chem. Soc.*, **48**, 876–879.

Dupuy, H.P., Brown, M.L., Legendre, M.G., Wadsworth, J.I. and Rayner, E.T. (1978). Instrumental analysis of volatiles in food products. In *Lipids as a Source of Flavor*, ed. M.K. Supran, ACS Symposium Series No. 75, American Chemical Society, Washington, DC, pp. 60–67.

Dupuy, H.P., Flick, Jr., G.J., Bailey, M.E., St. Angelo, A.J. and Legendre, M.G. (1985). Direct sampling capillary gas chromatography of volatiles in vegetable oils. *J. Amer. Oil Chem. Soc.*, **62**, 1690–1693.

Dupuy, H.P., Bailey, M.E., St. Angelo, A.J., Legendre, M.G. and Vercellotti, J.R. (1987). Instrumental analyses of volatiles related to warmed-over flavour of cooked meats. In *Warmed-Over Flavor of Meat*, ed. A.J. St. Angelo and M.E. Bailey, Academic Press, Orlando, Florida, pp. 165–191.

Green, B.E. (1969). Lipid oxidation and pigment changes in raw beef. *J. Food Sci.*, **34**, 110–113.

Johnsen, P.B. and Civille, G.V. (1986). A standardized lexicon of meat WOF descriptors. *J. Sens. Studies*, **1**, 99–104.

Kosugi, H., Kato, T. and Kikugawa, K. (1988). Formation of red pigment by a two-step 2-thiobarbituric acid reaction of alka-2,4-dienals. Potential products of lipid oxidation. *Lipids*, **23**, 1024–1031.

Legendre, M.G., Fisher, G.S., Schuller, W.H., Dupuy, H.P. and Rayner, E.T. (1979). Novel technique for the analysis of volatiles in aqueous and nonaqueous systems. *J. Amer. Oil Chem. Soc.*, **56**, 552–555.

Love, J. (1988). Sensory analysis of warmed-over flavor in meat. *Food Technol.*, **42**(6), 140–143.

Meilgaard, M.C., Civille, G.C. and Carr, B.T. (1987). *Sensory Evaluation Techniques*, Vol. II. CRC Press Inc., Boca Raton, Florida, pp. 47–49.

Milliken, G.A. and Johnson, D.E. (1989). *Analysis of Messy Data—Nonreplicated Experiments*, vol. 2. Van Nostrand Reinhold Ltd., New York.

Pangborn, R.M. (1967). Use and misuse of sensory methodology. *Food Qual. Control*, **15**, 7–12.

Sato, K. and Hegarty, G.R. (1971). Warmed-over flavor in cooked meat. *J. Food Sci.*, **36**, 1098–1102.

Sinnhuber, R.O. and Yu, T.C. (1977). The 2-thiobarbituric acid reaction, an objective measure of the oxidative deterioration occurring in fats and oils. *J. Japan Oil Chem. Soc.*, **26**, 259–267.

Spanier, A.M., McMillin, K.W. and Miller, J.A. (1990). Enzyme activity levels in beef: effect of postmortem aging and endpoint cooking temperature. *J. Food Sci.*, **55**, 318–322.

Spanier, A.M., Vercellotti, J.R. and James, C., Jr. (1992). Correlation of sensory, instrumental and chemical attributes of beef as influenced by meat structure and oxygen exclusion, *J. Food Sci.*, **56** (in press).

St. Angelo, A.J., Vercellotti, J.R., Legendre, M.G., Vinnett, C.H., Kuan, J.W., James, C. Jr., and Dupuy, H.P. (1987). Chemical and instrumental analysis of warmed-over flavor in beef. *J. Food Sci.*, **52**, 1163–1168.

St. Angelo, A.J., Koohmaraie, M., Crippen, K.L. and Crouse, J. (1991). Acceleration of post-mortem tenderization/inhibition of warmed-over flavor by calcium chloride–antioxidant infusion into lamb carcasses. *J. Food Sci.*, **56**, 359–362.

St. Angelo, A.H., Spanier, A.M. and Bett, K.L. (1992). In *Lipid Oxidation in Foods*, ed. A.J. St. Angelo. ACS Symposium Series, American Chemical Society, Washington, DC pp. 140–160.

Steel, R.G.D. and Torrie, J.H. (1980). *Principles and Procedures of Statistics—A Biometrical Approach*, Second Edition. McGraw-Hill Book Company, New York.

Tabachnick, B.G. and Fidell, L.S. (1983). *Using Multivariate Statistics*. Harper & Row Publishers, New York.

Tarladgis, B.G., Watts, B.M., Younathan, M.T. and Dugan, L. (1960). A distillation method for the quantitative determination of malonaldehyde in rancid foods. *J. Amer. Oil Chem. Soc.*, **37**, 44–48.

Tims, M.J. and Watts, B.M. (1958). Protection of cooked meats with phosphates. *Food Technol.*, **12**, 240–243.

Tsoukala, B. and Grosch, W. (1977). Analysis of fat deterioration: comparison of some photometric tests. *J. Amer. Oil Chem. Soc.*, **54**, 490–493.

Wampler, T.P., Bowe, W.A. and Levy, E.J. (1985). Splitless capillary GC analysis of herbs and spices using cryofocusing. *American Laboratory*, **October**, pp. 76–81.

Witty, V.C., Krause, G.F. and Bailey, M.E. (1970). A new extraction method for determining 2-thiobarbituric acid values of pork and beef during storage. *J. Food Sci.*, **35**, 582–585.

Index